Masoud Nikravesh
Ben Azvine
Ronald Yager
Lotfi A. Zadeh (Eds.)

Enhancing the Power of the Internet

Springer

Prof. Masoud Nikravesh
E-mail: nikravesh@cs.berkeley.edu

Prof. Dr. Lotfi A. Zadeh
E-mail: zadeh@cs.berkeley.edu

University of California
Dept. Electrical Engineering and Computer
Science - EECS
94720 Berkeley, CA
USA

Dr. Ben Azvine
BTexact Technologies
Adastral Park
Martlesham Heath
IP5 3RE Ipswich
United Kingdom
E-mail: ben.azvine@bt.com

Prof. Ronald Yager
Iona College
Hagan School Business
10801 New Rochelle, NY
USA
E-mail: ryager@iona.edu

ISSN 1434-9922
ISBN 3-540-20237-4 Springer-Verlag Berlin Heidelberg New York

Library of Congress Cataloging-in-Publication-Data
Enhancing the power of the Internet / Masoud Nikravesh ... [et al.] (eds.).
p. cm. -- (Studies in fuzziness and soft computing ; v. 139)
Includes bibliographical references.
ISBN 3-540-20237-4 (alk. paper)
1. Soft computing. 2. Fuzzy logic. 3. Internet research. 4. Internet searching.
I. Nikravesh, Masoud, 1959- II. Series.
QA76.9.S63E55 2004
006.3--dc22

Springer-Verlag is a part of Springer Science+Business Media
springeronline.com

Typesetting: Camera-ready by editors
Cover design: E. Kirchner, Springer-Verlag, Heidelberg
Printed on acid free paper   62/3020/M - 5 4 3 2 1 0

Preface

Under leadership of DARPA, ARPANET has been designed through close collaboration with UCLA during 1962-1969, 1970-1973, and 1974-1981. Initially designed to keep military sites in communication across the US. In 1969, ARPANET connected researchers from Stanford University, UCLA, UC Santa Barbara and the University of Utah. The Internet community formed in 1972 and the Email is started in 1977. While initially a technology designed primarily for needs of the U.S. military, the Internet grew to serve the academic and research communities. More recently, there has been tremendous expansion of the network both internationally and into the commercial user domain.

During last decade significant progress has been made in the Internet industry by using computational intelligence technology. Underlying this evolving technology there have been ideas transforming the very language we use to describe problems with imprecision, uncertainty and partial truth. These developments offer exciting opportunities, but at the same time it is becoming clearer that further advancements are confronted by fundamental problems. During the August 2001, BISC program hosted a workshop toward better understanding of the issues related to the Internet (Fuzzy Logic and the Internet-FLINT2001, Toward the Enhancing the Power of the Internet). The main purpose of the Workshop was to draw the attention of the fuzzy logic community as well as the Internet community to the fundamental importance of specific Internet-related problems. This issue is critically significant about problems that center on search and deduction in large, unstructured knowledge bases. The Workshop provided a unique opportunity for the academic and corporate communities to address new challenges, share solutions, and discuss research directions for the future. Following are the areas that were recognized as challenging problems and the new direction toward the next generation of the search engines and Internet.

The chapters of the book are evolved from presentations made by the participants at the Meeting. The papers include reports from the front of soft computing in the Internet industry and address the problems of the fields by considering a very important topic of search engine, fuzzy query, decision analysis and support system, e-business and e-commerce.

The book provides a collection of sixteen (16) articles including search engines, user modeling and personal information provision, e-commerce, e-business, e-health, semantic web/net, web-assistant and agents, knowledge representation for e-learning, content-based information retrieval, information organization, intrusion detection and network management.

We would like to take this opportunity to thank all the contributors and reviewers of the articles. We also wish to acknowledge our colleagues who have contributed to the area directly or indirectly related to the content of this book. Finally, we gratefully acknowledge the BTexact technologies -- specially, Dr. Nader Azarmi-- for the financial and technical support, which made the Meeting and book possible.

Masoud Nikravesh, Ben Azvine, Ron Yager and Lotfi A Zadeh
Berkeley Initiative in Soft Computing (BISC)
Berkeley-USA
May 2003

Contents

Enhancing the Power of the Internet

Masoud Nikravesh[1] and Tomohiro Takagi [2]
[1] BISC Program, Computer Sciences Division, EECS Department
University of California, Berkeley, CA 94720, USA
Email: nikravesh@cs.berkeley.edu,URL: http://www-bisc.cs.berkeley.edu
Tel: (510) 643-4522, Fax: (510) 642-5775
[2] Dept. of Computer Science, Meiji University, Email: takagi@cs.meiji.ac.jp

Abstract: World Wide Web search engines have become the most heavily-used online services, with millions of searches performed each day. Their popularity is due, in part, to their ease of use. The central tasks for the most of the search engines can be summarize as 1) query or user information request- do what I mean and not what I say!, 2) model for the Internet, Web representation-web page collection, documents, text, images, music, etc, and 3) ranking or matching function-degree of relevance, recall, precision, similarity, etc. Design of any new intelligent search engine should be at least based on two main motivations: 1) The web environment is, for the most part, unstructured and imprecise. To deal with information in the web environment what is needed is a logic that supports modes of reasoning which are approximate rather than exact. While searches may retrieve thousands of hits, finding decision-relevant and query-relevant information in an imprecise environment is a challenging problem, which has to be addressed and 2) Another, and less obvious, is deduction in an unstructured and imprecise environment given the huge stream of complex information.

1. Introduction

Under leadership of DARPA, ARPANET has been designed through close collaboration with UCLA during 1962-1969, 1970-1973, and 1974-1981. Initially designed to keep military sites in communication across the US. In 1969, ARPANET connected researchers from Stanford University, UCLA, UC Santa Barbara and the University of Utah. The Internet community formed in 1972 and the Email is started in 1977. While initially a technology designed primarily for needs of the U.S. military, the Internet grew to serve the academic and research communities. More recently, there has been tremendous expansion of the network both internationally and into the commercial user domain. ***Table 1*** shows the evolution of the Internet, World Wide Web, and Search Engines.

There are many publicly available Web search engines, but users are not necessarily satisfied with speed of retrieval (i.e., slow access) and quality of retrieved information (i.e., inability to find relevant information). It is important to remember that problems related to speed and access time may not be resolved by considering Web information access and retrieval as an isolated scientific problem. An August 1998 survey by Alexa Internet (<alexa.com>) indicates that 90% of all Web traffic is spread over 100,000 different hosts, with 50% of all Web traffic headed towards the top 900 most popular sites. Effective means of managing uneven concentration of information packets on the Internet will be needed in addition to the development of fast access and retrieval algorithms (Kabayashi and Takeda 2000).

During 80, most of the advances of the automatic document categorization and IR were based on knowledge engineering. The models were built manually using expert systems capable of taking decision. Such expert system has been typically built based on a set of manually defined rules. However, the bottleneck for such manual expert systems was the knowledge acquisition very similar to expert system. Mainly, rules needed to be defined manually by expert and were static. Therefore, once the database has been changed or updated the model must intervene again or work has to be repeated anew if the system to be ported to a completely different domain. By explosion of the Internet, these bottlenecks are more obvious today. During 90, new direction has been merged based on machine learning approach. The advantage of this new approach is evident compared to the previous approach during 80. In machine learning approach, most of the engineering efforts goes towards the construction of the system and mostly is independent of the domain. Therefore, it is much easier to port the system into a new domain. Once the system or model is ported into a new domain, all that is needed is the inductive, and updating of the system from a different set of new dataset, with no required intervention of the domain expert or the knowledge engineer. In term of the effectiveness, IR techniques based on machine learning techniques achieved impressive level of the performance and for example made it possible automatic document classification, categorization, and filtering and making these processes viable alternative to manual and expert system models.

During the recent years, applications of fuzzy logic and the Internet from Web data mining to intelligent search engine and agents for Internet applications have greatly increased (Nikravesh, 2002; Nikravesh et al., 2002, 2003a, 2003b, 2003c; Nikravesh and Choi 2003, Loia et al. 2002, 2003; Nikravesh and Azvine, 2001, 2002; Takagi et al., 2002a, 2002b). Martin (2001) concluded that semantic web includes many aspects, which require fuzzy knowledge representation and reasoning. This includes the fuzzification and matching of concepts. In addition, it is concluded that fuzzy logic can be used in making useful, human-understandable, deduction from semi-structured information available in the web. It is also presented issues related to knowledge representation focusing on the process of fuzzy matching within graph structure. This includes knowledge representation based on conceptual graphs and Fril++. Baldwin and Morton (1985) studied the use of fuzzy logic in conceptual graph framework. Ho (1994) also used fuzzy conceptual graph to be implemented in the machine-learning framework. Baldwin (2001) presented the basic concept of fuzzy Bayesian Nets for user modeling, message filtering and data mining. For message filtering the protoype model representation has been used. Given a context, prototypes represent different types of people and can be modeled using fuzzy rules, fuzzy decision tree, fuzzy Bayesian Net or a fuzzy conceptual graph. In their study, fuzzy set has been used for better generalization. It has been also concluded that the new approach has many applications. For example, it can be used for personalization of web pages, intelligent filtering of the Emails, providing TV programs, books or movie and video of interest. Cao (2001) presented the fuzzy conceptual graphs for the semantic web. It is concluded that the use of conceptual graph and fuzzy logic is complementary for the semantic web. While conceptual graph provide a structure for natural language sentence, fuzzy logic provide a methodology for computing with words. It has been concluded that fuzzy conceptual graphs is suitable language for knowledge representation to be used by Semantic web. Takagi and Tajima (2001a, 2001b) presented the conceptual matching of text notes to be used by search engines. An new search engine proposed which conceptually matches keywords and the web pages. Conceptual fuzzy set has been used for context-dependent keyword expansion. A new structure for search engine has been proposed which can resolve the context-dependent word ambiguity using fuzzy conceptual matching technique. Berenji (2001) used Fuzzy Reinforcement Learning (FRL) for text data mining and Internet search engine. Choi (2001) presented a new technique, which integrates document index with perception index. The techniques can be used for refinement of fuzzy queries on the Internet. It has been concluded that the use of perception index in commercial search engine provides a framework to handle fuzzy terms (perception-based), which is further step toward a human-friendly, natural language-based interface for the Internet. Sanchez (2001) presented the concept of Internet-based fuzzy Telerobotic for the WWW. The system receives the information from human and has the capability for fuzzy reasoning. It has be proposed to use fuzzy applets such as fuzzy logic propositions in the form of fuzzy rules that can be used for smart data base search. Bautista and Kraft (2001) presented an approach to use fuzzy logic for user profiling in Web retrieval applications. The technique can be used to expand the queries and knowledge extraction related to a group of users with common interest. Fuzzy representation of terms based on linguistic qualifiers has been used for their study. In addition, fuzzy clustering of the user profiles can be used to construct fuzzy rules and inferences in order to modify queries. The result can be used for knowledge extraction from user profiles for marketing purposes. Yager (2001) introduced fuzzy aggregation methods for intelligent search. It is concluded that the new technique can increase the expressiveness in the queries. Widyantoro and Yen (2001) proposed the use of fuzzy ontology in search engines. Fuzzy ontology of term relations can be built automatically from a collection of documents. The proposed fuzzy ontology can be used for query refinement and to suggest narrower and broader terms suggestions during user search activity. Presser (2001) introduced fuzzy logic for rule-based personalization and can be implemented for personalization of newsletters.

It is concluded that the use of fuzzy logic provide better flexibility and better interpretation which helps in keeping the knowledge bases easy to maintain. Zhang et al. (2001a) presented granular fuzzy technique for web search engine to increase Internet search speed and the Internet quality of service. The techniques can be used for personalized fuzzy web search engine, the personalized granular web search agent. While current fuzzy search engines uses keywords, the proposed technique provide a framework to not only use traditional fuzzy-key-word but also fuzzy-user-preference-based search algorithm. It is concluded that the proposed model reduces web search redundancy, increase web search relevancy, and decrease user's web search time. Zhang et al. (2001b) proposed fuzzy neural web agents based on granular neural network, which discovers fuzzy rules for stock prediction. Fuzzy logic can be used for web mining. Pal et al. (2002) presented issues related to web mining using soft computing framework. The main tasks of web mining based on fuzzy logic include information retrieval and generalization. Krisnapuram et al. (1999) used fuzzy c medoids and triimed medoids for clustering of web documents. Joshi and Krisnapuram (1998) used fuzzy clustering for web log data mining. Sharestani (2001) presented the use of fuzzy logic for network intruder detection. It is concluded that fuzzy logic can be used for approximate reasoning and handling detection of intruders through approximate matching; fuzzy rule and summarizing the audit log data. Serrano (2001) presented a web-based intelligent assistance. The model is an agent-based system which uses a knowledge-based model of the e-business that provide advise to user through intelligent reasoning and dialogue evolution. The main advantage of this system is based on the human-computer understanding and expression capabilities, which generate the right information in the right time.

2. Web Intelligence: Conceptual Search Engine and Navigation

World Wide Web search engines have become the most heavily-used online services, with millions of searches performed each day. Their popularity is due, in part, to their ease of use. The central tasks for the most of the search engines can be summarize as 1) query or user information request- do what I mean and not what I say!, 2) model for the Internet, Web representation-web page collection, documents, text, images, music, etc, and 3) ranking or matching function-degree of relevance, recall, precision, similarity, etc.

One can use clarification dialog, user profile, context, and ontology, into an integrated frame work to design a more intelligent search engine. The model will be used for intelligent information and knowledge retrieval through conceptual matching of text. The selected query doesn't need to match the decision criteria exactly, which gives the system a more human-like behavior. The model can also be used for constructing ontology or terms related to the context of search or query to resolve the ambiguity. The new model can execute conceptual matching dealing with context-dependent word ambiguity and produce results in a format that permits the user to interact dynamically to customize and personalized its search strategy.

It is also possible to automate ontology generation and document indexing using the terms similarity based on Conceptual-Latent Semantic Indexing Technique (CLSI). Often time it is hard to find the "right" term and even in some cases the term does not exist.

The ontology is automatically constructed from text document collection and can be used for query refinement. It is also possible to generate conceptual documents similarity map that can be used for intelligent search engine based on CLSI, personalization and user profiling. The user profile is automatically constructed from text document collection and can be used for query refinement and provide suggestions and for ranking the information based on pre-existence user profile.

In our perspective, one can use clarification dialog, user profile, context, and ontology, into an integrated frame work to design a more intelligent search engine. The model will be used for intelligent information and knowledge retrieval through conceptual matching of text. The selected query doesn't need to match the decision criteria exactly, which gives the system a more human-like behavior. The model can also be used for constructing ontology or terms related to the context of search or query to resolve the ambiguity. The new model can execute conceptual matching dealing with context-dependent word ambiguity and produce results in a format that permits the user to interact dynamically to customize and personalized its search strategy.

Given the ambiguity and imprecision of the "concept" in the Internet, which may be described by both textual and image information, the use of Fuzzy Conceptual Matching (FCM) (Nikravesh et al., 2003b; Takagi et al., 1995, 1996, 1999a, 1999b) is a necessity for search engines. In the FCM approach, the "concept" is defined by a series of keywords with different weights depending on the importance of each keyword. Ambiguity in concepts can be defined by a set of imprecise concepts. Each imprecise concept in fact can be defined by a set of fuzzy concepts. The fuzzy concepts can then be related to a set of imprecise words given the context. Imprecise words can then be translated into precise words given the ontology and ambiguity resolution through clarification dialog. By constructing the ontology and fine-tuning the strength of links (weights), we could construct a fuzzy set to integrate piecewise the imprecise concepts and precise words to define the ambiguous concept.

3. Challenges and Road Ahead

World Wide Web search engines have become the most heavily-used online services, with millions of searches performed each day. Their popularity is due, in part, to their ease of use. The central tasks for the most of the search engines can be summarize as 1) query or user information request- do what I mean and not what I say!, 2) model for the Internet, Web representation-web page collection, documents, text, images, music, etc, and 3) ranking or matching function-degree of relevance, recall, precision, similarity, etc.

Design of any new intelligent search engine should be at least based on two main motivations:

i· The web environment is, for the most part, unstructured and imprecise. To deal with information in the web environment what is needed is a logic that supports modes of reasoning which are approximate rather than exact. While searches may retrieve thousands of hits, finding decision-relevant and query-relevant information in an imprecise environment is a challenging problem, which has to be addressed.

ii· Another, and less obvious, is deduction in an unstructured and imprecise environment given the huge stream of complex information.

Search engines, with Google at the top, have many remarkable capabilities. But what is not among them is the deduction capability—the capability to synthesize an answer to a query by drawing on bodies of information which are resident in various parts of the knowledge base. It is this capability that differentiates a question-answering system, Q/A system for short, from a search engine.

Prof. L. A. Zadeh in his recent work and several of his presentation has addressed the key issues related to importance of the use of Search Engine and the construction of Q/A system.

From Search Engine to Q/A Systems: The Need for New Tools (Extracted text from Prof. Zadeh's presentation and abstracts; Nikravesh et al., Web Intelligence: Conceptual-Based Model, Memorandum No. UCB/ERL M03/19, 5 June 2003): *Construction of Q/A systems has a long history in AI. Interest in Q/A systems peaked in the seventies and eighties, and began to decline when it became obvious that the available tools were not adequate for construction of systems having significant question-answering capabilities. However, Q/A systems in the form of domain-restricted expert systems have proved to be of value, and are growing in versatility, visibility and importance.*

Search engines as we know them today owe their existence and capabilities to the advent of the Web. A typical search engine is not designed to come up with answers to queries exemplified by "How many Ph.D. degrees in computer science were granted by Princeton University in 1996?" or "What is the name and affiliation of the leading eye surgeon in Boston?" or "What is the age of the oldest son of the President of Finland?" or "What is the fastest way of getting from Paris to London?"

Upgrading a search engine to a Q/A system is a complex, effort-intensive, open-ended problem. Semantic Web and related systems for upgrading quality of search may be viewed as steps in this direction. But what may be argued, as is done in the following, is that existing tools, based as they are on bivalent logic and probability theory, have intrinsic limitations. The principal obstacle is the nature of world knowledge.

The centrality of world knowledge in human cognition, and especially in reasoning and decision-making, has long been recognized in AI. The Cyc system of Douglas Lenat is a repository of world knowledge. The problem is that much of world knowledge consists of perceptions. Reflecting the bounded ability of sensory organs, and ultimately the brain, to resolve detail and store information, perceptions are intrinsically imprecise. More specifically, perceptions are f-granular in the sense that (a) the boundaries of perceived classes are fuzzy; and (b) the perceived values of attributes are granular, with a granule being a clump of values drawn together by indistinguishability, similarity, proximity or functionality. What is not widely recognized is that f-granularity of perceptions put them well beyond the reach of computational bivalent-logic-based theories. For example, the meaning of a simple perception described as "Most Swedes are tall," does not admit representation in predicate logic and/or probability theory.

Dealing with world knowledge needs new tools. A new tool which is suggested for this purpose is the fuzzy-logic-based method of computing with words and perceptions (CWP), with the understanding that perceptions are described in a natural language. A concept which plays a key role in CWP is that of Precisiated Natural Language (PNL). It is this language that is the centerpiece of our approach to reasoning and decision-making with world knowledge.

A concept which plays an essential role in PNL is that of precisiability. More specifically, a proposition, p, in a natural language, NL, is PL precisiable, or simply precisiable, if it is translatable into a mathematically well-defined language termed precisiation language, PL. Examples of precisiation languages are: the languages of propositional logic; predicate logic; modal logic; etc.; and Prolog; LISP; SQL; etc. These languages are based on bivalent logic. In the case of PNL, the precisiation language is a fuzzy-logic-based language referred to as the Generalized Constraint Language (GCL). By construction, GCL is maximally expressive.

A basic assumption underlying GCL is that, in general, the meaning of a proposition, p, in NL may be represented as a generalized constraint of the form X isr R, where X is the constrained variable; R is the constraining relation, and r is a discrete-valued variable, termed modal variable, whose values define the modality of the constraint, that is, the way in which R constrains X. The principal modalities are; possibilistic (r=blank); probabilistic (r=p); veristic (r=v); usuality (r=u); fuzzy random set (r=rs); fuzzy graph (r=fg); and Pawlak set (r=ps). In general, X, R and r are implicit in p. Thus, precisiation of p, that is, translation of p into GCL, involves explicitation of X, R and r. GCL is generated by (a) combining generalized constraints; and (b) generalized constraint propagation, which is governed by the rules of inference in fuzzy logic. The translation of p expressed as a generalized constraint is referred to as the GC-form of p, GC(p). GC(p) may be viewed as a generalization of the concept of logical form. An abstraction of the GC-form is referred to as a protoform (prototypical form) of p, and is denoted as PF(p). For example, the protoform of p: "Most Swedes are tall" is Q A's are B's, where A and B are labels of fuzzy sets, and Q is a fuzzy quantifier. Two propositions p and q are said to be PF-equivalent if they have identical protoforms. For example, "Most Swedes are tall," and "Not many professors are rich," are PF-equivalent. In effect, a protoform of p is its deep semantic structure. The protoform language, PFL, consists of protoforms of elements of GCL.

With the concepts of GC-form and protoform in place, PNL may be defined as a subset of NL which is equipped with two dictionaries: (a) from NL to GCL; and (b) from GCL to PFL. In addition, PNL is equipped with a multiagent modular deduction database, DDB, which contains rules of deduction in PFL. A simple example of a rule of deduction in PFL which is identical to the compositional rule of inference in fuzzy logic, is: if X is A and (X, Y) is B then Y is A∘B, where A∘B is the composition of A and B, defined by

$$\mu_B(v) = \sup_u \left(\mu_A(u) \wedge \mu_B(u,v) \right),$$ *where μ_A and μ_B are the membership functions of A and B, respectively, and ∧ is min or, more generally, a T-norm. The rules of deduction in DDB are organized into modules and submodules, with each module and submodule associated with an agent who controls execution of rules of deduction and passing results of execution.*

In our approach, PNL is employed in the main to represent information in the world knowledge database (WKD). For example, the items:

If X/Person works in Y/City then it is likely that X lives in or near Y
If X/Person lives in Y/City then it is likely that X works in or near Y

are translated into GCL as:

Distance (Location (Residence (X/Person), Location (Work (X/Person) isu near,

where isu, read as ezoo, is th e usuality constraint. The corresponding protoform is:

F (A(B(X/C), A(E(X/C)) isu G.

A concept which plays a key role in organization of world knowledge is that of an epistemic (knowledge-directed) lexicon (EL). Basically, an epistemic lexicon is a network of nodes and weighted links, with node i representing an object in the world knowledge database, and a weighted link from node i to node j representing the strength of association between i and j. The name of an object is a word or a composite word, e.g., car, passenger car or Ph.D. degree. An object is described by a relation or relations whose fields are attributes of the object. The values of an attribute may be granulated and associated with granulated probability and possibility distributions. For example, the values of a granular attribute may be labeled small, medium and large, and their probabilities may be described as low, high and low, respectively. Relations which are associated with an object serve as PNL-based descriptions of the world knowledge about the object. For example, a relation associated with an object labeled Ph.D. degree may contain attributes labeled Eligibility, Length.of.study, Granting.institution, etc. The knowledge associated with an object may be context-dependent. What should be stressed is that the concept of an epistemic lexicon is intended to be employed in representation of world knowledge — which is largely perception-based—rather than Web knowledge, which is not.

As a very simple illustration of the use of an epistemic lexicon, consider the query "How many horses received the Ph.D. degree from Princeton University in 1996." No existing search engine would come up with the correct answer, "Zero, since a horse cannot be a recipient of a Ph.D. degree." To generate the correct answer, the attribute Eligibility in the Ph.D. entry in EL should contain the condition "Human, usually over twenty years of age."

In conclusion, the main thrust of the fuzzy-logic-based approach to question-answering which is outlined here, is that to achieve significant question-answering capability it is necessary to develop methods of dealing with the reality that much of world knowledge—and especially knowledge about underlying probabilities is perception-based. Dealing with perception-based information is more complex and more effort-intensive than dealing with measurement-based information. In this instance, as in many others, complexity is the price that has to be paid to achieve superior performance.

In this context, Berkeley Initiative in Soft Computing (BISC), University of California, Berkeley formed a Technical Committee to organize a Meeting entitled "Fuzzy Logic and the Internet: Enhancing the Power of the Internet" to understand the significance of the fields accomplishments, new developments and future directions. In addition, the Technical Committee selected and invited over 50 scientists (and industry experts as technical committee members) from the related disciplines to participate in the Meeting "State of the Art Assessment and New Directions for Research" which took place at the University of California, Berkeley, in August 14-18, 2001. The Workshop provided a unique opportunity for the academic and corporate communities to address new challenges, share solutions, and discuss research directions for the future. Following are the areas that were recognized as challenging problems and the new direction toward the next generation of the search engines and Internet. We summarize the challenges and the road ahead into four categories as follows:

Search Engine and Queries:	*Internet and the Academia:*
• *Deductive Capabilities* • *Customization and Specialization* • *Metadata and Profiling* • *Semantic Web* • *Imprecise-Querying* • *Automatic Parallelism via Database Technology* • *Approximate Reasoning* • *Ontology* • *Ambiguity Resolution through Clarification Dialog; Definition/Meaning & SpecificityUser Friendly* • *Multimedia* • *Databases* • *Interaction*	• *Ambiguity and Conceptual and Ontology* • *Aggregation and Imprecision Query* • *Meaning and structure Understanding* • *Dynamic Knowledge* • *Perception, Emotion, and Intelligent Behavior* • *Content-Based* • *Escape from Vector SpaceDeductive Capabilities* • *Imprecise-Querying* • *Ambiguity Resolution through Clarification Dialog* • *Precisiated Natural Languages (PNL)*
Internet and the Industry:	*Fuzzy Logic and Internet; Fundamental Research:*
• *XML=>Semantic Web* • *Workflow* • *Mobile E-Commerce* • *CRM* • *Resource Allocation* • *Intent* • *Ambiguity Resolution* • *Interaction* • *Reliability* • *Monitoring* • *Personalization and Navigation* • *Decision Support* • *Document Soul* • *Approximate Reasoning* • *Imprecise QueryContextual Categorization*	• *Computing with Words (CW)* • *Computational Theory of Perception (CTP)* • *Precisiated Natural Languages (PNL)*

The potential areas and applications of Fuzzy Logic for the Internet include:

- ***Potential Areas:***
  - *Search Engines*
  - *Retrieving Information*
  - *Database Querying*
  - *Ontology*
  - *Content Management*
  - *Recognition Technology*
  - *Data Mining*
  - *Summarization*
  - *Information Aggregation and Fusion*
  - *E-Commerce*
  - *Intelligent Agents*
  - *Customization and Personalization*

- ***Potential Applications:***
  - *Search Engines and Web Crawlers*
  - *Agent Technology (i.e., Web-Based Collaborative and Distributed Agents)*
  - *Adaptive and Evolutionary techniques for dynamic environment (i.e. Evolutionary search engine and text retrieval, Dynamic learning and adaptation of the Web Databases, etc)*
  - *Fuzzy Queries in Multimedia Database Systems*
  - *Query Based on User Profile*
  - *Information Retrievals*
  - *Summary of Documents*
  - *Information Fusion Such as Medical Records, Research Papers, News, etc*
  - *Files and Folder Organizer*
  - *Data Management for Mobile Applications and eBusiness Mobile Solutions over the Web*
  - *Matching People, Interests, Products, etc*
  - *Association Rule Mining for Terms-Documents and Text Mining*
  - *E-mail Notification*
  - *Web-Based Calendar Manager*
  - *Web-Based Telephony*
  - *Web-Based Call Centre*
  - *Workgroup Messages*
  - *E-Mail and Web-Mail*
  - *Web-Based Personal Info*
  - *Internet related issues such as Information overload and load balancing, Wireless Internet-coding and D-coding (Encryption), Security such as Web security and Wireless/Embedded Web Security, Web-based Fraud detection and prediction, Recognition, issues related to E-commerce and E-bussiness, etc.*

4. Conclusions

Intelligent search engines with growing complexity and technological challenges are currently being developed. This requires new technology in terms of understanding, development, engineering design and visualization. While the technological expertise of each component becomes increasingly complex, there is a need for better integration of each component into a global model adequately capturing the imprecision and deduction capabilities. In addition, intelligent models can mine the Internet to conceptually match and rank homepages based on predefined linguistic formulations and rules defined by experts or based on a set of known homepages. The FCM model can be used as a framework for intelligent information and knowledge retrieval through conceptual matching of both text and images (here defined as "Concept"). The FCM can also be used for constructing fuzzy ontology or terms related to the context of the query and search to resolve the ambiguity. This model can be used to calculate conceptually the degree of match to the object or query.

5. Future Works

5.1 TIKManD (Tool for Intelligent Knowledge Management and Discovery)

In the future work, we intent to develop and deploy an intelligent computer system is called *"TIKManD (Tool for Intelligent Knowledge Management and Discovery)"*.

The system can mine Internet homepages, Emails, Chat Lines, and/or authorized wire tapping information (which may include Multi-Lingual information) to recognize, conceptually match, and rank potential terrorist and criminal activities (both common and unusual) by the type and seriousness of the activities. This will be done automatically or semi-automatically based on predefined linguistic formulations and rules defined by experts or based on a set of known terrorist activities given the information provided through law enforcement databases (text and voices) and huge number of "tips" received immediately after the attack. Conceptual Fuzzy Set (CFS) model will be used for intelligent information and knowledge retrieval through conceptual matching of text, images and voice (here defined as "Concept"). The CFS can be also used for constructing fuzzy ontology or terms relating the context of the investigation (Terrorism or other criminal activities) to resolve the ambiguity. This model can be used to calculate conceptually the degree of match to the object or query. In addition, the ranking can be used for intelligently allocating resources given the degree of match between objectives and resources available.

5.2 Google™ and Yahoo! Concept-Based Search Engine

There are two type of search engine that we are interested and are dominating the Internet. First, the most popular search engines that are mainly for unstructured data such as Google ™ and Teoma which are based on the concept of Authorities and Hubs.   Second, search engines that are task spcifics such as 1) <u>Yahoo!</u>: manually-pre-classified, 2) <u>NorthernLight</u>: Classification, 3) <u>Vivisimo</u>: Clustering, 4) <u>Self-organizing Map</u>: Clustering + Visualization and 5) <u>AskJeeves:</u> Natural Languages-Based Search; Human Expert.

Google uses the PageRank and Teoma uses HITS (Ding et al. 2001) for the Ranking. To develop such models, state-of-the-art computational intelligence techniques are needed. These include and are not limited to:

- *Latent-Semantic Indexing and SVD for preprocessing,*
- *Radial-Basis Function Network to develop concepts,*
- *Support Vector Machine (SVM) for supervised classification,*
- *fuzzy/neuro-fuzzy clustering for unsupervised classification based on both conventional learning techniques and Genetic and Reinforcement learning,*
- *non-linear aggregation operators for data/text fusion,*
- *automatic recognition using fuzzy measures and a fuzzy integral approach*
- *self organization map and graph theory for building community and clusters,*
- *both genetic algorithm and reinforcement learning to learn the preferences,*
- *fuzzy-integration-based aggregation technique and hybrid fuzzy logic-genetic algorithm for decision analysis, resource allocation, multi-criteria decision-making and multi-attribute optimization.*
- *text analysis: next generation of the Text, Image Retrieval and concept recognition based on soft computing technique and in particular Conceptual Search Model (CSM). This includes*
    - *Understanding textual content by retrieval of relevant texts or paragraphs using CSM followed by clustering analysis.*
    - *Hierarchical model for CSM*
    - *Integration of Text and Images based on CSM*
    - *CSM Scalability, and*
    - *The use of CSM for development of*
        - *Ontology*
        - *Query Refinement and Ambiguity Resolution*
        - *Clarification Dialog*
        - *Personalization-User Profiling*

Acknowledgements

Funding for this research was provided by the British Telecommunication (BT) and the BISC Program of UC Berkeley.

References

J. Baldwin, Future directions for fuzzy theory with applications to intelligent agents, in M. Nikravesh and B. Azvine, FLINT 2001, New Directions in Enhancing the Power of the Internet, UC Berkeley Electronics Research Laboratory, Memorandum No. UCB/ERL M01/28, August 200.

J. F. Baldwin and S. K. Morton, conceptual Graphs and Fuzzy Qualifiers in Natural Languages Interfaces, 1985, University of Bristol.

M. J. M. Batista et al., User Profiles and Fuzzy Logic in Web Retrieval, in M. Nikravesh and B. Azvine, FLINT 2001, New Directions in Enhancing the Power of the Internet, UC Berkeley Electronics Research Laboratory, Memorandum No. UCB/ERL M01/28, August 2001.

H. Beremji, Fuzzy Reinforcement Learning and the Internet with Applications in Power Management or wireless Networks, in M. Nikravesh and B. Azvine, FLINT 2001, New Directions in Enhancing the Power of the Internet, UC Berkeley Electronics Research Laboratory, Memorandum No. UCB/ERL M01/28, August 2001.

T.H. Cao, Fuzzy Conceptual Graphs for the Semantic Web, in M. Nikravesh and B. Azvine, FLINT 2001, New Directions in Enhancing the Power of the Internet, UC Berkeley Electronics Research Laboratory, Memorandum No. UCB/ERL M01/28, August 2001.

D. Y. Choi, Integration of Document Index with Perception Index and Its Application to Fuzzy Query on the Internet, in M. Nikravesh and B. Azvine, FLINT 2001, New Directions in Enhancing the Power of the Internet, UC Berkeley Electronics Research Laboratory, Memorandum No. UCB/ERL M01/28, August 2001.

K.H.L. Ho, Learning Fuzzy Concepts by Example with Fuzzy Conceptual Graphs. In 1st Australian Conceptual Structures Workshop, 1994. Armidale, Australia.

A. Joshi and R. Krishnapuram, Robust Fuzzy Clustering Methods to Support Web Mining, in Proc Workshop in Data Mining and Knowledge Discovery, SIGMOD, pp. 15-1 to 15-8, 1998.

M. Kobayashi, K. Takeda, "Information retrieval on the web", ACM Computing Survey, Vol.32, pp.144-173 (2000)

R. Krishnapuram et al., A Fuzzy Relative of the K medoids Algorithm with application to document and Snippet Clustering , in Proceedings of IEEE Intel. Conf. Fuzzy Systems-FUZZIEEE 99, Korea, 1999.

V. Loia et al., "Fuzzy Logic an the Internet", to be published in the Series Studies in Fuzziness and Soft Computing, Physica-Verlag, Springer (August 2003)

V. Loia et al., *Journal of Soft Computing*, Special Issue; fuzzy Logic and the Internet, Springer Verlag, Vol. 6, No. 5; August 2002.

T. P. Martin, Searching and smushing on the Semantic Web – Challenges for Soft Computing, in M. Nikravesh and B. Azvine, FLINT 2001, New Directions in Enhancing the Power of the Internet, UC Berkeley Electronics Research Laboratory, Memorandum No. UCB/ERL M01/28, August 2001.

M. Nikravesh and B. Azvine, FLINT 2001, New Directions in Enhancing the Power of the Internet, UC Berkeley Electronics Research Laboratory, Memorandum No. UCB/ERL M01/28, August 2001.

M. Nikravesh et al., "Enhancing the Power of the Internet", to be published in the Series Studies in Fuzziness and Soft Computing, Physica-Verlag, Springer (August 2003) (2003a)

M. Nikravesh, et al., Perception-Based Decision processing and Analysis, UC Berkeley Electronics Research Laboratory, Memorandum No. UCB/ERL M03/21, June 2003 (2003b).

M. Nikravesh and D-Y. Choi, Perception-Based Information Processing, UC Berkeley Electronics Research Laboratory, Memorandum No. UCB/ERL M03/20, June 2003.

M. Nikravesh et al., Web Intelligence: Conceptual-Based Model, UC Berkeley Electronics Research Laboratory, Memorandum No. UCB/ERL M03/19, June 2003 (2003c).

M. Nikravesh et al., Fuzzy logic and the Internet (FLINT), Internet, World Wide Web, and Search Engines, *Journal of Soft Computing*, Special Issue; fuzzy Logic and the Internet, Springer Verlag, Vol. 6, No. 5; August 2002

M. Nikravesh, Fuzzy Conceptual-Based Search Engine using Conceptual Semantic Indexing, NAFIPS-FLINT 2002, June 27-29, New Orleans, LA, USA

M. Nikravesh and B. Azvine, Fuzzy Queries, Search, and Decision Support System, *Journal of Soft Computing*, Special Issue fuzzy Logic and the Internet, Springer Verlag, Vol. 6, No. 5; August 2002.

S. K. Pal, V. Talwar, and P. Mitra, Web Mining in Soft Computing Framework: Relevance, State of the Art and Future Directions, to be published in IEEE Transcations on Neural Networks, 2002.

G. Presser, Fuzzy Personalization, in M. Nikravesh and B. Azvine, FLINT 2001, New Directions in Enhancing the Power of the Internet, UC Berkeley Electronics Research Laboratory, Memorandum No. UCB/ERL M01/28, August 2001.

E. Sanchez, Fuzzy logic e-motion, in M. Nikravesh and B. Azvine, FLINT 2001, New Directions in Enhancing the Power of the Internet, UC Berkeley Electronics Research Laboratory, Memorandum No. UCB/ERL M01/28, August 2001.

A. M. G. Serrano, Dialogue-based Approach to Intelligent Assistance on the Web, in M. Nikravesh and B. Azvine, FLINT 2001, New Directions in Enhancing the Power of the Internet, UC Berkeley Electronics Research Laboratory, Memorandum No. UCB/ERL M01/28, August 2001.

S. Shahrestani, Fuzzy Logic and Network Intrusion Detection, in M. Nikravesh and B. Azvine, FLINT 2001, New Directions in Enhancing the Power of the Internet, UC Berkeley Electronics Research Laboratory, Memorandum No. UCB/ERL M01/28, August 2001.

T. Takagi, A. Imura, H. Ushida, and T. Yamaguchi, "Conceptual Fuzzy Sets as a Meaning Representation and their Inductive Construction," International Journal of Intelligent Systems, Vol. 10, 929-945 (1995).

T. Takagi, A. Imura, H. Ushida, and T. Yamaguchi, "Multilayered Reasoning by Means of Conceptual Fuzzy Sets," International Journal of Intelligent Systems, Vol. 11, 97-111 (1996).

T. Takagi, S. Kasuya, M. Mukaidono, T. Yamaguchi, and T. Kokubo, "Realization of Sound-scape Agent by the Fusion of Conceptual Fuzzy Sets and Ontology," 8th International Conference on Fuzzy Systems FUZZ-IEEE'99, II, 801-806 (1999).

T. Takagi, S. Kasuya, M. Mukaidono, and T. Yamaguchi, "Conceptual Matching and its Applications to Selection of TV Programs and BGMs," IEEE International Conference on Systems, Man, and Cybernetics SMC'99, III, 269-273 (1999).

T. Takagi, et al., Conceptual Fuzzy Sets as a Meaning Representation and their Inductive Construction, International Journal of Intelligent Systems, Vol. 10, 929-945 (1995).

T. Takagi and M.Tajima, Proposal of a Search Engine based on Conceptual Matching of Text Notes, IEEE International Conference on Fuzzy Systems FUZZ-IEEE'2001, S406- (2001a)

T. Takagi and M. Tajima, Proposal of a Search Engine based on Conceptual Matching of Text Notes, in M. Nikravesh and B. Azvine, FLINT 2001, New Directions in Enhancing the Power of the Internet, UC Berkeley Electronics Research Laboratory, Memorandum No. UCB/ERL M01/28, August 2001 (2001b).

T. Takagi, et al., Exposure of Illegal Website using Conceptual Fuzzy Sets based Information Filtering System, the North American Fuzzy Information Processing Society - The Special Interest Group on Fuzzy Logic and the Internet NAFIPS-FLINT 2002, 327-332 (2002a)

T. Takagi, et al., Conceptual Fuzzy Sets-Based Menu Navigation System for Yahoo!, the North American Fuzzy Information Processing Society - The Special Interest Group on Fuzzy Logic and the Internet NAFIPS-FLINT 2002, 274-279 (2002b)

Wittgenstein, "Philosophical Investigations," Basil Blackwell, Oxford (1953).

R. Yager, Aggregation Methods for Intelligent Search and Information Fusion, in M. Nikravesh and B. Azvine, FLINT 2001, New Directions in Enhancing the Power of the Internet, UC Berkeley Electronics Research Laboratory, Memorandum No. UCB/ERL M01/28, August 2001.

John Yen, Incorporating Fuzzy Ontology of Terms Relations in a Search Engine, in M. Nikravesh and B. Azvine, FLINT 2001, New Directions in Enhancing the Power of the Internet, UC Berkeley Electronics Research Laboratory, Memorandum No. UCB/ERL M01/28, August 2001.

L. A. Zadeh, The problem of deduction in an environment of imprecision, uncertainty, and partial truth, in M. Nikravesh and B. Azvine, FLINT 2001, New Directions in Enhancing the Power of the Internet, UC Berkeley Electronics Research Laboratory, Memorandum No. UCB/ERL M01/28, August 2001 (2001).

L.A. Zadeh, A Prototype-Centered Approach to Adding Deduction Capability to Search Engines -- The Concept of Protoform, BISC Seminar, Feb 7, 2002, UC Berkeley, 2002.

L. A. Zadeh and M. Nikravesh, Perception-Based Intelligent Decision Systems; Office of Naval Research, Summer 2002 Program Review, Covel Commons, University of California, Los Angeles, July 30th-August 1st, 2002.

Y. Zhang et al., Granular Fuzzy Web Search Agents, in M. Nikravesh and B. Azvine, FLINT 2001, New Directions in Enhancing the Power of the Internet, UC Berkeley Electronics Research Laboratory, Memorandum No. UCB/ERL M01/28, August 2001 (2001a).

Y. Zhang et al., Fuzzy Neural Web Agents for Stock Prediction, in M. Nikravesh and B. Azvine, FLINT 2001, New Directions in Enhancing the Power of the Internet, UC Berkeley Electronics Research Laboratory, Memorandum No. UCB/ERL M01/28, August 2001 (2001b).

Table 1. Understanding and History of Internet, World Wide Web and Search Engine;

Search Engine and Internet	Date	Developer	Affiliation	Comments
ARPANET	1962 –1969 1970-1973 1974-1981	UCLA	Under Leadership of DARPA	Initially designed to keep military sites in communication across the US. In 1969, ARPANET connected researchers from Stanford University, UCLA, UC Santa Barbara and the University of Utah. Internet community formed (1972). Email started (1977).
ALOHANET	1970		University of Hawaii	
USENET	1979	Tom Truscott & Jim Ellis Steve Bellovin	Duke University & University of North Carolina	The first newsgroup.
ARPANET	1982-1987	Bob Kahn & Vint Cerf	DARPA & Stanford University	ARPANET became "Internet". Vinton Cerf "Father of the Internet". Email and Newsgroups used by many universities.
CERT	1988-1990	Computer Emergency Response Team		Internet tool for communication. Privacy and Security. Digital world formed. Internet worms & hackers. The World Wide Web is born.
Archie through FTP	1990	Alan Ematage	McGill University	Originally for access to files given exact address. Finally for searching the archive sites on FTP server, deposit and retrieve files.
Gopher	1991	A team led by Mark MaCahill	University of Minnesota	Gopher used to organize all kinds of information stored on universities servers, libraries, non-classified government sites, etc. Archie and Veronica, helped Gopher (Search utilities).
World Wide Web "alt.hypertext	1991	Tim Berners-Lee	CERN in Switzerland	The first World Wide Web computer code. "alt.hypertext." newsgroup with the ability to combine words, pictures, and sounds on Web pages
Hyper Text Transfer Protocol (HTTP).	1991	Tim Berners-Lee	CERN in Switzerland	The 1990s marked the beginning of World Wide Web which in turn relies on HTML and Hyper HTTP. Conceived in 1989 at the CERN Physics Laboratory in Geneva. The first demonstration December 1990. On May 17, 1991, the World Wide Web was officially started, by granting HTTP access to a number of central CERN computers. Browser software became available-Microsoft Windows and

				Apple Macintosh
	1992			The first audio and video broadcasts:"MBONE." More than 1,000,000 hosts.
Veronica	1993	System Computing Services Group	University of Nevada	The search Device was similar to Archie but search Gopher servers for Text Files
Mosaic	1993	Marc Andeerssen	NCSA (the National Center for Supercomputing Applications); University of Illinois at Urbana Champaign	Mosaic, Graphical browser for the World Wide Web, were developed for the Xwindows/UNIX, Mac and Windows.
World Wide Web Wanderer; the first Spider robot	1993	Matthew Gary	MIT	Developed to count the web servers. Modified to capture URLs. First searchable Web database, the Wandex.
ALIWEB	1993	Martiijn Koster	Now with Excite	Archie-Like Indexing of the Web. The first META tag
JumpStation, World Wide Web Worm .	1993		NASA	Jump Station developed to gathere document titles and headings. Index the information by searching database and matching keywords. WWW worm index title tags and URLs.
Repository-Based Software Engineering (RBSE) Spider	1993		NASA	The first relevancy algorithm in search results, based on keyword frequency in the document. Robot-Driven Search Engine Spidered by content.
	1994			Broadcast over the M-Bone. Japan's Prime Minister goes online at www.kantei.go.jp. Backbone traffic exceeds 10 trillion bytes per month.
Netscape and Microsoft's Internet Explorer	1994-1998	Microsoft and Netscape	Microsoft and Netscape	Added a user-friendly point-and-click interface for browsing

Name	Year	Founder(s)	Institution/Company	Description
Netscape	1994	Dr. James H. Clark and Marc Andreessen		The company was founded in April 1994 by Dr. James H. Clark, founder of Silicon Graphics, Inc. and Marc Andreessen, creator of the NCSA Mosaic research prototype for the Internet. June 5, 1995 - change the character of the World Wide Web from static pages to dynamic, interactive multimedia.
Galaxy	1994	Administered by Microelectronics and computer Technology Corporation	Funded by DARPA and consortium of technologies companies and original prototype by MADE program.	Provided large-scale support for electronic commerce and links documents into hierarchical categories with subcategories. Galaxy merged into Fox/News in 1999.
WebCrawler	1994	Brian Pinkerton	University of Washington	Search text of the sites and used for finding information in the Web. AOL purchased WebCrawler in 1995. Excite purchased WebCrawler in 1996.
Yahoo!	1994	David Filo and Jerry Yang	Stanford University	Organized the data into searchable directory based on simple database search engine. With the addition of the Google, Yahoo! Is the top-referring site for searches on the Web. It led also the future of the internet by changing the focus from search retrieval methods to clearly match the user's intent with the database.
Lycous	1994	Michael Mauldin	Carnegie Mellon University	New features such as ranked relevance retrieval, prefix matching, and word proximity matching. Until June 2000, it had used Inktomi as its back-end database provide. Currently, FAST a Norwegian search provider, replaced the Inktomi.
Excite	1995	Mark Van Haren, Ryan McIntyre, Ben Lutch, Joe Kraus, Graham Spencer, and Martin Reinfried	Architext Sofware	Combined search and retrieval with automatic hypertext linking to document and includes subject grouping and automatic abstract algorithm. IT can electronically parse and abstract from the web.
Infoseek	1995	Steve Kirsch	Infoseek	Infoseek combined many functional elements seen in other search

		(now with Propel)		tools such as Yahoo! And Lycos, but it boasted a solid user-friendly interface and consumer-focused features such as news. Also speed in which indexed Web sites and then added them to its live search database.
AltaVista	1995	Lcuis Monier, with Mike Burrows	Digital Equipment Corporation	Speed and the first "Natural Language" queries and Boolean operators. It also proved a user-friendly interface and the first search engine to add a link to helpful search tips below search field to assist novice searchers.
MetaCrawler	1995	Erick Selberg and Cren Etizinoi	University of Washington	The first Meta search engine. Search several search engines and reformat the results into a single page.
SavvySearch	1995	Daniel Dreilinger	Colorado State University	Meta Search which was included 20 search engines. Today, it includes 200 search engine.
Inktomi and HotBot	1994-1996	Eric Brewer and Paul Gauthier	University of California-Berkeley Funded by ARPA	Cluster inexpensive workstation computers to achieve the same computing power as expensive super computer. Powerful search technologies that made use of the clustering of workstations to achieve scaleable and flexible information retrieval system. HotBot, powered by Inktomi and was able to rapidly index and spider the Web and developing a very large database within a very short time.
LookSmart	1996	Mr Evan Thornley	LookSmart	Delivers a set of categorized listing presented in a user-friendly format and providing search infrastructure for vertical portals and ISPs.
AskJeeves	1997	Davis Warthen and Garrett Gruener	AskJeeves	It is built based on a large knowledge base on pre-searched Web sites. It used sophisticated, natural-language semantic and syntactic processing to understand the meaning of the user's question and match it to a 'question template" in the knowledge base.
GoTo	1997	Bill Gross	Indealab!	Auctioning off search engine positions. Advertisers to attach a value to their search engine placement.

Snap	1997	Halsey Minor, CNET Founder	CNET, Computer Network	Redefining the search engine space with a new business model; "portal" as first partnership between a traditional media company and an Internet portal.
Google	1997-1998	Larry Page and Sergey Brin	Stanford University	PageRank™ to deliver highly relevant search results based on proximity match and link popularity algorithms. Google represent the next generation of search engines.
Northern Light	1997	Team of librarians, software engineers, and information industry	Northern Light	To Index and classify human knowledge and has two database 1) contains an index to the full text of millions of Web pages and 2) includes full-text articles from a variety of sources. It searches both Web pages and full-text articles and sorts its search results into folders based on keywords, source, and other criteria.
AOL, MSN and Netscape	1998	AOL, MSN and Netscape	AOL, MSN and Netscape	Search service for the users of services and software
Open Directory	1998	Rick Skrenta and Bob Truel	dmoz	Open directory
Direct Hit	1998	Mike Cassidy	MIT	Direct Hit is dedicated to providing highly relevant Internet search results. Direct Hit's highly scalable search system leverages the searching activity of millions of Internet searchers to provide dramatically superior search results. By analyzing previous Internet search activity, Direct Hit determines the most relevant sites for your search request.
FAST Search	1999	Isaac Elsevier	FAST; Norwegian Company- All the Web	High-capacity search and real-time content matching engines based on the All the Web technology. Using Spider technology to index pages very rapidly. FAST can index both Audio and Video files.

Soft Computing and User Modeling

T. P. Martin[1] and B. Azvine[2]

[1]University of Bristol, Bristol, BS8 1TR, UK
Trevor.Martin@bristol.ac.uk
[2]BTexact Technologies, Adastral Park, Ipswich, IP5 3RE, UK
Ben.Azvine@bt.com

Abstract. The next generation of consumer goods, including computers, will be much more sophisticated in order to cope with a less technologically literate user base. A user model is an essential component for "user friendliness", enabling the behavior of a system to be tailored to the needs of a particular user. Simple user profiles already personalise many software products and consumer goods such as digital TV recorders and mobile phones. A user model should be easy to initialise, and it must adapt in the light of interaction with the user. In many cases, a large amount of training data is needed to generate a user model, and adaptation is equivalent to completely retraining the system. This paper briefly outlines the user modelling problem and work done at BTexact on an Intelligent Personal Assistant (IPA) which incorporates a user profile. We go on to describe FILUM, a more flexible method of user modelling, and show its application to the Telephone Assistant and Email Assistant components of the IPA, with tests to illustrate its usefulness. An experimental testbed based on the iterated prisoner's dilemma, which allows the generation of unlimited data for learning or testing, is also proposed.

Introduction

User modeling is a key technology in increasing the effective use of computers and information appliances, and the next generation of consumer goods requires more sophistication in user-modeling and intelligent help systems to cope with a less technologically literate user base. For example, an integrated home information / entertainment system (computer, VCR, TV, hi-fi, etc) should be able to suggest TV/video choices based on past preferences, and automatically record programmes judged to be interesting to a user. With the increased access to information arising from the web and integration of digital TV and computer networking, this area of intelligent consumer goods is an extremely important next step.

We define user modeling to be the provision of a software sub-system able to observe and predict the actions of a user (from a limited set of possibilities), with the aim of improving the overall interaction between user and system. This is a relatively "soft" definition, as the quality of interaction is almost always a subjective judgment, and it is therefore difficult to discuss the success (or otherwise) of user modeling.

We can recognise a strongly growing strand of interest in user modelling arising from research into intelligent interfaces. In this context, we can identify three different outcomes of user modelling:

- Changing the way in which some fixed content is delivered to the user.
- Changing the content that is delivered to the user.
- Changing the way in which the device is used.

Each of these is discussed in turn below.

The first relates more to the device that is displaying content to a user. For example, a WAP browser must restrict graphical content. There is little room for user likes and dislikes, although [12] describe a system which implements different interfaces for different users on desktop systems. Those who have more difficulty navigating through the system use a menu-based interface whereas those with a greater awareness of the system contents are given an interface using a number of shortcut keys.

The second category—improving (or changing) information content—is perhaps the most common. Examples abound in internet-related areas, with applications to

- Deliver only "interesting" news stories to an individual's desktop. The pointcast news delivery systems are a first step (e.g. www.pointcast.com/products/pcn/ and cnn.com/ads/advertiser/pointcast2.0/); see also [11] and IDIoMS [13].
- Identify interesting web pages—for example Syskill &Webert [25] uses an information-theoretic approach to detect "informative" words on web pages. These are used as features, and user ratings of web pages (very interesting, interesting, not interesting, etc.) creates a training data set for a naive Bayesian classifier. A similar approach can be used for the retrieval of documents from digital libraries, using term frequency/inverse document frequency [31] to select keywords and phrases as features. A user model can be constructed in terms of these features, and used to judge whether new documents are likely to be of interest.
- Remove unwanted emails – see [18]or [32, 33] for example. (e.g. [5-7] as well as earlier work listed in [2])

The problem of "information overload" from email was identified as far back as 1982

"in current message systems, the message remains uninterpreted ... The system delivers the message but does not manage the messages .... In order to enhance their functionality, message systems have to interpret, at least partially, the messages they handle ..."

[8] quoted in [1]. The latter authors also noted that

"information inundation may cause information entropy, when "incoming messages are not sufficiently organized by topic or content to be easily recognized as important"

With the incorporation of powerful embedded computing devices in consumer products, there is a blurring of boundaries between computers and other equipment, resulting in a convergence to *information appliances* or *information devices*. Personalisation, which is equivalent to user modelling, is a key selling point of this technology—for example, to personalise TV viewing (www.tivo.com, 1999):

"With TiVo, getting your favorite programs is easy. You just teach it what shows you like, and TiVo records them for you automatically. As you're watching TV, press the Thumbs Up or Thumbs Down button on the TiVo remote to teach TiVo what you like As TiVo searches for shows you've told it to record, it will also look for shows that match your preferences and get those for you as well..."

Sony have implemented a prototype user modelling system [37] which predicts a viewing timetable for a user, on the basis of previous viewing and programme classification. Testing against a database of 606 individuals, 108 programme categories and 45 TV channels gave an average prediction accuracy of 60-70%. We will not discuss social or collaborative filtering systems here. These are used to recommend books (e.g. amazon.com), films, and so on, and are based on clustering the likes and dislikes of a group of users.

The third category - changing the way in which the device is used - can also be illustrated by examples. Microsoft's Office Assistant is perhaps the best known example of user modelling, and aims to provide appropriate help when required, as well as a "tip of the day" that is intended to identify and remedy gaps in the user's knowledge of the software. The Office Assistant was developed from the Lumiere [16] project, which aimed to construct Bayesian models for reasoning about the time-varying goals of computer users from their observed actions and queries. Although it can be argued that the Office Assistant also fits into the previous category (changing the content delivered to the user), its ultimate aim is to change the way the user works so that the software is employed more effectively.

The system described by [20] has similar goals but a different approach. User modelling is employed to disseminate expertise in use of software packages (such as Microsoft Word) within an organisation. By creating an individual user model and comparing it to expert models, the system is able to identify gaps in knowledge and offer individualised tips as well as feedback on how closely the user matches expert use of the package. The key difference from the Office Assistant is that this system monitors all users and identifies improved ways of accomplishing small tasks; this expertise can then be spread to other users. The Office Assistant, on the other hand, has a static view of best practice.

Hermens and Schlimmer [14] implemented a system which aided a user filling in an electronic form, by suggesting likely values for fields in the form, based on the values in earlier fields.

The change in system behaviour may not be obvious to the user. Lau and Horvitz [19] outline a system which uses a log of search requests from Yahoo, and classifies users' behaviour so that their next action can be predicted using a Bayesian net. If it is likely that a user will follow a particular link, rather than refining or reformulating their query, then the link can be pre-fetched to improve the perceived performance of the system. This approach generates canonical user models, describing the behaviour of a typical *group* of users rather than individual user models.

There are two key features in all these examples:

- the aim is to improve the interaction between human and machine. This is a property of the whole system, not just of the machine, and is frequently a subjective judgement that can not be measured objectively.
- the user model must adapt in the light of interaction with the user.
- Additionally, it is desirable that the user model
- be gathered unobtrusively, by observation or with minimal effort from the user.
- be understandable and changeable by the user - both in terms of the knowledge held about the user and in the inferences made from that knowledge.
- be correct in actions taken as well as in deciding when to act.

User models—Learning, Adaptivity and Uncertainty

The requirement for adaptation puts user modelling into the domain of machine learning (see [17] and [36]). A user model is generally represented as a set of attribute-value pairs—indeed the W3C proposals [34] on profile exchange recommend this representation. This is ideal for machine learning, as the knowledge representation fits conveniently into a propositional learning framework. To apply machine learning, we need to gather data and identify appropriate features plus the desired attribute for prediction. To make this concrete, consider a system which predicts the action to be taken on receiving emails, using the sender's identity and words in the title field. Most mail readers allow the user to define a *kill file*, specifying that certain emails may be deleted without the user seeing them. A set of examples might lead to rules such as

```
if title includes $ or money then action = delete
if sender = boss then action = read, and subsequently file
if sender = mailing list then action = read and subse-
quently delete
```

This is a conventional propositional learning task, and a number of algorithms exist to create rules or decision trees on the basis of data such as this [4, 5, 7, 27, 28]. Typically, the problem must be expressed in an attribute-value format, as above; some feature engineering may be necessary to enable efficient rules to be induced. Rule-based knowledge representation is better than (say) neural nets due to better understandability of the rules produced - the system should propose rules which the user can inspect and alter if necessary. See [24] for empirical evidence of the importance of allowing the user to remain in control.

One problem with propositional learning approaches is that it is difficult to extract relational knowledge. For example:

```
if several identical emails arrive consecutively from a
list server, then delete all but one of them
```

Also, it can be difficult to express relevant background knowledge such as:

```
if a person has an email address at acme.com then that per-
son is a work colleague
```

These problems can be avoided by moving to relational learning, such as inductive logic programming [23], although this is not without drawbacks as the learning process becomes a considerable search task.

Possibly more serious issues relate to the need to update the user model, and to incorporate uncertainty. Most machine learning methods are based on a relatively large, static set of training examples, followed by a testing phase on previously unseen data. New training examples can normally be addressed only by restarting the learning process with a new, expanded, training set. As the learning process is typically quite slow, this is clearly undesirable. Additionally in user modelling it is relatively expensive to gather training data - explicit feedback is required from the user, causing inconvenience. The available data is therefore more limited than is typical for machine learning.

A second problem relates to uncertainty. User modeling is inherently uncertain—as [15] observes, *"Uncertainty is ubiquitous in attempts to recognize an agent's goals from observations of behavior,"* and even strongly logic-based methods such as [26] acknowledge the need for "graduated assumptions." There may be uncertainty over the feature definitions. For example:

```
if the sender is a close colleague then action = read very
soon
```

where *close colleague* and *very soon* are fuzzily defined terms) or over the applicability of rules. For example:

```
if the user has selected several options from a menu and
undone each action, then it is very likely that the user
requires help on that menu
```

where the conclusion is not always guaranteed to follow.

It is an easy matter to say that uncertainty can be dealt with by means of a fuzzy approach, but less easy to implement the system in a way that satisfies the need for understandability. The major problem with many uses of fuzziness is that they rely on intuitive semantics, which a sceptic might translate as "no semantics at all." It is clear from the fuzzy control literature that the major development effort goes into adjusting membership functions to tune the controller. Bezdek [9, 10] suggests that membership functions should be *"adjusted for maximum utility in a given situation."* However, this leaves membership functions with no objective meaning—they are simply parameters to make the software function correctly. For a fuzzy knowledge based system to be meaningful to a human, the membership functions should have an interpretation which is independent of the machine operation—that is, one which does not require the software to be executed in order to determine its meaning. Probabilistic representations of uncertain data have a strictly defined interpretation, and the approach adopted here uses Baldwin's mass assignment theory and voting model semantics for fuzzy sets [3, 8].

The Intelligent Personal Assistant

BTexact's Intelligent Personal Assistant (IPA) [1, 2] is an adaptive software system that automatically performs helpful tasks for its user, helping the user achieve higher levels of productivity. The system consists of a number of assistants specialising in time, information, and communication management:

- The Diary Assistant helps users schedule their personal activities according to their preferences.
- Web and Electronic Yellow Pages Assistants meet the user's needs for timely and relevant access to information and people.
- The RADAR assistant reminds the user of information pertaining to the current task.
- The Contact Finder Assistant puts the user in touch with people who have similar interests.
- The Telephone and Email Assistants give the user greater control over incoming messages by learning priorities and filtering unwanted communication.

As with any personal assistant, the key to the IPA's success is an up-to-date understanding of the user's interests, priorities, and behaviour. It builds this profile by tracking the electronic information that a user reads and creates over time—for example, web pages, electronic diaries, e-mails, and word processor documents. Analysis of these information sources and their timeliness helps IPA understand the users personal interests. By tracking diaries, keyboard activity, gaze, and phone usage, the IPA can build up a picture of the habits and preferences of the user.

We are particularly interested in the Telephone and E-mail assistants for communication management, used respectively for filtering incoming calls and prioritising incoming e-mail messages. The Telephone Assistant maintains a set of priorities of the user's acquaintances, and uses these in conjunction with the caller's phone number to determine the importance of an incoming call. The Email Assistant computes the urgency of each incoming message based on its sender, recipients, size and content. Both assistants use Bayesian networks for learning the intended actions of the user, and importantly, the system continually adapts its behaviour as the user's priorities change over time.

The telephone assistant handles incoming telephone calls on behalf of the user with the aim of minimising disruption caused by frequent calls. For each incoming call, the telephone assistant determines whether to interrupt the user (before the phone rings) based on the importance of the caller and on various contextual factors such as the frequency of recent calls from that caller and the presence of a related entry in the diary (e.g. a meeting with the caller). When deciding to interrupt the user, the telephone assistant displays a panel indicating that a call has arrived; the user has the option of accepting or declining to answer the call. The telephone assistant uses this feedback to learn an overall user model for how the user weights the different factors in deciding whether or not to answer a call. Although this model has been effective, its meaning is not obvious to a user, and hence it is not adjustable. To address this issue, the FILUM [21, 22] approach has been applied to the telephone and email assistants.

Assumptions for FILUM

We consider any interaction between a user and a software or hardware system in which the user has a limited set of choices regarding his/her next action. For example, given a set of possible TV programmes, the user will be able to select one to watch. Given an email, the user can gauge its importance and decide to read it immediately, within the same day, within a week, or maybe classify it as unimportant and discardable. The aim of user modelling is to be able to predict accurately the user's decision and hence improve the user's interaction with the system by suggesting or making such decisions automatically. Human behaviour is not generally amenable to crisp, logical modelling. Our assumption is that the limited aspect of human behaviour to be predicted is based mainly on observable aspects of the user's context—for example, in classifying an email the context could include features such as the sender, other recipients of the message, previously received messages, current workload, time of day, and so on. Of course, there are numerous unobservable variables - humans have complex internal states, emotions, external drives, and so on. This complicates the prediction problem and motivates the use of uncertainty modelling—we can only expect to make correct predictions "most" of the time.

We define a set of possible output values

$$B = \{b^1, b^2, ..., b^j\},$$

which we refer to as the behaviour, and a set of observable inputs

$$I = \{i^1, i^2, ..., i^m\}.$$

Our assumption is that the $n+1$th observation of the user's behaviour is predictable by some function of the current observables and all previous inputs and behaviours.

$$b_{n+1} = f(I_1, b_1, I_2, b_2, ... I_n, b_n, I_{n+1})$$

The user model, including any associated processing, is equivalent to the function f. This is assumed to be relatively static; within FILUM, addition of new prototypes would correspond to a major change in the function.

We define a set of classes implemented as Fril++ [6, 29] or java programs.

$$C = \{c_1, c_2, ... c_k\},$$

A user model is treated as an instance that has a probability of belonging to each class according to how well the class behaviour matches the observed behaviour of the user. The probabilities are expressed as support pairs, and updated each time a new observation of the user's behaviour is made.

We aim to create a user model m, which correctly predicts the behaviour of a user. Each class c_i must implement the method *Behaviour*, giving an output in B (this may be expressed as supports over B). Let $S_n(m \in c_i)$ be the support for the user model m belonging to the ith class before the nth observation of behaviour. Initially,

$$S_1(m \in c_i) = [0, 1]$$

for all classes ci, representing complete ignorance.

Each time an observation is made, every class makes a prediction, and the support for the user model being a member of that class is updated according to the predictive success of the class :

$$S_{n+1}(m \in c_i) = \frac{n \times S_n(m \in c_i) + S(c_i.Behaviour_{n+1} == b_{n+1})}{n+1} \tag{1}$$

where $S(c_i.Behaviour_{n+1} == b_{n+1})$ represents the (normalised) support for class c_i predicting the correct behaviour (from the set B) on iteration $n+1$.

The overall model behavior is predicted by multiplying the support for each prototype by the support for each behavior prediction made by that prototype, and then taking the best support over all prototypes.
i.e.

$$PredictedBehaviour = \max_{c \in C, b \in B} \left(S(m \in c) \times S(c.Behaviour == b) \right)$$

where B is the set of possible behaviors and P is the set of user prototypes.

For example, take the problem of predicting the outcome of a biased coin which always lands on heads (this corresponds to a "user", with behaviors "heads" and "tails") and three prototypes P1, P2, P3 which are simple probabilistic programs predicting heads with probabilities 0.4, 0.5 and 0.6 respectively (we will work with probabilities rather than supports to simplify the example). After several iterations (tosses), their success rates in correctly predicting the outcome (i.e. support for m ∈ Ci) will be as shown in the table below,

prototype	$S(m \in C_i)$	prediction	weighted prediction
C1	0.4	heads 0.4	heads 0.16
		tails 0.6	tails 0.24
C2	0.5	heads 0.5	heads 0.25
		tails 0.5	tails 0.25
C3	0.6	heads 0.6	heads 0.36
		tails 0.4	tails 0.24

The next prediction would be heads, as this has the highest weighted support (0.36 from prototype C3). This is the prediction we would intuitively expect, and would also be the prediction of the weighted sum of supports (Heads : 0.16+0.25+0.36 = 0.77, Tails : 0.24+0.25+.0.24 = 0.73).

Note that if we add a fourth prototype which predicts heads 10% of the time, the user model behaves as before :

prototype	$S(m \in C_i)$	prediction	weighted prediction
C1	0.4	heads 0.4	heads 0.16
		tails 0.6	tails 0.24
C2	0.5	heads 0.5	heads 0.25
		tails 0.5	tails 0.25
C3	0.6	heads 0.6	heads 0.36
		tails 0.4	tails 0.24
C4	0.1	heads 0.1	heads 0.01
		tails 0.9	tails 0.09

The prototypes in the user model are more inclined towards "tails", so that taking a weighted combination of predictions would not be an accurate user model. In practice it may be impossible to determine whether a set of prototypes is biased towards any particular behavior – the whole point of the FILUM approach is that the prototypes provide knowledge-based estimations of the effects of external factors, because adequate statistics are not available.

Updating Support

The user model is treated as a partial instance of all prototype classes, with its degree of membership in each class determined by the accuracy with which the prototype predicts the correct behavior, as shown in Eq. 1. Note that this does not discriminate against a prototype which gives high support to all possible behaviors – to take an extreme case, allocating a support of (1 1) to every possibility. Thus in addition to requiring each prototype to give the support for each behavior, they must also predict a single behavior. If this is correct, support is updated as above; if it is wrong, the updating support is (0 0). Prototypes are allowed to return *FILUMPrototype.NOPREDICTION* in cases where they are not applicable.

To illustrate this process, consider three prototypes – one which always predicts correctly, one which always predicts wrongly and a third which is alternately correct and incorrect. The supports for these prototypes will evolve as shown in Fig 1.

Clearly as n becomes large, supports change relatively slowly. [30] discuss an alternative updating algorithm which is weighted in favour of more recent behaviour, particularly unexpected actions. The accuracy of the user model at any stage is the proportion of correct predictions made up to that point—this metric can easily be changed to use a different utility function, for example, if some errors are more serious than others.

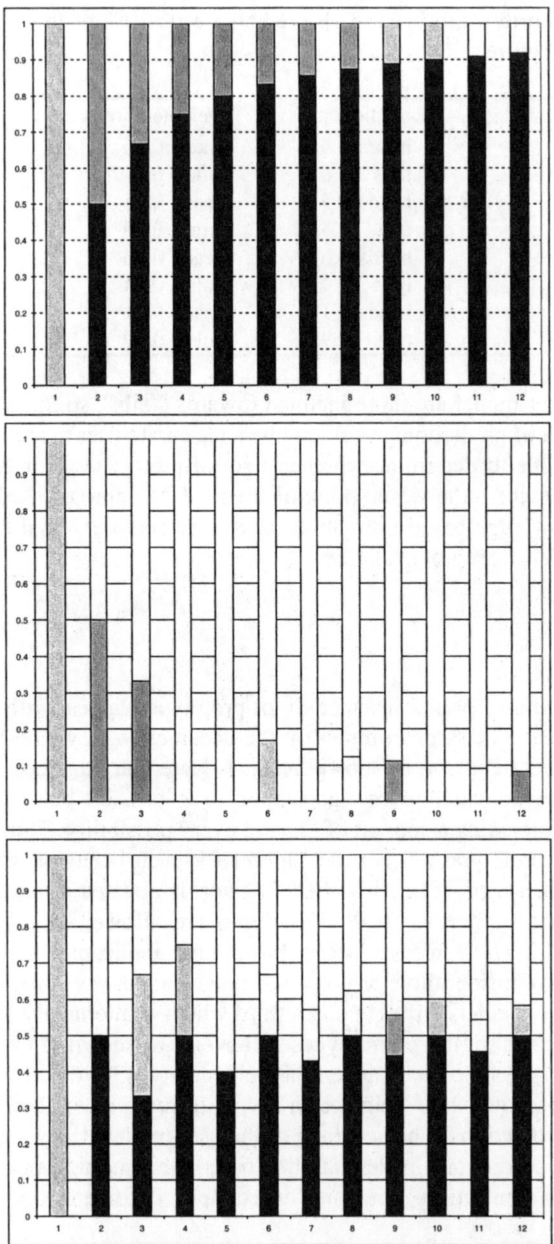

Fig.1 : evolution of support over 12 iterations for three user prototypes. The dark bars show the necessary support, diagonal stripes represent uncertainty and light horizontal shading represents support against the prototype predicting correctly. The top graph shows a prototype which always predicts correctly and the support evolves towards 1; the second graph shows a prototype which always predicts incorrectly, and the support evolves towards 0. The third shows a prototype which alternates between correct and incorrect predictions, and the support evolves towards 0.5

Testing

In order to test user modeling approaches, data is needed. This can either be gathered from a real situation or generated using an artificial model problem. The *n*-player iterated prisoner's dilemma described below provides a method of generating data for user modeling experiments. It should be emphasized that the game itself is of no interest here; it is used purely as a platform for generating data to test user modeling approaches.

The Prisoner's Dilemma is a well-known example of a non-zero sum game in game theory. As an illustration, consider two prisoners who are jointly charged with a crime for which the standard sentence is five years. They are separately offered a deal whereby they can stay silent (co-operate with their partner) or defect, providing evidence to implicate their partner. Each can choose to co-operate or defect, and the reduction in their sentences according to their joint behavior is given in the following table:

player1 / player2	co-operate	defect
co-operate	3 / 3	0 / 5
defect	5 / 0	1 / 1

From each individual's point of view, the rational strategy is to defect; collectively the best solution is for both to co-operate.

The iterated version extends the game to a sequence of interactions, where each player has access to the history of interactions. The n-player version considers more than two individuals. In each round, each pair of players participates in a pairwise interaction as above.

In all cases, each player aims to maximize their own score. There is an incentive to co-operate (e.g. the payoff from three co-operative interactions, c-c, c-c, c-c will be 9 whereas the payoff from one "exploitative" interaction and two mutual defections (d-c, d-d, d-d) will only be 7).

The iterated prisoner's dilemma or IPD [1] has been used to explain arms race escalation, the formation of economic cartels, and evolutionary biology, as well as acting as a test bed for multi-agent systems [10] and evolutionary programming. It can be shown that there is no optimum strategy, as much is dependent on the environment. For example if all other players co-operate irrespective of one's own actions, the optimum strategy is to defect. If all players respond by echoing one's last move (the "tit-for-tat" strategy) then the optimum strategy is to co-operate. The game was widely publicized in computer tournaments e.g. [13,20] in which co-operative strategies tended to do best. In particular successful strategies are

☐ "nice" in that they do not defect without provocation

☐ "responsive" in that they punish provocation (i.e. defection by the other player)

☐ "forgiving" in that they will attempt to co-operate again after punishing an opponent

n-player IPD as a Test Bed for User Modeling

The n-IPD is a good test bed for user modeling as it is possible to generate as much data as necessary and the true behavior can be obtained in each case, so that it is possible to get an objective evaluation of the predictive accuracy of the user model.

The aim is to replace a selected player in an n-player tournament by a user model, which behaves in the same way as the player without knowledge of or access to the internal structure of the player, i.e. without knowing the code governing the player's behavior. The user model has access to the interaction history of the selected player. There is a close analogy to the situation in user modeling applications, where the previous behavior of a user is known without any detailed understanding of the "algorithm" that led to the decision.

There is no intention to judge whether a given strategy is successful or to optimize a strategy. At each iteration, the user model of a player P simply makes a prediction of P's behavior in an interaction with each other player in the tournament.

Experiments

A Fril++[3,5,6] -based system was developed to run n-IPD tournaments, allowing a choice of strategies to be included in the environment. The number of players using each strategy could also be specified. Examples of strategies are:
- *trust* - always co-operates
- *defect* - always defects
- *tft* (tit-for-tat) initially co-operates, subsequently echoes whatever opponent did last time
- *rand* - random (50-50) choice
- *crand* - random (75-25) choice biased towards co-operation
- *drand* - random (25-75) choice biased against co-operation
- *tftt* - (tit-for-two-tats) co-operates unless two consecutive defections from opponent
- *stft* - sneaky tit-for-tat initially defects then echoes last opponent response
- *massret* (massive retaliation) - co-operates until a defection, then defects against that opponent forever
- other responsive strategies - a method must be supplied to determine initial behaviour, subsequently co-operate unless the number of defections by opponent in a given number of previous interactions exceeds some threshold, in which case defect (*tit for tat*, *tit for two tats*, *massret* are all subtypes of this strategy). Examples included in the tournaments are
- *respc* - co-operate unless all of the opponent's last 6 responses were *d*
- *respd* - defect unless all of the opponent's last 6 responses were *c*

Figure 2 shows the results from two sample tournaments, plotting the average reward per interaction for selected player in the population. If all interactions are co-operative, then the average will be 3; similarly if a player manages to exploit every other player then its average reward would be 5. In Figure 1 (a) there are too few responsive players, and too many that co-operate unconditionally. The "defect every time" strategy is able to exploit the other players and maintain a clearly better overall average.

In Figure 2 (b) the tournament contained a similar number of co-operative players but more were responsive, and withdrew co-operation from the the "defect every

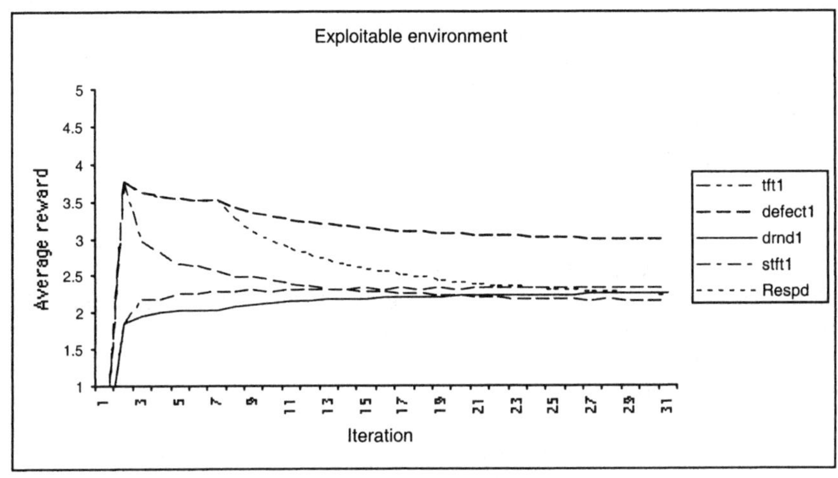

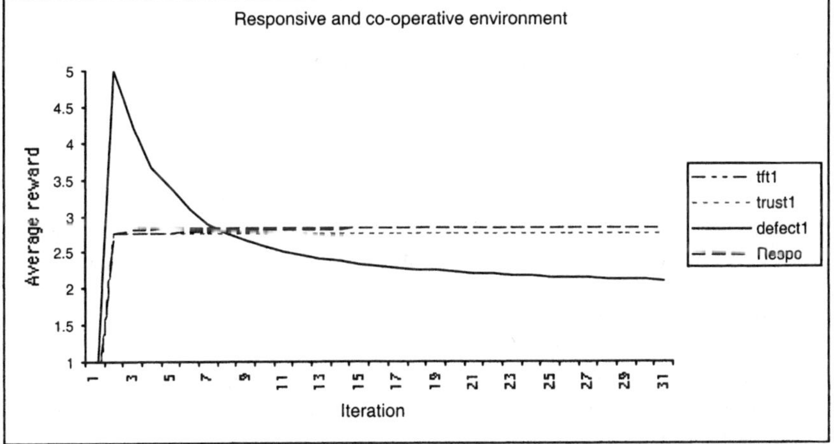

Fig. 2: Average reward per interaction for various players in two n-IPD tournaments

(a) exploitable environment, where the top line represents a player defecting on every move, and
(b) responsive environment where the defecting strategy scores highly at first but then performs poorly. Twenty strategies were represented in each tournament, drawn from those listed in Section 7. It is possible for the same strategy to be included several times as different "players" in the tournament.

time" strategy. The latter does worse than the others, despite its initial success.

The precise details are not important, other than to note that the average score is higher in more co-operative environments, and that the best strategy in one environment is not necessarily the same as the best strategy in another. Such observations are in line with other n-IPD tournaments and help to verify that the software performs correctly.

From the point of view of user modeling, we aim to reproduce the behavior of a player by means of some simple prototypes. The properties of successful players observed in tournaments are niceness, responsiveness, and forgiveness. In order to roughly model these characteristic behaviors, the four prototypes exhibit the following patterns:

- trusting - always co-operates
- defective - always defects
- responsive - identified by echoing the opponents last move
- provocative - identified by defecting when the opponent co-operated last time (this can be viewed as the negation of nice)

Note that there is no prototype to explicitly identify "forgiving" behavior, although the "responsive" prototype will effectively detect it.

Thus if we regard the interaction between two specified players P and Q as a sequence of pairs of actions (pi, qi), we are looking for instances of :

$(c, \_)$ to support P belonging to the co-operative prototype

$(d, \_)$ to support P belonging to the unco-operative prototype

$(c, d), (d, \_)$ or $(d, c), (c, \_)$ to support P belonging to the responsive prototype

$(c, c), (d, \_)$ to support P belonging to the provocative prototype

With a history of interactions generated by an n-IPD tournament, we can use these prototypes to model different players. Table 1 shows selected user models after 12 iterations of a 20 player tournament. The user models have converged after 12 iterations, i.e. the supports for the models belonging to each class do not change significantly after this iteration.

Table 1 : Protoype supports in user models derived from 12 iterations of an n-IPD tournament. Column headings refer to prototypes Trusting, Responsive, Provocative and Defective as defined above

Player	T	R	P	D
co-op	[0.9 1]	[0, 1]	[0. 0.09]	[0, 0.09]
uncoop	[0, 0.09]	[0, 1]	[0. 0.09]	[0.9 1]
tit-for-tat	[0.6,0.7]	[0.3,1]	[0,0.5]	[0.3,0.4]
random	[0.4,0.5]	[0.3,0.6]	[0.2,0.7]	[0.5,0.6]
respd	[0.2,0.3]	[0.1,0.8]	[0.1,0.9]	[0.7,0.8]

Prediction success rates vary between 40-60% for the random strategies, and 80-95% for the others.

The support for class membership is determined purely by the success of the class rule in predicting the behavior of the user. Thus a player showing a strategy of "always defect" appears as highly unco-operative but has [0, 1] support for membership in the "provocative" class, as the player never exhibits the provocative behavior pattern of mutual co-operation followed by a defection.

The models for the selected player are derived from all of the player's interactions; within each model, there are models for the other players derived from their interactions with the selected player.

Overall predictive success rates are good although the random strategies are difficult to predict, as would be expected. As a rule of thumb, if a user model is giving a success rate of less than 60% then a new prototype is required, either by a human expert or using an inductive logic-type approach to generate prototypes which predict the observed behavior more accurately. Note that these new prototypes need only explain the behavior in a subset of cases; they can give uncertain support for all outcomes when the prototype is not applicable.

The success rate of a user model can easily be calculated by comparing the predicted and observed behavior of the user. Clearly the user model changes with each new observation, and there is very little overhead in updating the user model. This approach depends on having a "good" set of prototypes, which are able to give a reasonable coverage of possible user behavior. It is assumed that a human expert is able to provide such a set; however, it is possible that new prototype behaviors could be generated by techniques such as inductive logic programming. This is an interesting avenue for future research.

4.2 User Models in the Telephone Assistant

The FILUM approach has also been applied to the prediction of user behaviour in the telephone assistant. The following assumptions have been made:

- The user model must decide whether to divert the call to voicemail or pass it through to be answered.
- The user is available to answer calls.
- Adaptive behaviour is based on knowing the correct decision after the call has finished.
- A log of past telephone activity and the current diary are available
- The identity of all callers is known.

A sample set of user prototypes is shown in Table 2.

Table 2. User Prototypes for the telephone assistant

Prototype	Identifying Characteristic	Behaviour
Talkative	none	always answer
Antisocial	none	always divert to voicemail
Interactive	recent calls or meetings involving this caller	answer
Busy	small proportion of free time in next working day (as shown by diary)	answer if caller is brief, otherwise divert to voicemail
Overloaded	small proportion of free time in next working day (as shown by diary)	divert to voicemail
Selective	none	answer if caller is a member of a selected group, else divert to voicemail
Regular	large proportion of calls answered at particular times of the day e.g. early morning	answer if this is a regular time

This approach assumes that all activities are planned and recorded accurately in an electronically accessible format. Other ways of judging a user's activity would be equally valid and may fit in better with a user's existing work pattern - for example the IPA system investigated the use of keyboard activity, gaze tracking and monitoring currently active applications on a computer. There is a need to model callers using a set of caller prototypes, since a user can react in different ways to different callers in a given set of circumstances. For example, the phone rings when you are due to have a meeting with the boss in five minutes. Do you answer if (a) the caller is the boss or (b) the caller is someone from the other side of the office who is ringing to talk about last night's football results while waiting for a report to print. The sample set of caller prototypes is shown in Table 3.

The user and caller prototypes are intended to illustrate the capabilities of the system rather than being a complete set; it is hoped that they are sufficiently close to real behaviour to make detailed explanation unnecessary.

Terms in italic are either fuzzy definitions that can be changed to suit a user. Note that support pairs indicate the degree to which a user or caller satisfies a particular prototype - this can range from uncertain (0 1) to complete satisfaction (1 1) or its opposite (0 0), through to any other probability interval.

Table 3. Caller Prototypes

Prototype	Identifying Characteristic
Brief	always makes short calls to user
Verbose	always makes long calls to user
Frequent	calls user frequently
Reactive	calls following a recent voicemail left by user
Proactive	calls prior to a meeting with user

A sample diary is shown in Figure 2. Note that the diary is is relatively empty at the beginning and end of the week but relatively full in the middle of the week. The *busy* and *overloaded* prototypes are written to be applicable when there is a small proportion of free time in the immediate future, that is, during the latter part of Tuesday and Wednesday.

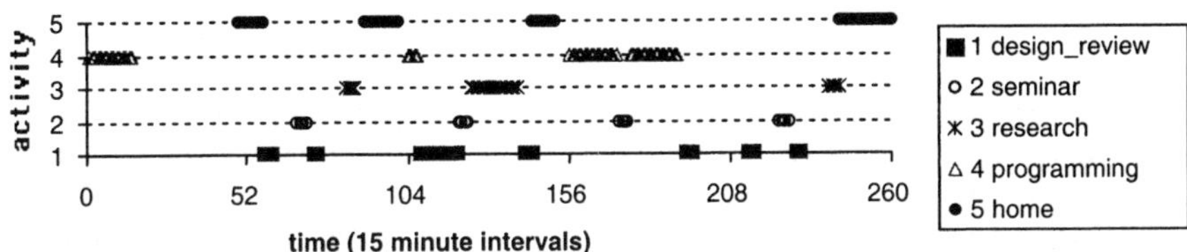

Fig. 2. Sample of diary. The window for the working day has been defined as 7:00 am - 8:00 pm, and diaried activities for each fifteen minute period within the window are shown; unassigned slots represent free time which can be used as appropriate at the time.

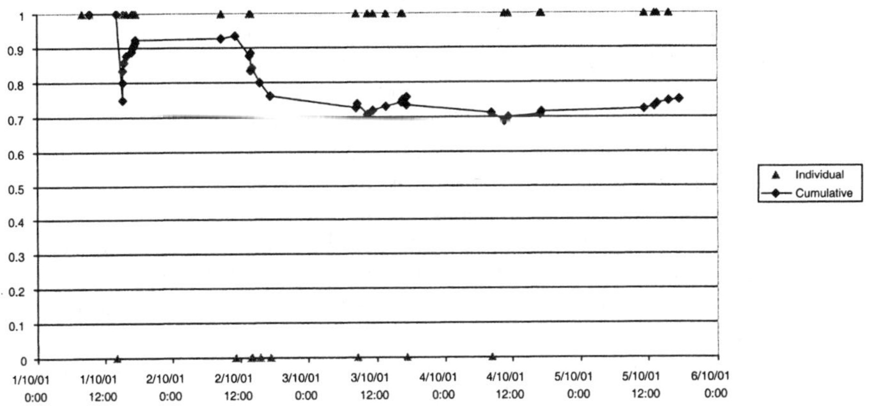

Fig. 3. Performance of the user model on individual calls (triangles, correct prediction =1, incorrect prediction = 0) and as a cumulative success rate (diamonds, continuous line).

Fig 3 shows the success rate of the user model in predicting whether a call should be answered or diverted to voicemail. The drop in performance on the second day

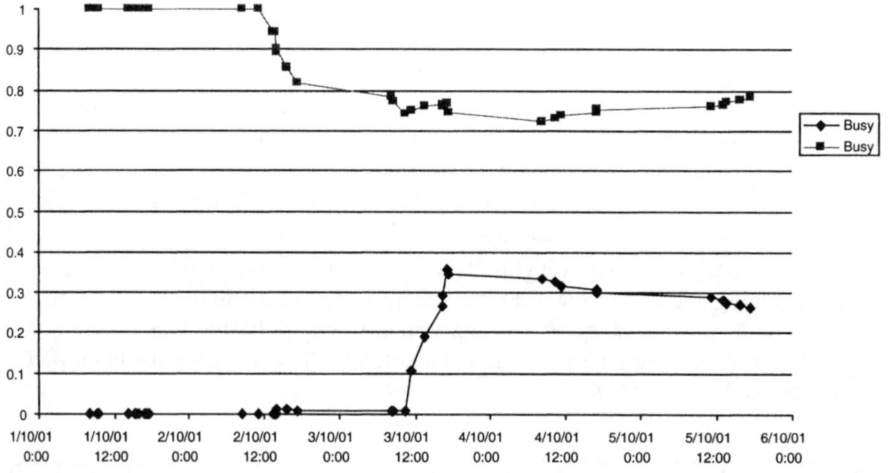

Fig. 4. Evolution of support pair for the "busy" prototype in the user model. The lower and upper lines show respectively the necessary and possible support for the observed behavior matching this prototype. At the start, the identifying characteristics (full diary) are not satisfied and support remains at unknown (0 1). In the middle of the week, conditions are satisfied. Initially the user does not behave as predicted by this prototype and possible support drops (i.e. support *against* increases); subsequently, the user behaves as predicted and necessary support increases. At the end of the week, once again the identifying characteristics are not satisfied and the prototype makes no predictions.

occurs because the *busy* and *overloaded* prototypes become active at this time, due to the full diary on Wednesday and Thursday. It takes a few iterations for the system to increase the membership of the user model in the *busy* and *overloaded* classes; once this has happened, the prediction rate increases again.

The necessary and possible supports for membership of the user model in the *busy* class is shown in Fig 4, where the evolution of support can be seen on the third and fourth days where this prototype is applicable. At the start, the identifying characteristics (full diary) are not satisfied and support remains at unknown (0 1). In the middle of the week, conditions are satisfied. Initially the user does not behave as predicted by this prototype and possible support drops (i.e. support *against* increases); subsequently, the user behaves as predicted and necessary support increases. At the end of the week, once again the identifying characteristics are not satisfied.

By repeating the week's data, there is relatively little change in the support pairs—this is suggestive that the learning has converged, although additional testing is necessary. Evolution of caller models can also be followed within the system, and good convergence to a stable caller model is observed. It should be emphasised that the supports for each prototype can be adjusted by the user at any stage. The user modelling software has been tested in several diary and call log scenarios, with good rates of prediction accuracy. Further testing is needed to investigate the user model evolution over longer periods.

User Models in the Email Assistant

The FILUM approach has also been also applied to the prediction of user behaviour in the email assistant. In many respects, this is a simpler problem than the telephone assistant – for example, in the latter there was a need to create caller models to predict the likely duration of a call. In the email assistant, this is no longer true as the required data is available from the fields of the email message, and only the user needs to be modeled. The aim is to predict the user's classification of emails into the following categories:

B = {READIMMEDIATE, READ24HOURS, READ5DAYS, READ4WEEKS, READNEVER}

on the basis of the message content and header fields i.e. to, from, date, time, subject and length.

The set of behaviors could easily be extended to more sophisticated actions such as "read this message now, then file in folder X" or "read now and answer within 24 hours". Rather than automatically moving emails to the predicted folder, a list of most likely destinations could be provided, with a facility for the user to enable automatic execution once s/he was happy with the accuracy of the system. Alternatively we could redirect mail to a different device, when appropriate. If a user is only available by telephone, a short and high priority (read immediately) message could be converted to speech and relayed to the user; if a handheld device had a greater capacity then all high priority messages could be redirected.

The set of behaviours was chosen to match those used in previous versions of the email assistant. The sample set of user prototypes is shown in Table 4.

Table 4. User Prototypes for the email assistant

Prototype	Identifying Characteristic	Behaviour
Always readNever	none	always classify as readNever
Always readImmed	none	always classify as readImmediate
Always 24Hr	none	always classify as read24Hours
Always 5Days	none	always classify as read5Days
Always 4Weeks	none	always classify as read4Weeks
Always uncertain	none	equal support (0 1) for all behaviours
PreviousBehav	previous emails received from this sender	repeat behaviour on last email from this sender
FixedTimes	current time matches set intervals	classify as readImmediate if current time is in the fuzzy intervals 8:00 – 10:00 or 16:00 – 18:00; otherwise inherit classifiaction method from PreviousBehav
Len/keywords	none	Uses a fuzzy "acceptable" message length and fixed fuzzy sets of words to identify mails to be read immediately ("good" keywords) or never ("bad" keywords). Otherwise returns (0 1) for all behaviours.

All testing was carried out on a data set of 94 email messages. Several different user behaviours were simulated on this set of emails, using different "verdict" files as shown in Table 5.

Table 5. Success rates of user model in email prediction against different scenarios

Short description	categorisation method	% Accuracy of user model after all emails processed
verdictAll24Hours	all messages categorised as "Read in 24 hours"	99
verdictAll4W	all messages categorised as "Read in 4 weeks"	99
verdictAll5Day	all messages categorised as "Read in 5 days"	99
verdictAllImmed	all messages categorised as "Read immediately"	99
verdictAllNo	all messages categorised as "Read never"	99
verdictByHand	manually assigned categories	80
verdictFromName	Sender contains "bt.co" -> immediate sales or market -> never yahoo or hotmail -> 5 days otherwise 24 hours	83
verdictFromTime	if hour = 8,9,16,17then immediate 7,10,18,19 then 24 hours otherwise random	81
verdictRandom	random categorisation	16
verdictRandom2	random categorisation	26

NB accuracy is given as the proportion of correct predictions. In some cases this involves a random choice, so results may vary slightly from run to run. Evolution of prediction accuracy as emails are presented is also shown below for the sixth and seventh cases in the table. The first five and last two cases are to test the user model's behavior in extremes; the remaining three cases are more realistic.

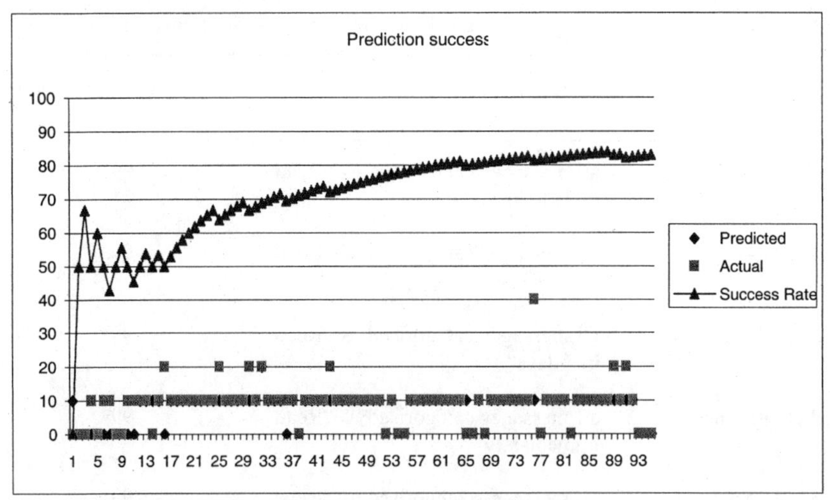

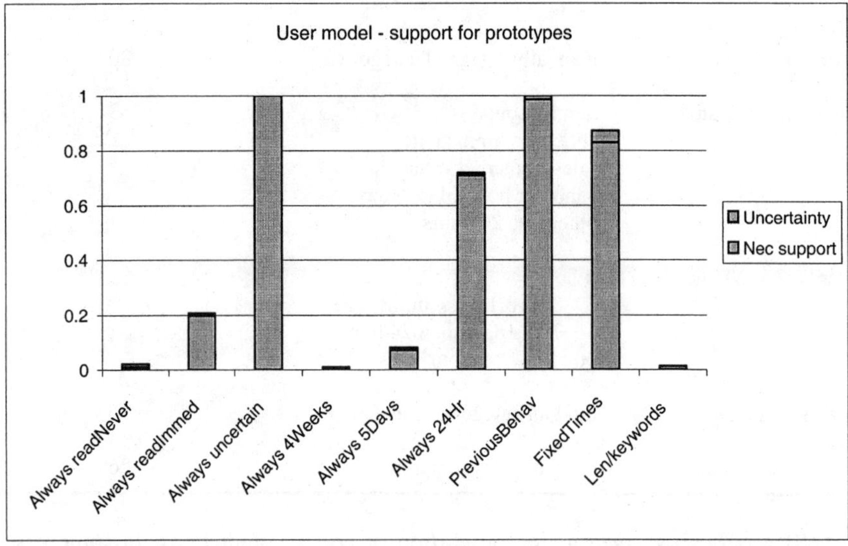

Fig 5   *Prediction success and user model for verdicts according to sender name*

It is interesting to note that whilst success rates are roughly the same in these cases, the user models are very different, e.g. the prototype based on message length and keywords (FILUMUserPrototypeLW) has very little support in the *ver-*

dictFromName case, but high support in the *verdictByHand* scenarios (and *verdictFromTime*, although this is not shown.

Because the prototypes almost always return a prediction, there is little uncertainty in the user models - with more sophisticated prototypes, there would be a greater difference between necessary and possible supports.

The predicted and actual behaviours are shown on the success plots - 0 corresponds to READIMMEDIATE, ... 40 corresponds to READNEVER.

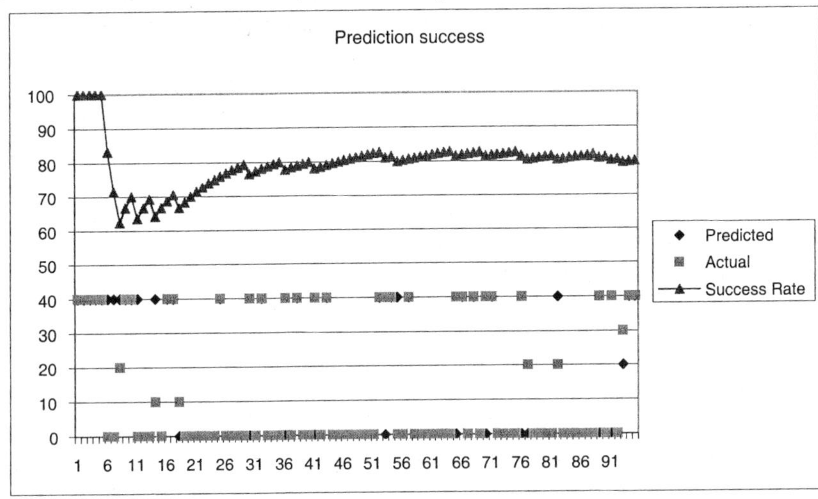

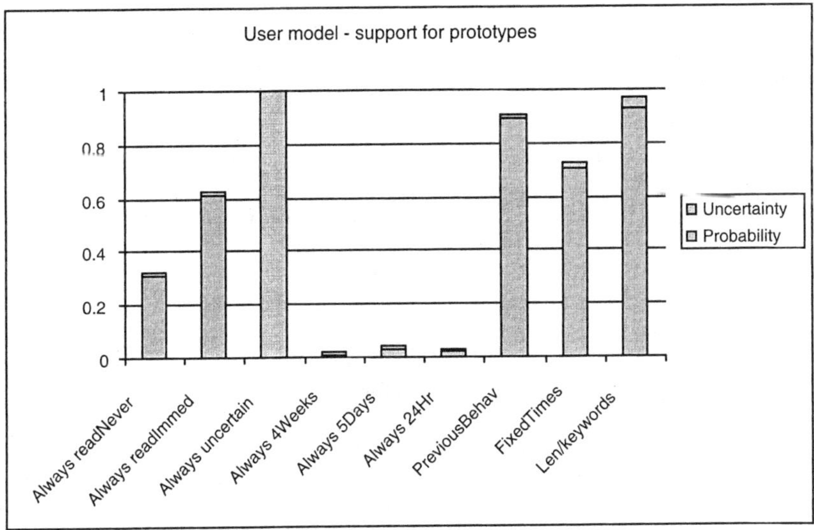

Fig 6   *Prediction success and user model for realistic verdicts (generated by hand)*

Summary

Many research projects have considered the use of machine learning for user modelling. In general, the problems identified include
- collection of training data
- length of time to train
- changes in profile cannot be incorporated without costly retraining

The work reported here extends previous work on FILUM and addresses the second and third issues listed above.

The aim of user modelling is to increase the quality of interaction—this is almost always a subjective judgement, and it can be difficult to discuss the success (or otherwise) of user modelling. We have developed an experimental testbed based on the iterated prisoner's dilemma, allowing generation of unlimited data. Prediction success rates vary between 80-95% for non-random behaviours in the testbed, and accuracy of over 80% has been obtained in a series of simulated tests of the telephone and email assistants.

The user model changes with each observation, and there is very little overhead in updating the user model. This approach depends on having a "good" set of prototypes, giving reasonable coverage of possible user behaviour. It is assumed that a human expert is able to provide such a set; however, it is possible that (for example) inductive logic programming could generate new prototype behaviours. This is an interesting avenue for future research.

Finally, there is a need for more rigorous testing – although it is difficult to obtain 'real" data due to privacy issue.

References

1. Azvine, B., D. Djian, K.C. Tsui, and W. Wobcke, *The Intelligent Assistant: An Overview*. Lecture Notes in Computer Science, 2000(1804): p. 215-238.
2. Azvine, B. and W. Wobcke, *Human-centred intelligent systems and soft computing*. Bt Technology Journal, 1998. **16**(3): p. 125-133.
3. Baldwin, J.F., *The Management of Fuzzy and Probabilistic Uncertainties for Knowledge Based Systems*, in *Encyclopedia of AI*, S.A. Shapiro, Editor. 1992, John Wiley. p. 528-537.
4. Baldwin, J.F., J. Lawry, and T.P. Martin, *A mass assignment based ID3 algorithm for decision tree induction*. International Journal of Intelligent Systems, 1997. **12**(7): p. 523-552.
5. Baldwin, J.F., J. Lawry, and T.P. Martin. *Mass Assignment Fuzzy ID3 with Applications*. in *Fuzzy Logic - Applications and Future Directions*. 1997. London: Unicom.
6. Baldwin, J.F. and T.P. Martin, *Fuzzy classes in object-oriented logic programming*, in *Fuzz-Ieee '96 - Proceedings of the Fifth Ieee International Conference On Fuzzy Systems, Vols 1-3*. 1996. p. 1358-1364.
7. Baldwin, J.F. and T.P. Martin. *A General Data Browser in Fril for Data Mining*. in *EUFIT-96*. 1996. Aachen, Germany.

8. Baldwin, J.F., T.P. Martin, and B.W. Pilsworth, *FRIL - Fuzzy and Evidential Reasoning in AI*. 1995, U.K.: Research Studies Press (John Wiley). 391.

9. Bezdek, J.C., *Fuzzy Models*. IEEE Trans. Fuzzy Systems, 1993. **1**(1): p. 1-5.

10. Bezdek, J.C., *What is Computational Intelligence?*, in *Computational Intelligence - Imitating Life*, C.J. Robinson, Editor. 1994, IEEE Press. p. 1-12.

11. Billsus, D. and M.J. Pazzani. *A Hybrid User Model for News Story Classification*. in *User Modeling 99*. 1999. Banff, Canada.

12. Bushey, R., J.M. Mauney, and T. Deelman. *The Development of Behavior-Based User Models for a Computer System*. in *User Modeling 99*. 1999. Banff, Canada.

13. Case, S.J., N. Azarmi, M. Thint, and T. Ohtani, *Enhancing e-Communities with Agent-Based Systems*. IEEE Computer, 2001. **33**(7): p. 64.

14. Hermens, L.A. and J.C. Schlimmer, *A Machine-Learning Apprentice for The Completion of Repetitive Forms*. IEEE Expert-Intelligent Systems & Their Applications, 1994. **9**(1): p. 28-33.

15. Horvitz, E., *Uncertainty, action, and interaction: in pursuit of mixed-initiative computing*. Ieee Intelligent Systems & Their Applications, 1999. **14**(5): p. 17-20.

16. Horvitz, E., et al. *The Lumière Project: Bayesian User Modeling for Inferring the Goals and Needs of Software USers*. in *Fourteenth Conference on Uncertainty in Artificial Intelligence*. 1998. Madison, WI.

17. Langley, P. *User Modeling in Adaptive Interfaces*. in *User Modeling 99*. 1999. Banff, Canada.

18. Lashkari, Y., M. Metral, and P. Maes, *Collaborative Interface Agents*. Proceedings of the National Conference on Artificial Intelligence, 1994: p. 444.

19. Lau, T. and E. Horvitz. *Patterns of Search: Analyzing and Modeling Web Query Refinement*. in *User Modeling 99*. 1999. Banff, Canada.

20. Linton, F., A. Charron, and D. Joy, *Owl: A Recommender System for Organization-Wide Learning*. 1998, MITRE Corporation. p. http://www.mitre.org/technology/tech_tats/modeling/owl/Coaching_Software_Skills.pdf.

21. Martin, T.P. *Incremental Learning of User Models - an Experimental Testbed*. in *IPMU 2000*. 2000. Madrid.

22. Martin, T.P. and B. Azvine. *Learning User Models for an Intelligent Telephone Assistant*. in *IFSA-NAFIPS 2001*. 2001. Vancouver: IEEE Press.

23. Muggleton, S., *Inductive Logic Programming*. 1992: Academic Press. 565.

24. Nunes, P. and A. Kambil, *Personalization? No Thanks*. Harvard Business Review, 2001. **79**(4): p. 32-34.

25. Pazzani, M. and D. Billsus, *Learning and revising user profiles: The identification of interesting Web sites*. Machine Learning, 1997. **27**(3): p. 313-331.

26. Pohl, W., *Logic-based representation and reasoning for user modeling shell systems*. User Modeling and User-Adapted Interaction, 1999. **9**(3): p. 217-282.

27. Quinlan, J.R., *Induction of Decision Trees*. Machine Learning, 1986. **1**: p. 81-106.

28. Quinlan, J.R., *C4.5: Programs for Machine Learning*. 1993: Morgan Kaufmann.

29. Rossiter, J.M., T. Cao, T.P. Martin, and J.F. Baldwin. *A Fril++ Compiler for Soft Computing Object-Oriented Logic Programming*. in *IIZUKA2000, 6th International Conference on Soft Computing*. 2000. Japan.

30. Rossiter, J.M., T.H. Cao, T.P. Martin, and J.F. Baldwin. *User Recognition in Uncertain Object Oriented User Modelling*. in *Tenth IEEE International Conference On Fuzzy Systems (Fuzz-IEEE 2001)*. 2001. Melbourne.

31. Salton, G. and G. Buckley, *Term-weighting Approaches in Automatic Text Retrieval.* Information Processing and Management, 1988. **24**: p. 513-523.

32. Segal, R.B. and J.O. Kephart, *MailCat: An Intelligent Assistant for Organizing E-Mail.* Proceedings of the National Conference on Artificial Intelligence, 1999: p. 925-926.

33. Segal, R.B. and J.O. Kephart. *Incremental Learning in SwiftFile.* in *International conference on machine learning.* 2000. Stanford, CA: Morgan Kaufmann Publishers.

34. W3C, *Composite Capability/Preference Profiles (CC/PP): A user side framework for content negotiation,* http://www.w3.org/TR/NOTE-CCPP/#2. 1999.

35. W3C, *Metadata and Resource Description,* http://www.w3.org/Metadata/. 1999.

36. Webb, G., *Preface to Special Issue on Machine Learning for User Modeling.* User Modeling and User-Adapted Interaction, 1998. **8**(1): p. 1-3.

37. Yiming, Z. *Fuzzy User Profiling for Broadcasters and Service Providers.* in *Computational Intelligence for User Modelling.* 1999. Bristol, UK.

Intelligent Web Searching Using Hierarchical Query Descriptions

Ronald R. Yager

Machine Intelligence Institute, Iona College, New Rochelle, NY 10801.
E-Mail: yager@panix.com

Abstract. We describe a document retrieval language which enables a user to better represent their requirements with respect to the desired documents to be retrieved. This language allows for a specification of the interrelationship between the desired attributes using linguistic quantifiers. This framework also supports a hierarchical formulation of queries. These features allow for an increased expressiveness in the queries that be handled by a retrieval system.

Keywords: document retrieval, fuzzy sets, web searching, OWA aggregation

1. Introduction

With the explosive growth of the Internet the need to effectively retrieve documents satisfying user requirements has emerged as one of the most important technological problems we are facing. We use the term document to generically indicate a broad spectrum of objects which include web–pages, reports, books, movie titles etc. At the heart of the current problem with retrieval systems is the ability to effectively express search requirements in a way that can be "understood" by the computer. Here we need a bridge to translate between the linguistic type expression favorite by humans and the formal representations needed for computer manipulation. Fuzzy logic and the related disciplines of approximate reasoning [1] and computing with words [2] have been primarily developed to provide such a bridge.

For the most part, the current retrieval paradigm involves a situation in which a document is "represented." Essentially this representation consists of a decomposition of a document into attributes or features on which the document can be scored. These attributes can be based upon ideas as simple as the appearance of a word or phase in the document or can involve complex processing of the document using notions like frequency of occurrence. When searching, a user expresses their requirements in terms of a subset of these primary attributes. The overall evaluation of a document is obtained by an aggregation of the scores of these specified attributes. The method of aggregation used can be seen to inherently reflect an expression of a desired interrelationship between the specified attributes. Thus the aggregation can be seen as a kind of *recomposition* of the document from its attributes. Typical examples of aggregation are the simple average and those based upon logical connections, *anding* and *oring*.

One way to improve retrieval systems is to provide a wide class of aggregation operations to enable the system to implement sophisticated interactions and thereby allow the user increased expressiveness in specifying their desires. In addition any extension of aggregation options would greatly benefit from a strong correspondence between formal methods of aggregation and natural language specification. One goal of the agenda of computing with words is to provide such a capability.

Here we shall describe a query language for evaluating user specifications, which we call this language Hi-RET. This language will make considerable us of the **O**rdered **W**eighted **A**veraging (OWA) operators [3]. These OWA operators provide a large class of aggregation methods, which have a natural correspondence between a linguistic specification, and a formal method of aggregation. In addition the expressive capability of the query language will be enhanced by the use of a hierarchical structure to represent queries.

As will become clear this approach is extremely generic and can be used as an integral part of many different types search and retrieval systems.

2. OWA Aggregation Operators

Central to any document retrieval system is the need to aggregate scores. In order to provide a very general framework to implement aggregations, we shall use the OWA operators [3].

Definition: An Ordered Weighted Averaging, OWA, operator of dimension n is a mapping which has an associated weighting vector W in which $w_j \in [0,1]$ and $\Sigma_j\, w_j = 1$ and where

$$F(a_1, a_2,..., a_n) = \Sigma_j\, w_j\, b_j$$

with b_j being the jth largest of the a_i.

By appropriately selecting the vector W we can implement various different types of aggregation.

We now consider a basic application of the OWA operator in a document retrieval system. Assume $A_1, A_2, \ldots, A_n$ is a collection of attributes of interest to a user of a retrieval system. For any document d, let $A_i(d) \in [0,1]$ indicate the degree to which document d satisfies the property associated with attribute A_i. Using the OWA operator, we can obtain an overall valuation of document d as

$$Val(d) = F_w(A_i(d), A_2(d), \ldots, A_n(d)).$$

A number of different approaches have been suggested for obtaining the weighting vector W to use in any given application [4]. For our purposes, that of developing a user friendly document retrieval system, we shall describe an approach based upon the idea of linguistic quantifiers.

The concept of linguistic quantifiers was originally introduced by Zadeh [5]. A proportional linguistic quantifier is a term corresponding to a proportion of objects. Examples of this are *most, at least half, all, about 0.3.* Let Q be a proportional linguistic quantifier we can represent this as a fuzzy subset Q over I = [0, 1] in which for any proportion $r \in I$, Q(r) indicates the degree to which r satisfies the concept indicated by the quantifier Q.

In [3] Yager showed how we could use a linguistic quantifier to obtain a weighting vector W associated with an OWA aggregation. For our purposes we shall restrict ourselves to regularly increasing monotonic (RIM) quantifiers. A fuzzy subset Q: I $\rightarrow$ I is said to represent a RIM linguistic quantifier if: **1.** Q(0) = 0, **2.** Q(1) = 1 and **3.** If $r_1 > r_2$ then $Q(r_1) \geq Q(r_2)$ (monotonic). These RIM quantifiers model the class in which an increase in proportion results in an increase in compatibility to the linguistic expression being modeled. Examples of these types of quantifiers are *at least one, all, at least α %, most, more than a few, some.*

Assume Q is a RIM quantifier. Then we can associate with Q an OWA weighting vector W such that for j = 1 to n

$$w_j = Q(j/n) - Q((j - 1)/n)$$

Using this approach we can obtain a weighting vector directly from the linguistic expression of the quantifier

Let us look at the situation for some prototypical quantifiers. The quantifier *for all* is shown in Fig 1.

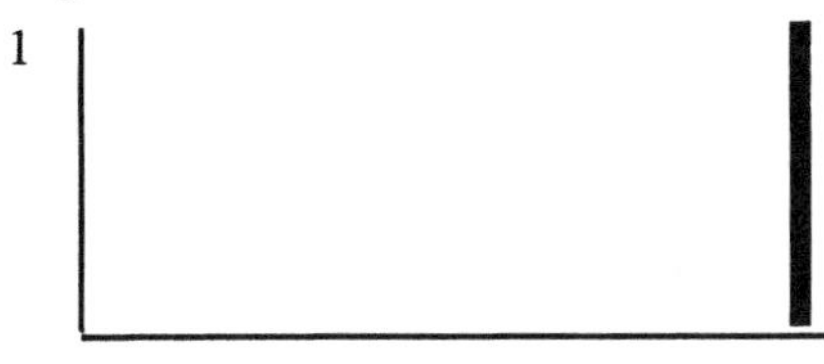

Fig. 1. Linguistic quantifier "for all"

In this case we get that $w_j = 0$ for $j \neq n$, and $w_n = 1$, $W = W_*$ In this case we get as our aggregation the minimum of the aggregates. We also recall that the quantifier *for all* corresponds to the logical "anding" of all the arguments

In Fig. 2 we see the existential quantifier, *at least one*, this is the same as *not none*.

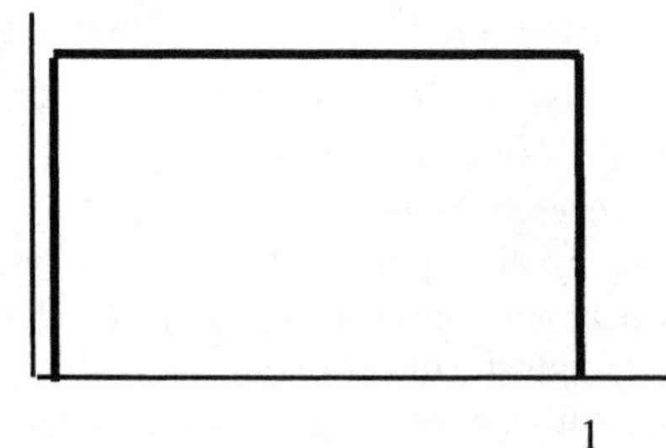

Fig. 2. Linguistic quantifier "not none"

In this case $w_1 = 1$ and $w_j = 0$ for $j > 1$, $W = W^*$. This can be seen as inducing the maximum aggregation. It is recalled this quantifier corresponds to a logical *oring* of the arguments

Figure #3 is seen as corresponding to the quantifier *at least α*. For this quantifier $w_j = 1$ for j such that $(j - 1)/n < \alpha \leq j/n$ and $w_i = 0$ for all other.

Another quantifier is one in which $Q(r) = r$ for $r \in [0, 1]$. For this quantifier we get $w_j = \dfrac{j}{n} - \dfrac{j-1}{n} = \dfrac{1}{n}$. for all j. This gives us the simple average. We shall denote this quantifier as *some*.

As discussed by Yager [6] one can consider parameterized families of quantifiers. For example consider the parameterized family $Q(r) = r^\rho$, where $\rho \in [0, \infty]$. Here if $\rho = 0$, we get the existential quantifier; when $\rho = \infty$, we get the quantifier *for all* and when $\rho = 1$, we are get the quantifier *some*. In addition for the case in which $\rho = 2$, $Q(r) = r^2$, we get one possible interpretation of the quantifier *most*.

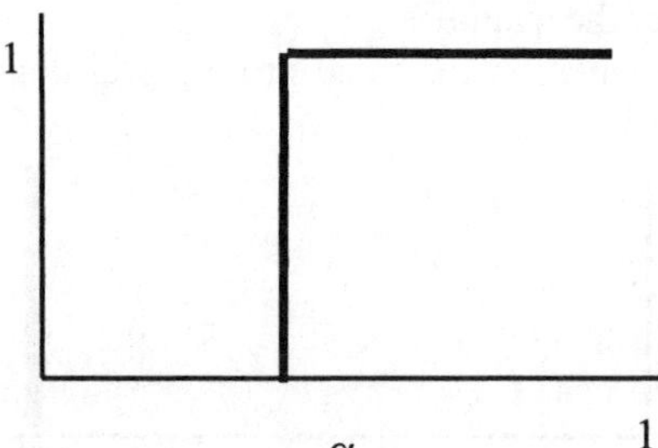

Fig. 3. Linguistic quantifier "at least α"

We are now in a position to address the issue of obtaining the OWA weighting vector to be used in a search in a user-friendly document retrieval system. In constructing such a user-friendly system we shall make available to the user a vocabulary, $Q = \{Q_1, Q_2, ..., Q_q\}$, of linguist expressions, each corresponding to a linguistic quantifier. When posing a query, the user, after specifying a collection of attributes of interest $(A_1, A_2, ... , A_n)$, will be prompted to also specify one of the linguistic quantifiers in Q as guiding the query formation. Transparent to the user is the association of each of the linguistic terms Q_i in Q with a representative fuzzy subset, $Q_i \Leftrightarrow Q_i$, and the process of converting this fuzzy subset into an OWA weighting vector based on the formulation

$$w_j = Q_i(\frac{j}{n}) - Q_i(\frac{j-1}{n}).$$

One of the elements in Q should be designated as the default quantifier; this is the one that is to be used when the user specifies no selection. Perhaps the most appropriate choice for this is the average quantifier $w_j = 1/n$, which corresponds to the linguistic expression *some*. Thus thus the implicit interpretation of the user providing a collection of attributes is the statement "I want *some* of these attributes satisfied by a retrieved document."

The process of actually determining the set of available quantifiers, Q, while of great importance is beyond our scope here and should benefit from some empirical research trying to match users perceptions and vocabulary with fuzzy sets. It should also deal the issue of finding a selection of terms to cover a spectrum wide enough to allow users to appropriately express their desires.

Based on the ideas so far presented here we can introduce the idea of a ***query module***: $\langle A_1, A_2,.... A_n: Q\rangle$, consisting of a collection of attributes of interest and a linguistic quantifier indicating the proportion of the relevant attributes we desire. Implicit in this query module is the fact the linguistic expression Q is essentially defining a weighting vector W for an OWA aggregation. For any document d we can find its satisfaction to the user query, Val(d), by calculating

$$F_Q(A_1(d), A_2(d), ..., A_n(d))$$

which indicates the OWA aggregation of the documents satisfaction to the criteria, $A_i(d)$, using a weighting vector W determined from the quantifier Q.

3. Introducing Importances In Queries

In the preceding we have indicated a query object to be a module consisting of a collection of attributes of interest and a quantifier Q indicating a mode of interaction between the attributes. Implicit in the preceding is the equal treatment of all desired attributes. Often a user may desire to ascribe different weights or importances to the different attributes. In the following we shall consider the introduction of importance weights into our procedure.

Let $\alpha_i \in [0, 1]$ be a value associated with an attribute indicating the importance associated with the attribute. We shall assume the larger α_i the more important and let $\alpha_i = 0$ stipulate zero importance. With the introduction of these weights we can consider a more general query object, $<A_1, A_2,...., A_n: M: Q>$. Here as before, the A_i are a collection of attributes and Q is a linguistic quantifier, however, here M is an n vector whose component $m_j = \alpha_j$, the importance associated with A_j.

Our goal now is to calculate the overall score $Val(d)$ associated with a document d, we shall denote this

$$Val(d) = F_{Q/M}(A_1(d), A_2(d),...., A_n(d))$$

Here $F_{Q/M}$ indicates an OWA operator. Our agenda here will be to first find an associated OWA weighting vector, $W(d)$, based upon Q and M. Once having obtained this vector we calculate $Val(d)$ by the usual OWA process

$$Val(d) = W(d)^T B(d) = \sum_j w_j(d)\, b_j(d)$$

Here $b_j(d)$ is denoting the jth largest of the $A_i(d)$ and $w_j(d)$ is the jth component of the associated OWA vector, $W(d)$

We now describe a procedure, developed in [7], that shall be used to calculate the weighting vector, $w_j(d)$. The first step is to calculate the ordered argument vector $B(d)$ such that $b_j(d)$ is the jth largest of the $A_i(d)$. Furthermore, we shall let μ_j denote the importance weight associated with the attribute that has the jth largest value. Thus if $A_5(d)$ is the largest of the $A_i(d)$, then $b_1(d) = A_5(d)$ and $u_1 = \alpha_5$. Our next step is to calculate the OWA weighting vector $W(d)$. We obtain the associated weights as

$$w_j(d) = Q\left(\frac{S_j}{T}\right) - Q\left(\frac{S_{j-1}}{T}\right)$$

where $S_j = \sum_{k=1}^{j} u_k$ and $T = S_n = \sum_{k=1}^{n} u_k$. Thus T is the sum of all the importances and S_j is the sum of the importances of the jth most satisfied attributes. Once having obtained these weights we can then obtain the aggregated value by the usual method, $B^T W$. The following example will illustrate the use of this technique.

Example: We assume four criteria of interest to the user: A_1, A_2, A_3 and A_4. Their importances are $\alpha_1 = 1$, $\alpha_2 = 0.6$, $\alpha_3 = 0.5$ and $\alpha_4 = 0.9$. From this we get

$$T = \sum_{k=1}^{4} \alpha_k = 3.$$ We shall assume the quantifier guiding this aggregation is *most*

which is defined by $Q(r) = r^2$. Consider document x whose satisfaction to each of the attributes is: $A_1(x) = 0.7$, $A_2(x) = 1$, $A_3(x) = 0.5$, $A_4(x) = 0.6$

Our objective here is to obtain the valuation of the document with respect to this query structure. In this case the ordering of the criteria satisfactions gives us:

	b_j	u_j
A_2	1	0.6
A_1	0.7	1
A_4	0.6	0.9
A_3	0.5	0.5

Calculating the weights associated with x, which we denote $w_i(x)$, we get

$$w_1(x) = Q(\frac{0.6}{3}) - Q(0) = 0.04$$

$$w_2(x) = Q(\frac{1.6}{3}) - Q(\frac{0.6}{3}) = 0.24$$

$$w_3(x) = Q(\frac{2.5}{3}) - Q(\frac{1.6}{3}) = 0.41$$

$$w_4(x) = Q(\frac{2}{3}) - Q(\frac{1.6}{3}) = 0.31$$

Using this we calculate Val(x)

$$\text{Val}(x) = \sum_{j=1}^{4} w_j(x)\, b_j = (.04)(1) + (.24)(.7) + (.41)(.6) +$$

$(.31)(.5) = 0.609$

More details with respect to the properties of this methodology can be found in [7], however here we shall point one special case. Consider the case of the quantifier *some*, $Q(r) = r$. For this quantifier $w_j = \dfrac{\alpha_j}{T}$ and hence $\text{Val}(d) = \dfrac{1}{T}$

$\sum_{j=1}^{n} \alpha_j\, a_j$ This is simply the weighted average of the attributes. Thus the

weighted average of the criteria satisfaction is a special case of our approach.

4. Concepts and Hierarchies

In the preceding we have considered a retrieval system in which we have a collection of primary attributes A_i, $i = 1$ to n. These attributes are characterized by the fact that for any $d \in D$ we we can directly access $A_i(d) \in [0,1]$. We shall now associate with our retrieval system a slightly more general idea called a *concept*. We define a concept as **an object whose measure of satisfaction can be obtained for any document in D**. It is clear that the attributes are examples of concepts, they are special concepts in that their values are directly accessible from the document base.

Consider now a query object of the type we have previously introduced. This is an object of the form $<A_1, A_2, ..., A_q: M: Q>$. As we have indicated, the satisfaction of this object for any $d \in D$ can be obtained by our aggregation process. In the light of this observation we can consider this query object to be a concept, with

$$Con = <A_1, A_2, ..., A_q: M: Q>$$

then

$$Con(d) = F_{Q/M}(A_1(d), A_2(d), ..., A_q(d)).$$

Thus a query object is a concept. A special concept is an individual attribute,

$$Con = <A_j : M: Q> = A_j,$$

we shall call these atomic concepts. These atomic concepts require no Q or M, but just need A_j specification.

Let us look at the query object type concept in more detail. The basic components in these objects are the attributes, the A_j. However, from a formal point of view, the ability to evaluate the query objects-concept is based upon the fact that for each A_j, we have a value for any d, $A_j(d)$. As we have just indicated, a concept also has this property, for any d we can obtain a measure of its satisfaction. This observation allows us to extend our idea of query object-concept to allow for concepts whose evaluation depends upon other concepts. Thus we can consider concepts of the form

$$Con = <Con_1, Con_2, ..., Con_1: M: Q>.$$

Here each of the Con_j are concepts used to determine the satisfaction of Con by an aggregation process where M determines the weight of each of the participating concepts and Q is the quantifier guiding the aggregation of the component concepts.

The introduction of concepts into the query objects results in a hierarchical structure for query formation. Essentially, we unfold until we end up with queries made up of just attributes, which we can directly evaluate. The following simple examples illustrate the structure.

Example: Consider here the query

$$(A_1 \text{ and } A_2 \text{ and } A_3) \text{ or } (A_3 \text{ and } A_4).$$

We can consider this as a concept

$$<Con_1, Con_2 : M: Q>.$$

Here Q is the existential quantifier and $M = \begin{bmatrix} 1 \\ 1 \end{bmatrix}$. In addition

$$Con_1 = <A_1, A_2, A_3: M_1: Q_1>$$

$$Con_2 = <A_3, A_4 : M_2: Q_2>$$

Here $Q_1 = Q_2 = all$ and $M_1 = \begin{bmatrix} 1 \\ 1 \\ 1 \end{bmatrix}$ and $M_2 = \begin{bmatrix} 1 \\ 1 \end{bmatrix}$. This query can be expressed in a hierarchical fashion as shown in Fig. 4.

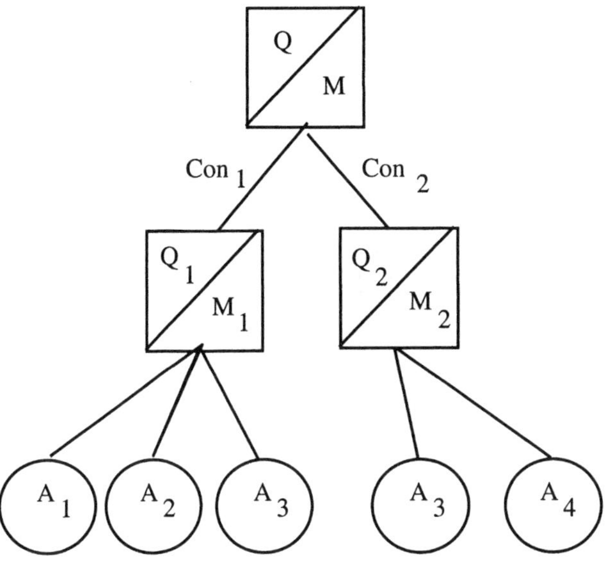

Fig. 4. Hierarchical Formulation of Query

5. Hierarchical Querying in Information Retrieval

Using the ideas discussed in the preceding we shall describe a hierarchical querying framework that can be used for information retrieval, we shall call this the Hierarchical Document Retrieval Language and use the acronym HI-RET. This language can be used to retrieve documents from the Internet or intranet type environment or any other computer-based environment.

Associated with any implementation of this language is a set $A = \{A_1, A_2, \ldots, A_n\}$ of atomic attributes, words or concepts. These atomic concepts are such that for any document d in D and any concept A_j in A we have directly available the value $A_j(d) \in [0,1]$, the satisfaction of attribute A_j by document d. This information can be stored in a database such that each record is a tuple consisting of the values $A_j(d)$ for $j = 1$ to n and the address of document d. Essentially each object can be viewed as a n vector whose components are the $A_j(d)$.

In addition to the attributes we also assume associated with any implementation of HI-RET is a vocabulary of linguistic quantifiers, $Q = \{Q_1, Q_2, ..., Q_q\}$ available to the searcher. Within this set of quantifiers we should surely have the quantifiers *all*, *any*. and *some*. One quantifier should be designated as the default quantifier. Perhaps a best choice for this is the quantifier *some*. Transparent to the user is a fuzzy subset Q_i on the unit interval associated with each linguistic quantifier Q_i. This fuzzy subset is used to generate the associated weights used in the aggregation.

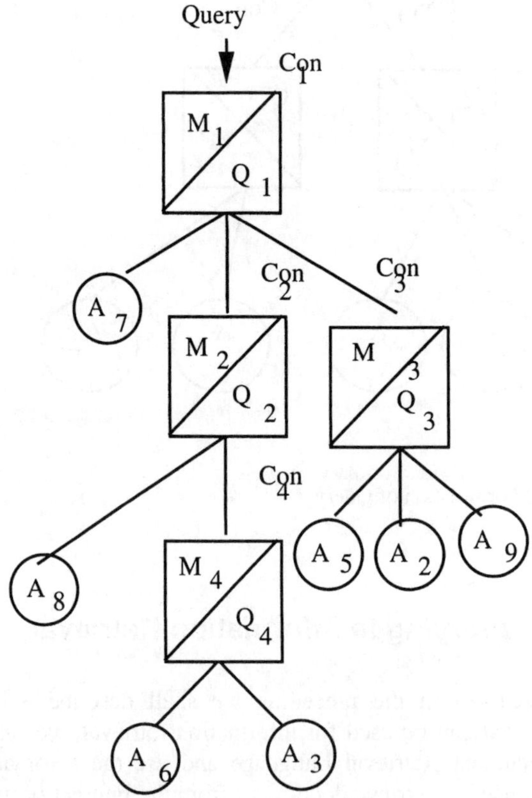

Fig. 5. Prototypical Query in HI-RET

The user indicates a query to the document retrieval system by the specification of a "concept" that the user desires satisfied. The user is asked to "define" this concept by expressing it in terms of a query object, $\langle C_1, C_2,...,C_n: M: Q\rangle$, consisting of a group of components C_j, an importance weight associated with each of the components, M, and a quantifier, Q, expressing the imperative for aggregating the components. The specifications of the importance weights as well as quantifier are optional. If the weights are not expressed, then by default they are assumed to have importance one, if the quantifier is not expressed, then the designated default quantifier is assumed. For each of the components of the query that are not an atomic object the searcher is asked to provide a definition. This process is continued until the complete hierarchy defining the query is formulated. It is noted that this hierarchy is a tree like structure in the leaves are atomic components. Figure #5 shows a prototypical example of such a query.

Once having obtained the HI-RET expansion of a query as in Figure#5, we can then use our aggregation methods to evaluate the query for each document. For example, in the case of figure #5 for document d we have

$$Con_4(d) = F_{M4/Q4}(A_6(d), A_3(d))$$

$$Con_3(d) = F_{M3/Q3}(A_2(d), A_5(d), A_9(d))$$

$$Con_2(d) = F_{M2/Q2}(Con_4(d), A_8(d))$$

$$Con_1(d) = F_{M1/Q1}(A_7(d), Con_2(d), Con_3(d))$$

6. Thesaurus

Modern document retrieval systems use thesaurus [8]. The function of a thesaurus is to allow for the consideration of satisfaction to synonyms when trying to evaluate the satisfaction of an attribute by a document. This functioning of a thesaurus can be modeled within the framework which we presented. Let $A = \{A1, A_2, ... A_3\}$ be a collection of concepts corresponding to our atomic attributes. A thesaurus can be represented as a relationship T on $A \times A$ such that for each pair $A_i, A_j \in A$, $T(A_i, A_j) \in [0, 1]$ indicates the degree of similarity that A_j has to the A_i. The basic properties of the thesaurus T are:

1. Identity: $T(A_i, A_i) = 1$ for all A_i,

2. Symmetry: $T(A_i, A_j) = T(A_j, A_i)$.

If A_j is an atomic attribute we shall let $\hat{A_j}$ indicate the concept corresponding to the extended definition of the original attribute A_j. We define $\hat{A_j}$ as follows

$$\hat{A_j} = <EX[A_j|A_1], EX[A_j|A_2], ..., EX[A_j|A_n]: -: "Or">,$$

where $EX[A_j|A_i]$, the extension of A_j by A_i, is a concept defined as

$$EX[A_j|A_i] = <T(A_j, A_i), A_i : -:"And">.$$

The concept $EX[A_j|A_i]$ corresponds to the statement A_i is similar to A_j and A_i is satisfied by the document. We see that the concept $EX[A_j|A_i]$ has two components, A_i and $T(A_j, A_i)$, it uses equal importances and uses the quantifier *and*. The component $T(A_j, A_i)$ can be directly accessible from the definition of the thesaurus and as such can be viewed as an atomic component. We further note that $\hat{A}_j$, a concept whose components are the n concepts $EX[A_j|A_i]$, uses the default importance and uses the quantifier *or*.

Let us look at the form of $\hat{A}_j(d)$ resulting from this structure. First we see that $EX[A_j|A_i]$ evaluate for document d, $EX[A_j|A_i](d) = F_Q[T(A_j, A_i), A_i(d)]$ where Q = "and." The importances are assumed to be equal. This gives us that
$$EX[A_j|A_i](d) = Min[T(A_j, A_i), A_i(d)] = T(A_j, A_i) \wedge A_i(d)$$
Next we see that
$$\hat{A}_j(d) = F_{"any"}(EX[A_j|A_1](d), EX[A_j|A_2](d),...., EX[A_j|A_n](d))$$
this gives us
$$\hat{A}_j(d) = \underset{i=1}{\overset{n}{Max}} (EX[A_j|A_i](d)) = \underset{i=1}{\overset{n}{Max}} (T(A_j, A_i) \wedge A_i(d)]).$$

We note since $T(A_j, A_j) = 1$, then
$$\hat{A}_j(d) = A_j(d) \vee \underset{i \neq j}{Max} [T(A_j, A_i) \wedge A_i(d)]$$

Having defined $\hat{A}_j(d)$ we can now explain how to include the thesaurus in the document evaluation process. Given a query addressed to a retrieval system we expand it as usual which results in a hierarchical query structure such as that shown in figure #5. To include the thesaurus we now replace the terminal attributes, the A_j, by their extended versions, $\hat{A}_j$.

When using the HI-RET query language the user can indicate whether they want to use the thesaurus or not.

7. Enhancing the Use of Importances

We have described a method for evaluating the overall score of documents based upon a query object of the form $<A_1, A_2, ..., A_n: M: Q>$. In this query object the jth component of the vector M, α_j, indicates the importance weight associated with the attribute A_j. Implicit in our preceding discussion was the assumption that the weight α_j was explicitly provided by the user. This assumption is not necessarily required. It is possible for the weight associated with an attribute A_j be determined by some property of the document being evaluated. Thus let B_j be some measurable attribute associated with the document, and let $B_j(d)$ be the degree to which document d satisfies this attribute. Then without any additional complexity we can allow $\alpha_j(d) = B_j(d)$. Thus here the weight associated with attribute A_j depends upon the document itself via the value $B_j(d)$. Thus within this framework we have the option of specifying the importance weights conditionally or non-conditionally or not at all. It should be noted we could of course let $\alpha_j(d) = \alpha_j \, B_j(d)$, that is some proportion of $B_j(d)$. Additional sophistication with respect to the determination of the importance value can be had if allow the the determination of the importance be obtained using a fuzzy systems model [9]. Thus we can obtain $\alpha_j(d)$ from a rule base whose antecedents are based on some measurable features of the document being evaluated.

Typically the association of importance weights with attributes indicates some measure of trade-off between the worth of the attributes. For example, consider the averaging quantifier where $\mathrm{Val}(d) = \sum_{j=1}^{n} A_j(d) \, \alpha_j$. Here we see that a gain of Δ in $A_j(d)$ results in an increase in overall evaluation of $\alpha_j \, \Delta$, while a gain of Δ in $A_i(d)$ is worth an increase of $\alpha_i \, \Delta$. This of course manifests itself in the ordering of the documents. In particular, if $\alpha_j = 2$ and $\alpha_i = 1$, then we are willing to trade a gain of Δ in A_j for a loss of less than 2Δ in A_i. In some cases where we desire two attributes, we may not be willing to trade-off one of for the other. For example, in designing a car, while we would like both safety and low-cost, we are not willing to give up safety for low cost. Such a situation is characterized as one in which a **priority** exists between the attributes, safety has priority over cost.

In some situations in document retrieval systems we might desire to include a priority relationship between attributes. In the following we shall suggest a mechanism that allows for the inclusion of a priority type effects.

Assume A_1 and A_2 are two attributes for which there exists a priority relationship; A_1 has priority over A_2. One way to manifest this relationship is to allow the importance associated with A_2 to be dependent upon the satisfaction of attribute A_1. Here then for example we have $\alpha_2(d) = A_1(d)$. Let us initially investigate this for the case of the simple weighted average. Assuming α_1 is fixed, we get

$$\mathrm{Val}(d) = \alpha_1 \, A_1(d) + A_1(d) \, A_2(d) = (\alpha_1 + A_2(d))$$

We see that if $A_1(d)$ is low, the weight associated with $A_2(d)$, becomes small and it is not possible for a high value of $A_2(d)$ to compensate for the low value of $A_1(d)$. On the other hand if $A_1(d)$ is high then the good score for A_2 can help improve the valuation of d.

More generally, consider the quantifier Q and assume A_i has priority over A_j. To implement this priority we make the importance associated with A_j related to the satisfaction of A_i. In particular, we let $\alpha_j = \alpha \, A_i$, where $\alpha \in [0, 1]$. Using this we get for the weight w_j associated with A_j that

$$w_j = Q(S_{k-1} + \frac{\alpha A_j(d)}{T}) - Q(S_{k-1})$$

where $S_{k-1} = \dfrac{\sum\limits_{k=1}^{j-1} \alpha_k}{T}$. We see that as $A_i(d)$ gets smaller, the value w_j will decrease.

Thus here we have suggested a method for including priorities in the process of formulating queries.

8. Granular Satisfactions

In the preceding we have assumed that the satisfaction of a document to an attribute, $A_i(d)$, was a precise value lying the unit interval. In some cases rather then being able to preciously provide these values we may be only able to specify them linguistically with terms such as *high* or *about .0.5*. More generally this information may in some kind of granular form. [10]. In this environment the formal representation of $A_i(d)$ can be made in terms of a fuzzy subset of the unit interval. One problem we now become faced with is the implementation of the OWA aggregation operator in the case where the arguments are fuzzy subsets. We shall briefly look at this problem.

We note that Mitchell and Estrakh [11] considered issue of the OWA operator in which the ranking was fuzzy ranks. In their approach they use a pairwise comparison between the objects to be ordered. Here we shall suggest an alternative approach. Although related to the interesting suggestion in [11].

Consider the situation when we desire to calculate $F_w(a_1, ..., a_n)$ where the a_j are fuzzy subsets over the unit interval. We recall the OWA aggregation involves a two-step process. The first step is an ordering of the arguments. This determines the assignment of the weights to the arguments. The second step is a weighted aggregation of the arguments. In particular if we have a index function such that index(j) is the index of the jth largest of the a_i then $F(a_1, ..., a_n) = \Sigma_j w_j a_{index(j)}$.

For the present let us assume this index function is available. Let us denote index(j) = b_j then $F(a_1, ..., a_n) = \Sigma_j w_j b_j$. In this case the calculation $F(a_1, ..., a_n)$ became the weighted aggregation of fuzzy numbers. This operation is well established in the literature. It was originally introduced by Zadeh [12] who provided a definition in terms of the extension principle. Dubois and Prade [13] studied this in considerable detail and suggested some methods for its efficient implementation. In particular performance of this weighted aggregation of fuzzy numbers results in a fuzzy subset b over its unit interval such that for each $x \in I$

$$b(x) = Min_j[a_j(x_j)]$$

The Min is taken over all x_j such that $\Sigma_j \, w_j \, x_j = x$.

In order to get to this step of calculating the weighted average, we must first address the problem of ordering to the index function.. Here we are faced with the problem of ordering of fuzzy numbers. Thus our problem is to order the collection $(a_1, \ldots, a_n)$ of fuzzy numbers. In [14] we suggested an approach to ordering fuzzy numbers. In particular if a_i^α is the α-level set of a_i, the subset of numbers such that $a_i(x) \geq \alpha$ and then we define $\mathrm{Ave}(a_i^\alpha)$ as the average of the elements in a_i^α. Using this we associate with a value a_i a unique value $\mathbf{Ave(a_i)}$ defined as

$$\mathrm{Ave}(a_i) = \int_0^1 a_i^\alpha \, d\alpha$$

We can then order the a_i using the $\mathbf{Ave(a_i)}$ values to order the arguments.. Thus $\mathbf{Ave(a_i)}$ induces an ordering on the a_i which is expressed by an index function and we get $F(a_1, \ldots, a_n) = \Sigma_j \, w_j \, a_{\text{index}(j)}$.

However we note there are many ways to transform fuzzy numbers into scaler values. This process is closely related to the issue of defuzzification. In [15] Yager and Filev discuss numerous ways to defuzzify a fuzzy number. The problem is that different defuzzification procedures have different properties and hence there is no approach that clearly dominates all the others. Each method may bring with it a desirable property. Here we suggest another approach that tries to accommodate the multiplicity of possible defuzzification methods.

Let $D_1, \ldots, D_q$ be a collection of defuzzification methods and let $u_1, \ldots, u_q$ be a collection of associated weights which sum to one and lie in the unit interval. Here u_k indicates the credibility we desire to give to defuzzification procedure D_k. The application of D_k to the collection $a_1, \ldots, a_n$ of fuzzy numbers induces an ordering over the a_i such that k-index(j) is the index of the jth largest of the a_i under defuzzification procedure D_k. If we use D_k and they perform the OWA aggregation indicate this order we get $V_k = \Sigma_j \, w_j \, a_{\text{k-index}(j)}$. Using these V_k we can obtain $F(a_1, \ldots, a_n) = \Sigma_k \, u_k \, v_k$. In this way we can allow the user to include any preference they have with respect to the defuzzification procedure by appropriately assigning the u_k.

We shall note pursue this any further here but only to make one last observation.

In [15] Yager and Filev introduced the idea of induced-OWA (IOWA) aggregation. With this aggregation operator they generalized the OWA operator by using an additional variable, called an order-inducing variable, to order the objects to aggregate instead of ordering them by the arguments being aggregated. We observe that in the proceeding each of the defuzzification procedures, D_k, can be seen as generating an order indicating variable. Here $D_k(a_j)$ is the value of a_j

for the kth order inducing variable. Implicit in the preceding is the idea of providing an aggregation based upon a weighting of the results obtained using the different order inducing variables. Thus u_k is the weight assigned the kth order-inducing variable.

9. References

[1]. Yager, R. R., "Approximate reasoning as a basis for computing with words," in Computing with Words in Information/Intelligent Systems 1, edited by Zadeh, L. A. and Kacprzyk, J., Springer-Verlag: Heidelberg, 50-77, 1999.

[2]. Zadeh, L. A., "Fuzzy logic = computing with words," IEEE Transactions on Fuzzy Systems 4, 103-111, 1996.

[3]. Yager, R. R., "On ordered weighted averaging aggregation operators in multi-criteria decision making," IEEE Transactions on Systems, Man and Cybernetics 18, 183-190, 1988.

[4]. Yager, R. R. and Kacprzyk, J., The Ordered Weighted Averaging Operators: Theory and Applications, Kluwer: Norwell, MA, 1997.

[5]. Zadeh, L. A., "A computational approach to fuzzy quantifiers in natural languages," Computing and Mathematics with Applications 9, 149-184, 1983.

[6]. Yager, R. R., "Families of OWA operators," Fuzzy Sets and Systems 59, 125-148, 1993.

[7]. Yager, R. R., "On the inclusion of importances in OWA aggregations," in The Ordered Weighted Averaging Operators: Theory and Applications, edited by Yager, R. R. and Kacprzyk, J., Kluwer Academic Publishers: Norwell, MA, 41-59, 1997.

[8]. Larsen, H. L. and Yager, R. R., "The use of fuzzy relational thesauri for classifactory problem solving in information retrieval and expert systems," IEEE Transactions on Systems, Man and Cybernetics 23, 31-41, 1993.

[9]. Yager, R. R. and Filev, D. P., Essentials of Fuzzy Modeling and Control, John Wiley: New York, 1994.

[10]. Zadeh, L. A., "A new direction in AI - toward a computational theory of perceptions," AI Magazine Vol 22, No 1, Spring, 73-84, 2001.

[11]. Mitchell, H. B. and Estrakh, D. D., "An OWA operator with fuzzy ranks," International Journal of Intelligent Systems 13, 69-81, 1998.

[12]. Zadeh, L., "The concept of a linguistic variable and its application to approximate reasoning: Part 1," Information Sciences 8, 199-249, 1975.

[13]. Dubois, D. and Prade, H., "Operations on fuzzy numbers," International Journal of Systems Science 9, 613 - 626, 1978.

[14]. Yager, R. R., "A procedure for ordering fuzzy subsets of the unit interval," Information Sciences 24, 143-161, 1981.

[15]. Yager, R. R. and Filev, D. P., "On the issue of defuzzification and selection based on a fuzzy set," Fuzzy Sets and Systems 55, 255-272, 1993.

[16]. Yager, R. R. and Filev, D. P., "Induced ordered weighted averaging operators," IEEE Transaction on Systems, Man and Cybernetics 29, 141-150, 1999.

Relationships at the Heart of Semantic Web: Modeling, Discovering, and Exploiting Complex Semantic Relationships

Amit Sheth[1,3], I. Budak Arpinar[1], and Vipul Kashyap[2]

[1] LSDIS Lab, Computer Science Department, University of Georgia
[2] National Library of Medicine, [3] Semagix, Inc.
amit@cs.uga.edu, budak@cs.uga.edu, kashyap@nlm.nih.gov

Abstract. The primary goal of today's search and browsing techniques is to find relevant documents. As the current web evolves into the next generation termed the Semantic Web, the emphasis will shift from finding documents to finding facts, actionable information, and insights. Improving ability to extract facts, mainly in the form of entities, embedded within documents leads to the fundamental challenge of discovering relevant and interesting relationships amongst the entities that these documents describe. Relationships are fundamental to semantics—to associate meanings to words, terms and entities. They are a key to new insights. Knowledge discovery is also about discovery of heretofore new relationships. The Semantic Web seeks to associate annotations (i.e., metadata), primarily consisting of based on concepts (often representing entities) from one or more ontologies/vocabularies with all Web-accessible resources such that programs can associate "meaning with data". Not only it supports the goal of automatic interpretation and processing (access, invoke, utilize, and analyze), it also enables improvements in scalability compared to approaches that are not semantics-based. Identification, discovery, validation and utilization of relationships (such as during query evaluation), will be a critical computation on the Semantic Web.

Based on our research over the last decade, this paper takes an empirical look at various types of simple and complex relationships, what is captured and how they are represented, and how they are identified, discovered or validated, and exploited. These relationships may be based only on what is contained in or directly derived from data (direct content based relationships), or may be based on information extraction, external and prior knowledge and user defined computations (content descriptive relationships). We also present some recent techniques for discovering indirect (i.e., transitive) and virtual (i.e., user-defined) yet meaningful (i.e., contextually relevant) relationships based on a set of patterns and paths between entities of interest. In particular, we will discuss modeling, representation and computation or validation of three types of complex semantic relationships:

(a) using predefined multi-ontology relationships for query processing and corresponding the issue of "loss of information" investigated in the OBSERVER project, (b) ρ (Rho) operator for semantic associations which seeks to discover contextually relevant and relevancy ranked indirect relationships or paths between entities using semantic metadata and relevant knowledge, and (c) IScapes which allows interactive, human-directed knowledge validation of hypothesis involving user-defined relationships and operations in a multi-ontology, and multi-agent InfoQuilt system.

Representing, identifying, discovering, validating and exploiting complex relationships are important issues related to realizing the full power of the Semantic Web, and can help close the gap between highly separated information retrieval and decision-making steps.

1. Introduction

Most Internet users today find information in one of two ways – either by browsing the information space or through the use of a search engine. Browsing is completely under the control of the user but requires choosing a good directory that has organized the document space, combined with user's constant attention and decision-making. Systems based on search engines perform essentially the task of delivering a document based on keywords or key phrases. Some search engines, such as Google, use heuristics and statistics to improve ranking for a generic user, but that only seeks to improve document retrieval for most users. None of these approaches attempts to get at the user's underlying intentions or information goals. And none give new insights related to user's information needs. This is readily evident from their results – most of the retrieved documents are either irrelevant unless the search objective is relatively straightforward (e.g., home page of a person or specific document posted at a well respected source), or contain the information buried in a morass of other data. A user must decide which of the retrieved documents are relevant or within his information need context, and then use his mental model of the information sought to "process" the documents to obtain the relevant information. This is a very serious and as yet unsolved problem, as evidenced by the fact that practically all of today's technical efforts in search engine, content management, and other technologies are geared towards dealing with data overload, which leads to information starvation (the inability to find useful and actionable information from massive amounts of data).

Significant past research has been conducted in managing heterogeneous data, and providing interoperability and integration of information systems so that data can be shared, collectively accessed, and processed [Sheth98]. This has been a long process, with earlier research dating to the late 1970, going through the architectures for federated databases [Sheth90], mediators [Weiderhold92] and information brokering systems [Kashyap00]. With the ability to access and share all forms of data, now we have the familiar challenge of data overload.

We believe the a more fundamental challenge is to make decisions or take actions based on data than finding relevant documents – an objective that a new generation of content management systems subscribe to, and the one most of today's search and browsing techniques fail to address. One step towards gaining this capability is to discover relevant and interesting relationships amongst the entities that these documents describe. These relationships are the basis of analysis, and underpin the semantics of the data. We face several challenges in meeting this task. One reason is that the data retrieval (i.e., "search") phase is not geared towards dealing with relationships. For instance, if a search for "data" results in a large numbers of irrelevant documents, any technique for finding relationships will generate a correspondingly much larger (perhaps by an order of magnitude) number of irrelevant, and useless relationships. As the adage says, every one is related by only six degrees of separation!

For computing (identifying, discovering or validating) relationships, what we need is very different from data mining, at least as it has been traditionally understood in terms of grouping or market basket type analysis through the discovery of association rules. Data mining techniques are typically based on statistics and look for patterns that are already present in the data. Moreover, the patterns are sought at a syntactic level, and do not take into consideration the meaning of the data. They are typically not easily extendable to look for the types of relationships that are meaningful to humans or to the software agent performed target information processing tasks, and they are not based on the semantics of the underlying data. The clustering and machine learning techniques in themselves will similarly not be sufficient.   However, computing complex relationships require new forms of processing data and relevant knowledge, and associated techniques of creating and maintaining a variety of relationships. Instead of relying on data alone, they utilize a broad variety of domain knowledge, and context, which enables scalability by ignoring irrelevant information, and knowledge.

Developing a system focused around finding semantic relationships rather than documents is challenging for several reasons. Each document may describe (and hence be annotated with) many entities. The number of relationships or paths connecting entities directly or through a Knowledge Base (KB), however, is vastly larger. Whether seen as a graph theoretic or deductive logic problem, many approaches for computation are not tractable, let along scalable. Furthermore, imposing constraints that only relevant or interesting relationships are discovered may add to the complexity.

This chapter significantly borrows from our benefits from past efforts including:

- Research in semantic interoperability and integration of heterogeneous data [Kashyap96], partially performed in InfoHarness [Shah99], and its follow on VisualHarness [Shah97], and VideoAnywhere projects [Bertram98],
- Semagix's Semantic Content Organization and Semantic Engine (SCORE) technology [Sheth02a, Hammond02] partially based on technology licensed from UGA, and based on above projects,

- UGA's research on human-directed knowledge discovery in InfoQuilt project [Sheth02b], multi-ontology query processing in OBSERVER project [Mena00], and the on-going project on Semantic Association discovery [Anyanwu02].

In this paper, we do not attempt to present a comprehensive taxonomy of relationships, nor do we survey all relevant literature. Rather our treatment is empirical and involves a review of semantic relationship computation and use in various research systems we have worked on during the last decade. Section 2 provides and overview and a partial classification of challenges in dealing with relationships. In Section 3, we start with *identification of simple semantic relationships* based on a large knowledge base in a state of the art commercial system SCORE based on technology transfer from our academic research. Section 4 discusses as examples of *semantic relationship discovery*. It introduces the concept of complex relations called Semantic Associations, and some preliminary thoughts on computing a ranked list of these associations using a context. In Section 5, we discuss IScapes, user-defined complex relationships, and their *validation* in the Info-Quilt system as a way to support user directed knowledge discovery. Section 6 provides an example of query *evaluation involving semantic relationships*. We discuss use of inter-ontology relationships in OBSERVER's multi-ontology query processing, and the corresponding effort in computing information loss. We conclude with Section 7.

2. Classification of Complex Relationships

The questions of if and how two or more entities relate to each other are both technical and philosophical questions. Yet, these are the essential questions to exploit to discover new, interesting, and useful relations across entities in diverse domains including national security, life sciences, and economics. On what dimensions should a study of different kinds of relationships be organized? One dimension of relationship is whether it is based on explicit, precise or exact knowledge, or that it is based on imprecise or approximate knowledge (such as one based statistical and probabilistic measures). As an enhancement of this perspective, we propose three dimensions along which it might be useful to organize such a study: (a) the information content captured by a relationship; (b) various ways of representing a relationship; and (c) methodologies for computing (i.e., identifying, discovering, and validating), and exploiting the various relationships.

2.1 A Taxonomy of Relationships Based on the Information Content

Metadata has been used to describe data, document or content [Boll98]. Patterned after the classification used for metadata [Kashyap95], we classify the relationships as follows:

- Content Independent Relationships: These types of relationships are typically independent of the content and are an artifact of the organization of content on a computer system due to reasons of organization, performance, scalability, etc., e.g., two documents may be related to each other by virtue of them being stored on the same server or file system, or the relationship between a document and it's date of modification, etc.
- Content Dependent Relationships: These capture the relationships between two entities based on the either the information content they refer to in the real world or based on some representation of it thereof. Various types of content dependent relationships are as follows:
  - Direct Content Dependent Relationships: These types of relationships typically depend on some representation of the information content to which the entities refer to and are directly computed from them. It may be noted that some of these relationships might be fuzzy in nature. For example, the relations between two entities being mentioned in the same paragraph and spatial locations of two objects in an image suggest crisp relations, whereas the similarity between two documents in a vector space is a fuzzy measure.
  - Content Descriptive Relationships: These types of relationships are based on the information content, which the entities refer to in the real world. These are typically not computable directly from the representation of the information content and help of additional resources such as taxonomies, and ontologies along with heuristic algorithms may be used to compute these relationships. For example, the fact that an entity X is the CEO of a company Y is computed based on the existence of an ontology that models businesses (which specifies the relationship "CEO") and heuristic document processing algorithms (which discover the relationship) applied to relevant documents. These relationships are typically viewed as crisp as some thresholding techniques are applied to the heuristic algorithms, whereas they are in reality fuzzy and reflect a probability of the person X being the CEO of a company Y. These relationships might associate entities within a domain (intra-domain relationships) or across multiple domains (inter-domain relationships). An informal (and incomplete) (sub-) classification of this type of relationship is as follows:
  - Direct Semantic Relationships: These are direct intra-domain relationships between two documents or entities, e.g., an HREF link annotated with semantic information (Figure 1.a), Intel *is-a-competitor-of* Motorola (Figure 1.b). Examples of these are discussed in the SCORE system in Section 3.
  - Complex Transitive Relationships: Remzi and Dick are associated with each other because they are linked to the same terrorist organization through their financial transaction (Figure 1.c). This . These type of intra-domain relationships are captured using the ρ operator discussed in Section 4.
  - Inter-domain Multi-ontology Relationships: Some relationships span across multiple domains and are typically represented as inter-ontology relationships across multiple ontologies. This type of relationships is discussed in the context of the OBSERVER System in Section 6.
  - Semantic Proximity Relationships: Two entities may have a semantic proximity or similarity that cannot be completely represented using crisp relationships. They may either be represented using a semantic proximity function associated with a relationship or depend on fuzzy predicates such as "close-enough" (Figure 1.e illustrates a similarity relations between two events)." Furthermore, they may be user defined

(Figure 1.d). These types of relationships are discussed in the context of IScapes in Section 5.

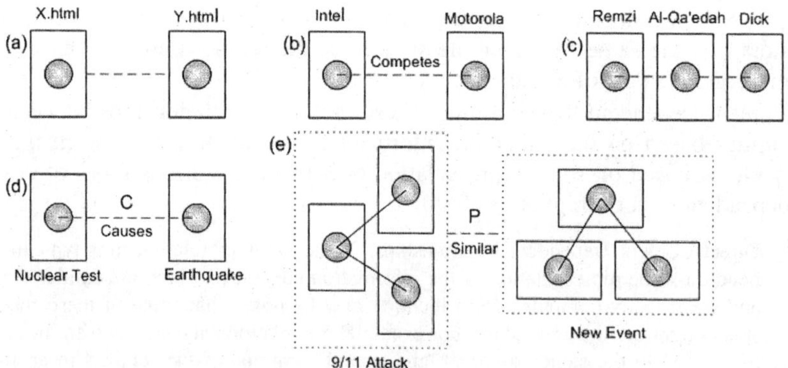

Fig. 1. Different types of structural composition of relationships

2.2 Representation of Relationships

A fundamental representation of a relationship between two concepts is a mathematical structure denoting it as a set mapping between the instances belonging to the two concepts. These mappings might be characterized along the following dimensions:

- Arity: Typically binary relationships are of most interest, but relationships can be of arbitrary arity, i.e., we could have 3 or more concepts participating in a relationship.
- Cardinality: These constraints are characterized in one of the following ways: 1–1, many–1, 1-many, or many-many. A more generalized way of representing these cardinality constraints is using a pair of numbers that specify the minimum and maximum number of times an instance of a concept can participate in a relationship. This is a very useful technique for n-ary relationships and also captures partial participation of concepts in relationships. 1–1 and many–1 relationships are functions which can be exploited in various ways.
- Direct v/s Transitive Relationships: Some entities might be directly related to each other via their participation in a common relationship, or might be related transitively to each other via a chain of relationships.
- Crisp vs. Fuzzy: Most of the current modeling approaches view relationships as crisp, i.e., for an n-ary relationship, instances of n concepts are either part of a relationship or not (e.g., is-a, part-of relations). In the case of fuzzy knowledge [Zadeh65], the extension of a relationship may be viewed as a joint probability distribution on the concepts participating in a relationship. For example semantic similarity (i.e., proximity) between two entities is an example for fuzzy relations.

- Properties vs. Relations: Properties are special relationships where the ranges of a relationship are values of a data type (e.g., dates, age) as opposed to instances of a concept.
- Structural Composition: Relationships can either be composed (if they are functional in nature) or combined using join operations to create new relationships and associations based on existing relationships.

Most frequently occurring relationship is that of hypertext link (HREF). One attempt to make it more meaningful was the proposal for MetadataMetadata Reference Link (MREF) [Shah98] that associated metadata represented in RDF to HREF. This metadata provided further semantics to otherwise a hypertext link without any information that a machine can use to understand what it is about (Figure 1.a).

Most modeling approaches whether they are graphical in nature, e.g., EER, UML diagrams or use object models and XML markup models, e.g., OMG object model, OKBC, DAML+OIL represent the fundamental structures described above using various modeling (graphical or markup) primitives which can be combined together using various (graphical, hierarchical or symbolic) constructors.

2.3 Computation and Exploitation of Relationships

Four main computations that can be performed to manage and exploit relationships are as follows.

- **Identify**: This is the process by which a relationships whose semantics is known and understood (e.g., via its representation in a domain specific ontology), and computation is directed towards identifying the presence of the relationship within a document or any other piece of data. We present an example of this in the discussion of the SCORE System (Section 3).
- **Discover**: This is the process by which we search for patterns among content or resources, within a semantic model or an ontology to discover new relationships. Other approaches of discovering new relationships might involve text mining operations. We present Rho operators that can search for patterns in an ontology and propose new relationships (Section 4).
- **Validate**: This is the process by which IScapes representing knowledge discovery hypothesis, possibly involving complex relationships and fuzzy operators (e.g., near to, same as), are validated by information gathering and analysis over a collection of heterogeneous data sources (Section 5).
- **Evaluate**: In the process of computing a given relationship, it may be noted that it may only be possible to estimate it, giving rise to uncertainty and confidence intervals. We discuss multi-ontology query processing in the OBSERVER System (Section 6), which computes the equivalence relationship between an information request and the answer (possibly spanning multiple ontologies), with the associated precision and recall measures.

3. Ontology Driven Relationship Identification: Example of the SCORE Technology

In this section, we discuss an example of identifying an instance of relationship based on a document analysis. The existence of a relationship is already know to the system, for example as part of an ontology, so the relationship is identified based on occurrence of entities in the relevant context in the document.

Identification of such a relationship is exemplified by a commercial semantic technology based on prior academic research. SCORE is a commercial Semantic Content Organization and Retrieval Engine [Sheth02a, Hammond02]. Semantic underpinning in SCORE is provided by an ontology with a definitional component (called World Model) and assertional component (called Knowledge Base – KB). In SCORE, through the use of automatic classification and contextually relevant ontology (i.e., relevant part of ontology including the assertions), domain specific metadata can be extracted from a document, enhancing the meaning of the original and allowing it to be linked with contextually heterogeneous content from multiple sources. In this way, relations between the entities, which are not explicitly evident in a single document, can be revealed. We call these types of one-to-one relations between the entities simple indirect relations.

The identification of indirect semantic relations between the entities and its use in document enhancement is illustrated in Figure 2. First, the classification technology determines the category for a document. This determines the domain of discourse, or relevant ontology, e.g., business ontology (or a relevant part of an ontology, e.g., equity market part of entertainment ontology). Then semantic metadata particular to the domain is targeted and extracted. This includes specific named entity types of interest in the category (such as "CEO" in "Business," "Downgrade" in "Equity Markets", or "SideEffects" in "Pharmacology") as well as category specific, regular expression-based knowledge extraction. This domain-specific metadata can be regarded as semantic metadata, or metadata within context. The automatic extraction of semantic metadata from documents which have not been previously associated with a domain is a unique feature of SCORE. In essence, this transports the document from the realm of text and mere syntax to a world of knowledge and semantics in a form that can be used for semantic computation.

An example is illustrated through a Web document in Figure 3. In the Figure, BEA Systems, Microsoft and PeopleSoft all engage in the "competes with" relationship with Oracle. When entities found within a document have relationships based on a known ontology, we refer to the relationships as "direct relationships." Some of the direct relationships found in this example include: HPQ identifies Hewlett-Packard Co.; HD identifies The Home Depot; Inc.; MSFT identifies Microsoft Corp.; ORCL identifies Oracle Corp.; Salomon Smith Barney's headquarters is in New York City; and MSFT, ORCL, PSFT, BEAS are traded on Nasdaq.

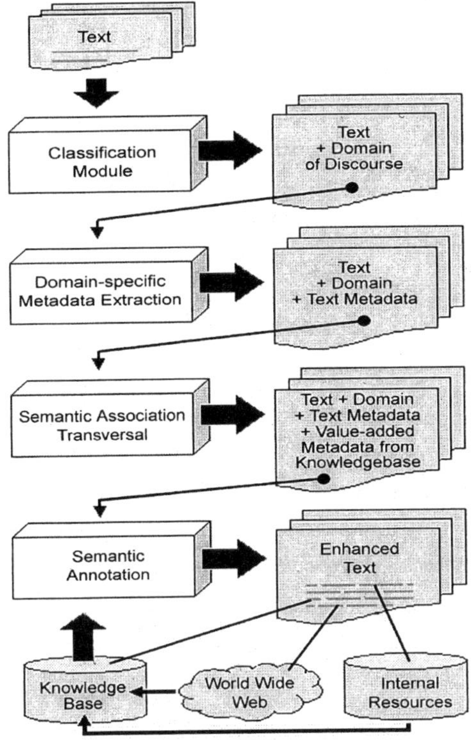

Fig. 2. Semantic Document Enhancement in SCORE System

Not all of the associated entities for an entity found in the text will appear in the document. Often, the entities mentioned will have one or more relationships with another common entity. In this case, some examples include: HPQ and HD are traded on the NYSE; BEAS, MSFT, ORCL and PSFT are components of the Nasdaq 100 Index; Hewlett-Packard and PeopleSoft invested in Marimba, Inc., which competes with Microsoft; BEA, Hewlett-Packard, Microsoft and People-Soft compete with IBM, Sun Microsystems and Apple Computer.

The use of semantic associations allows entities not explicitly mentioned in the text to be inferred or linked to a document. This one-step-removed linking is referred to as "indirect relationships." The relationships that are retained are application specific and are completely customizable. Additionally, it is possible to traverse relationship chains to more than one level. It is possible to limit the identification of relationships between entities within a document, within a corpus across documents or allow indirect relationships by freely relating an entity in a document with any known entity in the SCORE KB.

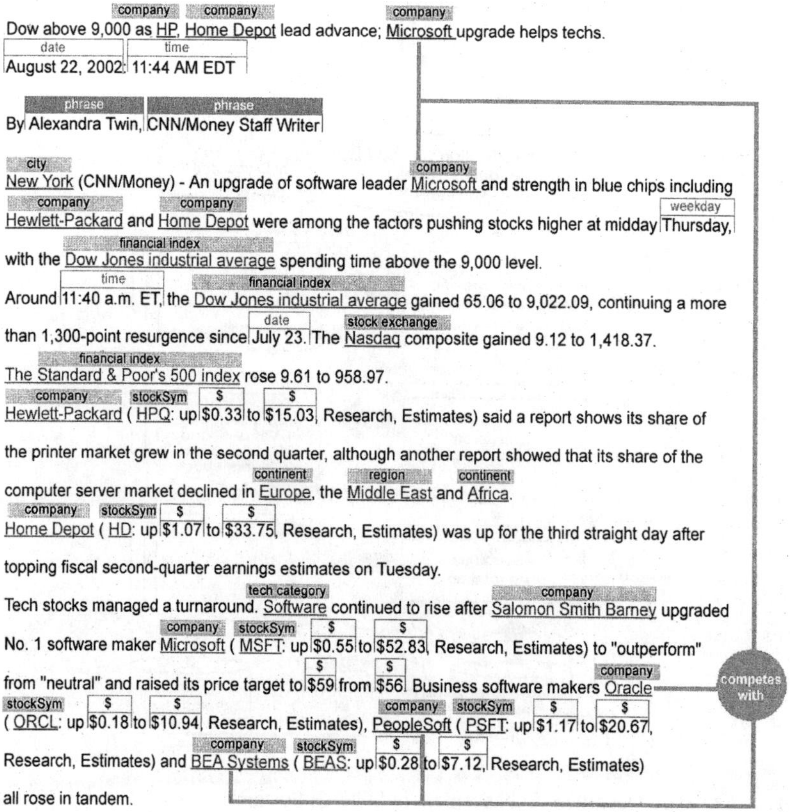

Dow above 9,000 as HP, Home Depot lead advance; Microsoft upgrade helps techs.

August 22, 2002: 11:44 AM EDT

By Alexandra Twin, CNN/Money Staff Writer

New York (CNN/Money) - An upgrade of software leader Microsoft and strength in blue chips including Hewlett-Packard and Home Depot were among the factors pushing stocks higher at midday Thursday, with the Dow Jones industrial average spending time above the 9,000 level.

Around 11:40 a.m. ET, the Dow Jones industrial average gained 65.06 to 9,022.09, continuing a more than 1,300-point resurgence since July 23. The Nasdaq composite gained 9.12 to 1,418.37.

The Standard & Poor's 500 index rose 9.61 to 958.97.

Hewlett-Packard (HPQ: up $0.33 to $15.03, Research, Estimates) said a report shows its share of the printer market grew in the second quarter, although another report showed that its share of the computer server market declined in Europe, the Middle East and Africa.

Home Depot (HD: up $1.07 to $33.75, Research, Estimates) was up for the third straight day after topping fiscal second-quarter earnings estimates on Tuesday.

Tech stocks managed a turnaround. Software continued to rise after Salomon Smith Barney upgraded No. 1 software maker Microsoft (MSFT: up $0.55 to $52.83, Research, Estimates) to "outperform" from "neutral" and raised its price target to $59 from $56. Business software makers Oracle (ORCL: up $0.18 to $10.94, Research, Estimates), PeopleSoft (PSFT: up $1.17 to $20.67, Research, Estimates) and BEA Systems (BEAS: up $0.28 to $7.12, Research, Estimates) all rose in tandem.

Fig. 3. When SCORE recognizes an entity, knowledge about its entity relationships to other entities becomes available through relevant (parts of) ontology based on context provided by automatic clas-sification

Indirect relationships provide a mechanism for producing value-added semantic metadata. Each entity in the KB provides an opportunity for rich semantic associations. As an example, consider the following:

Oracle Corp.	
Sector:	Computer Software and Services
Industry:	Database and File Management Software
Symbol:	ORCL
CEO:	Ellison, Lawrence J.
CFO:	Henley, Jeffrey O.
Headquartered in:	RedWood City, California, USA

Manufactured by:	8i Standard Edition, Application Server, etc.
Subsidiary of:	Liberate Technologies and OracleMobile
Competes with:	Agile, Ariba, BEA Systems, Informix, IBM, Microsoft, People-Soft and Sybase

This represents only a small sample of the sort of knowledge in the SCORE KB. Here, the ability to extract from disparate resources can be seen clearly. The "Redwood City" listed for the "Headquartered in" relationship above, has the relationship "located within" to "California," which has the same relationship to the "United States of America." Each of the entities related to "Oracle" are also related to other entities radiating outward. Each of the binary relationships has a defined *directionality (some may be bi-directional)*. In this example, *Manufactured by* and *Subsidiary of* are marked as *right-to-left* and should be interpreted as "8i Standard Edition, Application Server, etc. are *manufactured by* Oracle" and "Liberate Technologies and OracleMobile are *subsidiaries of* Oracle." SCORE can use these relationships to put entities within context.

When a document mentions "Redwood City," SCORE can add "California," "USA," and "North America." Thus, when a user looks for stories that occur in the United States or California, a document containing "Redwood City" can be returned, even though the more generalized location is not explicitly mentioned. This is one of the capabilities a keyword-based search cannot provide, where the information implicit in the text is revealed and can then be linked with other sources of content.

4. ρ Operator for Semantic Associations: Example of Semantic Relationship Discovery and Ranking

In this example, we will discuss an ongoing research on discovery of complex semantic relationships in the Semantic Web. Many applications in analytical domains such as national security and business intelligence require a more complex notion of relationships than the simple direct relationships between the entities, of the types discussed in Section 3. For example, in the light of the recent breach of flight security, it has become pertinent to enable airport security agents are able to ask questions like, what *important relationships* exist between Passenger X and Passenger Y? A new relationship may emerge because of complex transitive relations connecting these two persons. Furthermore, the notion of importance depends primarily on the context, which in this case is the assessment the risk of flight based on passenger associations. In this scenario, it is not possible to encode all the relevant relationships as rules, because they are not usually known; yet they can be discovered through an analytical process. In general, the relevant relationships emerge as a set of connections or various interesting patterns of connections between the entities. As an example, consider some passengers who are the nationals of the same country, and purchased their tickets using the same credit card, even though they do not have a known family relationship, and furthermore one of them is on the FBI watch-list. Because different domains may have different notions of relationships, in other words, what kind of connections constitute a relationship, it may be useful to use domain-specific ontology to guide the search for semantic relations.

Semantic relations in the most basic sense involve evaluating a set of contextually relevant paths of relations from one entity to another. By evaluating such paths we may identify relations based on connectivity or similarity of paths. This allows us to analyze sequences of binary relationships instead of just single binary relationships, and manipulate these sequences to find similar entities as well as entities that may be connected, albeit not directly. This technique is different from data mining that uses statistical techniques to find co-occurrence relationships between predicates based on patterns in data.

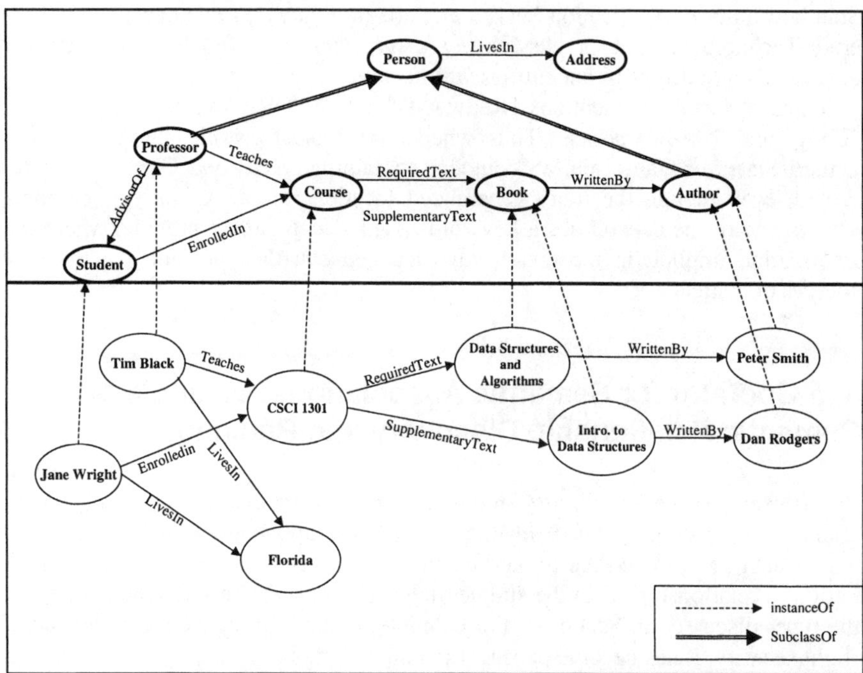

Fig. 4. An Example Ontology and Knowledge Base

We will illustrate the notion of complex semantic relations, termed semantic associations through a pedagogical example [Anyanwu02]. Figure 4 shows a simple ontology containing information about Professors, Students, Courses, Books, and Book Authors. The top part of the figure shows the descriptional part of ontology which contains the entity types (i.e., classes) depicted as nodes, and the domain specific relationships between entity types are illustrated by single-lined arcs. Entity types may also be related by special relationships such as a *subclassOf* relationship denoted by a double-lined arc. The bottom part of the Figure shows assertional component of the ontology, i.e., instances of the classes, and dotted lines illustrate *instanceOf* relations. In this simplified example, semantic relations include the following: Tim Black can be said to be associated with Peter Smith because he *Teaches* a course CSCI1301 that has as its text a book *WrittenBy* by Peter Smith. Also, Peter Smith and Dan Rodgers are associated in that they both are authors of the books that are used as textbooks in a particular course. These two relations are slightly different because the first involves a directed path between entities, while the second involves an undirected path. Discovery of more complex relations between two entities may require checking the semantic similarity between the sub-graphs of a knowledge base involving these entities; furthermore, the similarity checking may require custom defined computations (e.g., two Professors can be related because they use similar investigative methods in two different scientific experiments). Another dimension is aggregation of entities and associations to find more meaningful group associations than individual links connecting the entities of interest (i.e., discovery of association structures vs. individual associations). Some example association types we have been addressing are illustrated in Figure 5.

The associations 1, and 2–4 are examples of direct and transitive links between two entities, respectively. For example, 3 may represent a semantic relation between two Professors whose books are used for the same course. Entities that have a common successor and predecessor can be represented by 3 and 4 respectively. The arbitrary combinations of these link types may result in more complex relations as illustrated in 5. An example might be two Professors whose projects are funded by two different agencies having a common manager. In general, two entities having an un-directed path between them can be associated in varying degrees according to the path length (and possibly path strength).

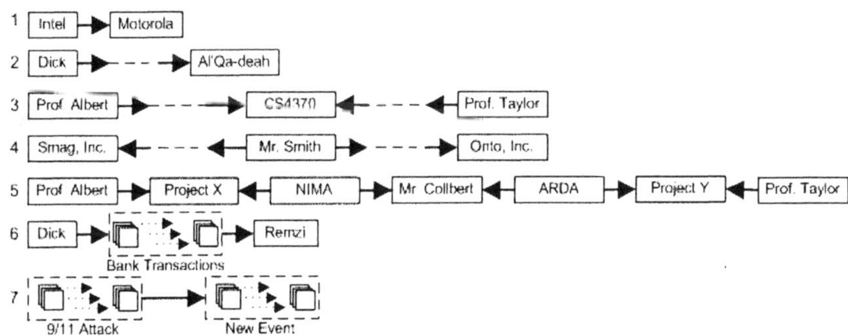

Fig. 5. Some Complex Semantic Association Types

Association 6 represents an aggregation of several associations, which is more meaningful and interesting than the individual member associations. For example if a person makes some periodic deposits to another person's account in an overseas bank the aggregation of the links for individual transactions may provide a clue for a money laundering operation. Similarly aggregation of certain entities into groups (i.e., *spheres of semantics*) and investigating group associations may yield more interesting results. In 7, a semantic similarity relation between two events exists, because both of them contain a "similar" set of associations. In another example, two terrorist organizations can be related if the set of associations representing their operation styles resemble each other.

Assigning more weights to certain entities and relations and favoring discovery process for visiting these entities and associations can improve the efficiency of the semantic association discovery. For example, if the entity of interest is a certain person, it can be given more weight and relationship discovery may focus on the paths passing through this person. Another technique involves specification of relevant context by identifying certain regions in the ontologies and knowledge base to limit the discovery in traversing transitive links.

If there are too many associations between the entities of interest, then analyzing them and deciding which ones are actually useful might be a burden on a user. Therefore ranking these new relations in accordance of the user's interest is an essential task. In general, a relation can be ranked higher if it is a relatively original (e. g., previously unknown), more trustworthy, and useful in a certain context.

4.1 A Comparative Analysis of Semantic Relation Discovery and Indexing

As the emergence of the Semantic Web gathers momentum, it is imperative to propagate the novel ideas of representing, correlating, and presenting the wealth of available semantic information. A traditional search engine with the associated inverted keyword index (or similar) has served the Web community quite well to a certain point. However, to make searching more precise, a typical search engine must evolve to incorporate a new query language, capable of expressing semantic relationships and conditions imposed on them.

Our KB contains *entities* as well as *relationships* connecting the entities. An entity has a name and a classification (type). A relationship has a name and a vector of entity classifications, specifying the types of entities allowed to participate in the relationship. Both entity classifications and relationships will be organized into their respective hierarchies. The *entity classification hierarchy* represents the similarities among the entity classifications. For example, a general entity class "terrorist" may have subtypes of "planner", "assassin", or "liaison". The *relationship hierarchy* is intended to represent the similarities among the existing relationships (following the "is-a" semantics). For example, "*supports*" is a relationship linking people and terrorist organizations (in the context of terrorism). It is the parent of several other relationships, including "*funds*", "*trains*", "*shelters*", etc.

A semantic query language can be used to express various semantic queries outlined below (the first two represents existing technology, third represents emerging technology, and the remaining represent novel research):

1. *Keyword queries*, as offered by traditional, search engines today. The query is a Boolean combination of search keywords and the result is the set of documents satisfying the query.
2. *Entity queries.* The query is a Boolean combination of entity names and the result is the set of documents satisfying the query. Note, that a given entity may be identified by different names (or different forms of the same name), as for example *"Usama bin Laden," "Osama bin Laden,"* and *"bin Laden, Osama,"* all identify the same entity.
3. *Relationship queries.* This type of queries involves using a specific relationship (for example, *sponsoredBy*) from the KB to find related entity(ies). A secondary result may include a set of documents matching the identified entities, and if possible, supporting the used relationship, as stored in the KB.
4. *Path queries.* Queries of this type involve using a sequence (path) of specific relationships in order to find connected entities. In addition, in order to take into account the relationship hierarchy, a query involving the relationship *supports* (as one of the relationships in the path) will result in entities linked by this and any of the sub-relationships (such as *"funds"*, *"trains"*, *"shelters"*, etc.). The secondary result may include a set of documents matching the identified entities, and if possible, supporting the relationships used in the path and stored in the KB.
5. *Path discovery queries.* This is the most powerful and arguably the most interesting form of semantic queries. This type of query involves a number of entities (possibly just a pair of entities) and attempts to return a set of paths (including relationships and intermediate entities) that connect the entities in the query. Each computed path represents a semantic association of the named entities.

Semantic query processing involves the construction of a specialized Semantic Index (SI). We view the structure of the SI as a three-level index, involving the "traditional" keywords (at level 1), entities and/or concepts (at level 2), as well relationships (at level 3) existing among the entities. The SI is shown in the Figure 6.

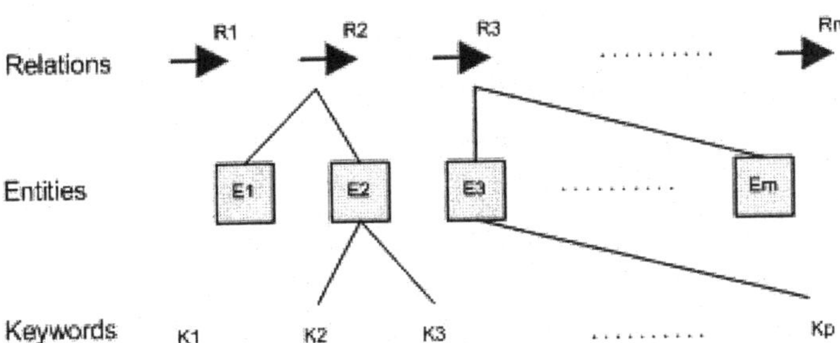

Fig. 6. Semantic Index

The SI constitutes a foundation for the design of a suitable semantic query engine. We must note that the most general of the semantic queries (of type 6 above) in an unconstrained form may be computationally prohibitive. However, when the length of the path is limited to a relatively small fixed number, the computation of the result set is possible.

4.2 ρ Operator

In this section, we highlight an approach for computing complex semantic relations using an operator we call ρ (Rho) [Anyanwu02]. The ρ operator is intended to facilitate complex path navigation in KBs. It permits the navigation of metadata (e.g., resource descriptions in RDF) as well as schema/taxonomies (e.g., ontologies in RDFS, DAML+OIL, or OWL [Heflin02]).

More specifically, the operator ρ provides the mechanism for reasoning about semantic associations that exist in KBs. The binary form of this operator, $\rho_T(a, b)$ [C, K], will return a set of semantic relations between entities a and b. Since semantic relations include not just single relationships but also associations that are realized as a sequence of relationships in a KB or based on certain patterns in such sequences, a mechanism that attempts to find possible paths, and in some cases makes comparisons about similarity of paths/sub-graphs is need. Of course this may be computationally very expensive. The parameters C and K allow us to focus and speed up the computation. C is the context (e.g., a relevant ontology) given by the user, which helps to narrow the search for associations to a specific region in the KB. K is a set of constraints that includes user given restrictions, heuristics and some domain knowledge that is used to limit the search and prioritize the results.

$\rho_T(a , b)$ [C, K] represents the generic form of the ρ operator where the subscript T represents the type of the operator. The types are as follows:

$\rho_{PATH}(a, b)$ [C, K]	Given the entities a, and b, ρ_{PATH} looks for directed paths from a to b and returns a subset of possible paths.
$\rho_{INTERSECT}(a, b)$ [C, K]	Given entities a, and b, $\rho_{INTERSECT}$ looks to see if there are directed paths from a and b that intersect at some node, say c. In other words, it checks to see if there exists a node c such that: $\rho_{PATH}(a , c) \& \rho_{PATH}(b , c)$. Thus, this query returns a set of path pairs where the paths in each pair are intersecting paths.
$\rho_{CONNECT}(a, b)$ [C, K]	Given entities a, and b, $\rho_{CONNECT}$ treats the graph as an undirected graph and looks for a set of edges forming an undirected path between a, and b. This query returns a subset of possible paths.
$\rho_{ISO}(a, b)$ [C, K]	Given entities a, and b, ρ_{ISO} looks for a pair of directed subgraphs rooted at a, and b, respectively, such that the 2 subgraphs are $\rho_{ISOMORPHIC}$. $\rho_{ISOMORPHISM}$ represents the notion of semantic similarity between the 2 sub-graphs.

5. Human-Assisted Knowledge Discovery Involving Complex Relations

In this section, we discuss the concept of IScape in the InfoQuilt system which allows a hypothesis involving complex relationships and its validation over heterogeneous, distribution content.

A great deal of research into enabling technologies for the Semantic Web and semantic interoperability in information systems has focused on domain knowledge representation through the use of ontologies. Current state-of-the-art ontological representational schemes represent knowledge as a hierarchical taxonomy of concepts and relationships such as is-a/role-of, instance-of/member-of and part-of. Fulfilling information requests on systems based on such representation and associated "crisp logic" based reasoning or inference mechanisms [Dec] allow for supporting queries of limited complexity [DHM+01], and additional research in query languages and query processing is rapidly continuing. For example, SCORE allows combining querying of metadata and ontology. An alternative approach has been taken in the InfoQuilt system that supports human-assisted knowledge discovery [Sheth02b]. Here users are able to pose questions that involve exploring complex hypothetical relationships amongst concepts within and across domains, in order to gain a better understanding of their domains of study, and the interactions between them. Such relationships across domains, e.g., causal relationships, may not necessarily be hierarchical in nature and such questions may involve complex information requests involving user defined functions and fuzzy or approximate match of objects, therefore requiring richer environment in terms of expressiveness and computation. For example, a user may want to know "Does Nuclear Testing *cause* Earthquakes?" Answering such a question requires correlation of data from sources of the domain Natural-Disasters.Earthquake with data from sources of Nuclear-Weapons.Nuclear-Testing domain. Such a correlation is only possible if, among other things, the user's notion of "cause" is clearly understood and exploited. This involves the use of ontologies of the involved domains for shared understanding of the terms and their relationships. Furthermore, the user should be allowed to express their meaning (or definition) of the causal relationship. In this case it could be based on the proximity in time and distance between the two events (i.e., nuclear tests and earthquakes), and this meaning should be exploited when correlating data from the different sources. Subsequent investigation of the relationship by refining and posing other questions based on the results presented, may lead the user to a better understanding of the nature of the interaction between the two events. This process is what we refer to as Human-Assisted KNowledge Discovery (HAND). Note that this approach is fundamentally different than the relationship types discussed earlier in the sense that a non-existent new relationship is named, and its precise semantic is defined through a computation. If that computation verifies the existence of this hypothetical relationship it can be placed permanently in an ontology.

InfoQuilt uses ontologies to model the domains of interest. Ontology captures useful semantics of the domain such as the terms and concepts of interest, their meanings, relationships between them and the characteristics of the domain. Ontology provides a structured, homogeneous view over all the available data sources. It is used to standardize the meaning, description and the representation of the attributes across the sources (we call it semantic normalization). All the resources are mapped to this integrated view and this helps to resolve the source differences and makes schema integration easier. An example of "disaster" ontology is shown in Figure 7.

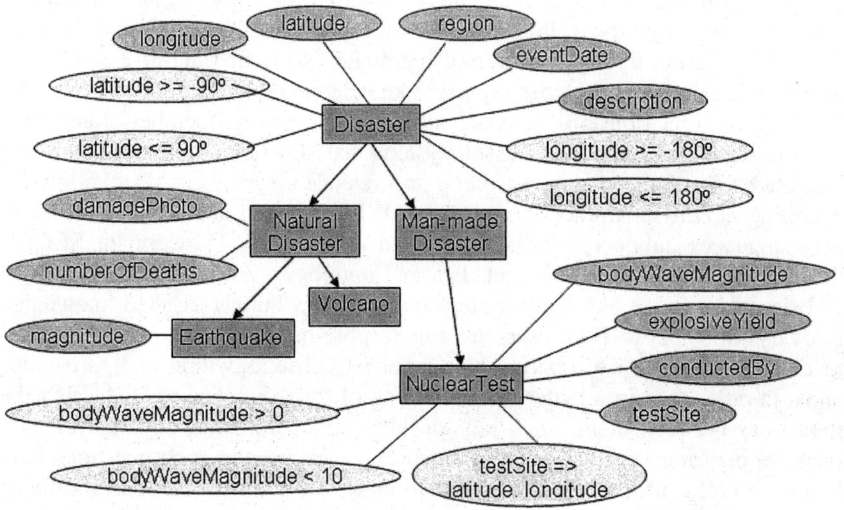

Fig. 7. Disaster Ontology

5.1 User-Defined Functions

A distinguishing feature of InfoQuilt is its framework to support user-defined operations. The user can use them to specify additional constraints in their information requests. For example, consider the information request:

"Find all earthquakes with epicenter in a 5000 mile radius area of the location at latitude 60.790 North and longitude 97.570 East"

The system needs to know how it can calculate the distance between two points, given their latitudes and longitudes, in order to check which earthquakes' epicenters fall in the range specified. The function distance can again be used here.

These user-defined functions are also helpful for supporting a context-specific fuzzy matching of attribute values. For example, assume that we have two data sources for the domain of earthquakes. It is quite possible that two values of an attribute testSite retrieved from the two sources may be syntactically unequal but refer to the same location. For example, the value available from one source could be "Nevada Test Site, Nevada, USA" and that from another source could be "Nevada Site, NV, USA". The two are semantically equal but syntactically unequal [KS96]. Fuzzy matching functions can be useful in comparing the two values.

Another important advantage of using operations is that the system can support complex post-processing of data. An interesting form of post-processing is the use of simulation programs. For instance, researchers in the field of Geographic Information Systems (GIS) use simulation programs to forecast characteristics like urban growth in a region based on a model. InfoQuilt supports the use of such simulations like any other operation.

5.2 Information Scapes (IScapes)

InfoQuilt uses IScape, a paradigm for information request which is *"a computing paradigm that allows users to query and analyze the data available from a diverse autonomous sources, gain better understanding of the domains and their interactions as well as discover and study relationships."*

Consider the following information request.

"Find all earthquakes with epicenter in a 5000 mile radius area of the location at latitude 60.790 North and longitude 97.570 East and find all tsunamis that they might have caused."

In addition to the obvious constraints, the system needs to understand what the user means by saying *"find all tsunamis that might have been caused due to the earthquakes"*. The relationship that *an earthquake caused a tsunami* is a complex inter-ontological relationship.

Any system that needs to answer such information requests would need a comprehensive knowledge of the terms involved and how they are related. An IScape is specified in terms of relevant ontologies, inter-ontological relationships and operations. Additionally, this abstracts the user from having to know the actual sources that will be used by the system to answer it and how the data retrieved from these sources will be integrated, including how the results should be grouped, any aggregations that need to be computed, constraints that need to be applied to the grouped data, and the information that needs to be returned in the result to the user.

The ontologies in the IScape identify the domains that are involved in the IScape and the inter-ontological relationships specify the semantic interaction between the ontologies. The preset constraint and the runtime configurable constraint are filters used to describe the subset of data that the user is interested in, similar to the WHERE clause in an SQL query. For example, a user may be interested in earthquakes that occurred in only a certain region and had a magnitude greater than 5. The difference between the preset constraint and the runtime constraint is that the runtime constraint can be set at the time of executing the IScape. The results of the IScape can be grouped based on attributes and/or values computed by functions.

5.3 Human <u>A</u>ssisted K<u>n</u>owledge <u>D</u>iscovery (HAND) Techniques

InfoQuilt provides a framework that allows users to access data available from a multitude of diverse autonomous distributed resources and provide tools that help them to analyze the data to gain a better understanding of the domains and the inter-domain relationships as well as help users to explore the possibilities of new relationships.

Existing relationships in the knowledgebase provide a scope for discovering new aspects of relationships through transitive learning. For example, consider the ontologies Earthquake, Tsunami and Environment. Assume that the relationships "Earthquake affects Environment", "Earthquake causes Tsunami" and "Tsunami affects Environment" are defined and known to the system. We can see that since Earthquake causes a Tsunami and Tsunami affects the environment, effectively this is another way in which an earthquake affects the environment (by causing a tsunami). If this aspect of the relationship between an earthquake and environment was not considered earlier, it can be studied further.

Another valuable source of knowledge discovery is studying existing IScapes that make use of the ontologies, their resources and relationships to retrieve information that is of interest to the users. The results obtained from IScapes can be analyzed further by post processing of the result data. For example, the Clarke UGM model forecasts the future patterns of urban growth using information about urban areas, roads, slopes, vegetation in those areas and information about areas where no urban growth can occur.

For the users that are well-versed with the domain, the InfoQuilt framework allows exploring new relationships. The data available from various sources can be queried by constructing IScapes and the results can be analyzed by using charts, statistical analysis techniques, etc. to study and explore trends or aspects of the domain. Such analysis can be used to validate any hypothetical relationships between domains and to see if the data validates or invalidates the hypothesis. For example, several researchers in the past have expressed their concern over nuclear tests as one of the causes of earthquakes and suggested that there could be a direct connection between the two. The underground nuclear tests cause shock waves, which travel as ripples along the crust of the earth and weaken it, thereby making it more susceptible to earthquakes. Although this issue has been addressed before, it still remains a hypothesis that is not conclusively and scientifically proven. Suppose we want to explore this hypothetical relationship.

Consider the NuclearTest and Earthquake ontologies again. We assume that the system has access to sufficient resources for both the ontologies such that they together provide sufficient information for the analysis. However, note that the user is not aware of these data sources since the system abstracts him from them. To construct IScapes, the user works only with the components in the knowledgebase. If the hypothesis is true, then we should be able to see an increase in the number of earthquakes that have occurred after the nuclear testing started.

An example IScape for testing this hypothesis is given below:

"Find nuclear tests conducted after January 1, 1950 and find any earthquakes that occurred not later than a certain number of days after the test and such that its epicenter was located no farther than a certain distance from the test site."

Note the use of *"not later than a certain number of days"* and *"no farther than a certain distance"*. The IScape does not specify the value for the time period and the distance. These are defined as runtime configurable parameters, which the user can use to form a constraint while executing the IScape. The user can hence supply different values for them and execute the IScape repeatedly to analyze the data for different values without constructing it repeatedly from scratch. Some of the interesting results that can be found by exploring earthquakes occurring that occurred no later than 30 days after the test and with their epicenter no farther than 5000 miles from the test site are listed below.

- China conducted a nuclear test on October 6, 1983 at Lop Nor test site. USSR conducted two tests, one on the same day and another on October 26, 1983, both at Easter Kazakh or Semipalitinsk test site. There was an earthquake of magnitude 6 on the Richter scale in Erzurum, Turkey on October 30, 1983, which killed about 1300 people. The epicenter of the earthquake was about 2000 miles away from the test site in China and about 3500 miles away from the test site in USSR. The second USSR test was just 4 days before the quake.
- USSR conducted a test on September 15, 1978 at Easter Kazakh or Semipalitinsk test site. There was an earthquake in Tabas, Iran on September 16, 1978. The epicenter was about 2300 miles away from the test site.

More recently, India conducted a nuclear test at its Pokaran test site in Rajasthan on May 11, 1998. Pakistan conducted two nuclear tests, one on May 28, 1998 at Chagai test site and another on May 30, 1998. There were two earthquakes that occurred soon after these tests. One was in Egypt and Israel on May 28, 1998 with its epicenter about 4500 miles away from both test sites and another in Afghanistan, Tajikistan region on May 30, 1998, with a magnitude of 6.9 and its epicenter about 750 miles away from the Pokaran test site and 710 miles from Chagai test site.

6. Evaluations involving Semantic Relationships: Example of Multi-ontology Query Processing

Our last section deals with some issues in evaluating complex relationships across information domains, potentially spanning multiple ontologies. Most practical situations in the Semantic Web will involve multiple overlapping or disjoint but related ontologies. For example, an information request might be formulated using terms in one ontology but the relevant resources may be annotated using terms in other ontologies. Computations such as query processing in such cases will involve complex relationships spanning multiple ontologies. This raises several difficult problems, but perhaps the key problem is that of impact on quality of results or the change in query semantics when the relationships involves are not synonyms. In this chapter, we present the case study of multi-ontology query processing in the OBSERVER project..

A user query formulated using terms in domain ontology is translated by using terms of other (target) domain ontologies. Mechanisms dealing with incremental enrichment of the answers are used. The substitution of a term by traversing inter-ontological relationships like synonyms (or combinations of them [Mena96]) and combinations of hyponyms (specializations) and hypernyms (generalizations) provide answers not available otherwise by using only a single ontology. This, however, changes the semantics of the query. We discuss with the help of examples, mechanisms to estimate loss of information (based on intensional and extensional properties) in the face of possible semantic changes when translating a query across different ontologies. This measure of the information loss (whose upper limit is defined by the user) guides the system in navigating those ontologies that have more relevant information; it also provides the user with a level of confidence in the answer that may be retrieved. Well-established metrics like precision and recall are used and adapted to our context in order to measure the change in semantics instead of the change in the extension, unlike techniques adopted by classical Information Retrieval methods.

6.1 Query Processing in OBSERVER

The idea underlying our query processing algorithm is the following: give the first possible answer and then enrich it in successive iterations until the user is satisfied. Moreover, certain degree of imprecision (defined by each user) in the answer could be allowed if it helps to speed up the search of the wanted information. We use ontologies, titled **WN** and **Stanford-I** (see [Mena00]) and the following example query to illustrate the main steps of our query expansion approach.

User Query: `Get title and number of pages of books written by Carl Sagan'`
The user browses the available ontologies (ordered by knowledge areas) and chooses a user ontology that includes the terms needed to express the semantics of her/his information needs. Terms from the user ontology are chosen, to express the constraints and relationships that comprise the query. In the example, the WN ontology is selected since it contains all the terms needed to express the semantics of the query, i.e., terms that store information about titles (`NAME'`), number of pages (`PAGES'`), books (`BOOK'`) and authors (`CREATOR'`).

Q = [NAME PAGES] for (**AND** BOOK (**FILLS** CREATOR "Carl Sagan"))

Syntax of the expressions is taken from CLASSIC [BBMR89], the system based on Description Logics (DL) that we use to describe ontologies.

Controlled and Incremental Query Expansion to Multiple Ontologies

If the user is not satisfied with the answer, the system retrieves more data from other ontologies in the Information System to "enrich" the answer in an incremental manner. In doing so, a new component ontology, the target ontology, whose concepts participate in inter-ontological relationships with the user ontology is selected. The user query is then expressed/translated into terms of that target ontology. The user and target ontologies are integrated by using the inter-ontology relationships defined between them.

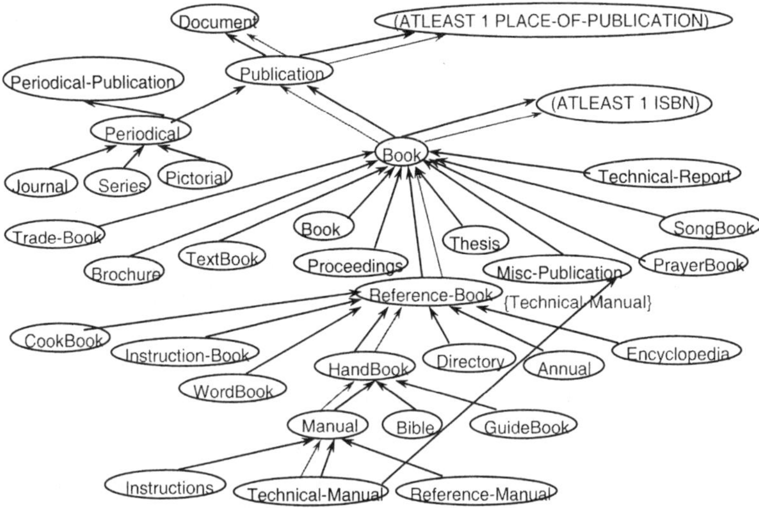

Fig. 8. Use of inter-ontological relationships to integrate multiple ontologies

- All the terms in the user query may have been rewritten by their corresponding synonyms in the target ontology. Thus the system obtains a semantically equivalent query (**full translation**) and no loss of information is incurred.
- There exist terms in the user query that can not be translated into the target ontology - they do not have synonyms in the target ontology (we called them conflicting terms). This is called a **partial translation**.

Each conflicting term in the user query is replaced by the intersection of its immediate parents (hypernyms) or by the union of its immediate children (hyponyms), recursively, until a translation of the conflicting term is obtained using only the terms of the target ontology. This could lead to several candidate translations, leading to change in semantics and loss of information. The query Q discussed above has to be translated into terms of the Stanford-I ontology [Mena00]. After the process of integrating the WN and Stanford-I ontologies (Figure 8), Q is redefined as follows:

Q = [title number-of-pages] for (**AND BOOK** (**FILLS** doc-author-name "Carl Sagan"))

The only conflicting term in the query is `BOOK' (it has no translation into terms of Stanford-I). The process of computing the various plans for the term `BOOK' results in four possible translations: `document', `periodical-publication', `journal' or `UNION(book, proceedings, thesis, misc-publication, technical-report)'. Details of this translation process can be found in [MKIS98]. This leads to 4 possible translations of the query:

Plan 1: (**AND** document (**FILLS** doc-author-name "Carl Sagan"))

Plan 2: (**AND** periodical-publication (**FILLS** doc-author-name "Carl Sagan"))

Plan 3: (**AND** journal (**FILLS** doc-author-name "Carl Sagan"))

Plan 4: (**AND** UNION(book, proceedings, thesis, misc-publication, technical-report) (**FILLS** doc-author-name "Carl Sagan"))

6.2 Estimating the Loss of Information

We use the Information Retrieval analogs of soundness (precision) and completeness (recall), which are estimated based on the sizes of the extensions of the terms. We combine these two measures to compute a composite measure in terms of a numerical value. This can then be used to choose the answers with the least loss of information.

Loss of information based on intensional information

The loss of information can be expressed like the terminological difference between two expressions, the user query and its translation. The terminological difference between two expressions consists of those constraints of the first expression that are not subsumed by the second expression. The loss of information for Plan 1 is as follows:

Plan 1: (**AND** document (**FILLS** doc-author-name "Carl Sagan"))

Taking into account the following term definitions[1]:

BOOK = (**AND** PUBLICATION (**ATLEAST** 1 ISBN)),

PUBLICATION = (**AND** document (**ATLEAST** 1 PLACE-OF-PUBLICATION))

The terminological difference is, in this case, the constraints not considered in the plan:

(**AND** (**ATLEAST** 1 ISBN) (**ATLEAST** 1 PLACE-OF-PUBLICATION))

The intensional loss of information of the 4 plans can thus be enumerated as:

- Plan = (**AND** document (**FILLS** doc-author-name "Carl Sagan"))
  Loss = "Instead of books written by Carl Sagan, all the documents written by Carl Sagan are retrieved, even if they do not have an ISBN and place of publication".

- Plan = (**AND** periodical-publication (**FILLS** doc-author-name "Carl Sagan"))
  Loss = "Instead of books written by Carl Sagan, all periodical publications written by Carl Sagan are retrieved, even if they do not have an ISBN and place of publication".

- Plan = (**AND** journal (**FILLS** doc-author-name "Carl Sagan"))

---

[1] The terminological difference is computed across extended definitions.

Loss = "Instead of books written by Carl Sagan, all journals written by Carl Sagan are retrieved, even if they do not have an ISBN and place of publication".
- Plan = (**AND** UNION(book, proceedings, thesis, misc-publication, technical-report)
(**FILLS** doc-author-name "Carl Sagan"))
Loss = "Instead of books written by Carl Sagan, book , proceedings, theses, misc-publication and technical manuals written by Carl Sagan are retrieved".

An intensional measure of the loss of information can make it hard for the system to decide between two alternatives, in order to execute first plan with less loss. Thus, some numeric way of measuring the loss should be explored.

Loss of information based on extensional information

The loss of information is based on the number of instances of terms involved in the substitutions performed on the query and depends on the sizes of the term extensions. A composite measure combining measures like *precision* and *recall* [Sal89] used to estimate the information loss is described, which takes into account the bias of the user ("is precision more important or recall ?").

The extension of a query expression is a combination of unions and intersections of concepts in the target ontology since and is estimated with an upper ($|Ext(Expr)|.high$) and lower ($|Ext(Expr)|.low$) bound. It is computed as follows:

$|Ext(Subexpr_1) \cap Ext(Subexpr_2)|.low = 0$

$|Ext(Subexpr_1) \cap Ext(Subexpr_2)|.high = min [|Ext(Subexpr_1)|.high, |Ext(Subexpr_2)|.high]$

$|Ext(Subexpr_1) \cup Ext(Subexpr_2)|.low = max [|Ext(Subexpr_1)|.high, |Ext(Subexpr_2)|.high]$

$|Ext(Subexpr_1) \cup Ext(Subexpr_2)|.high = |Ext(Subexpr_1)|.high + |Ext(Subexpr_2)|.high$

A composite measure combining precision and recall

Precision and Recall have been very widely used in Information Retrieval literature to measure loss of information incurred when the answer to a query issued to the information retrieval system contains some proportion of irrelevant data [Sal89]. The measures are adapted to our context, as follows:

$$\Pr ecision = \frac{|\ Ext\ (Term\ ) \cap Ext\ (Translatio\ n)\ |}{|\ Ext\ (Translatio\ n)\ |}$$

$$\mathrm{Re}\ call = \frac{|\ Ext\ (Term\ ) \cap Ext\ (Translatio\ n)\ |}{Ext\ (Term\ )}$$

We use a composite measure [vR] which combines the precision and recall to estimate the loss of information. We seek to measure the extent to which the two sets do not match. This is denoted by the shaded area in Figure 9. The area is, in fact, the symmetric difference:

RelevantSet Δ RetrievedSet = RelevantSet $\cup$ RetrievedSet - RelevantSet $\cap$ RetrievedSet

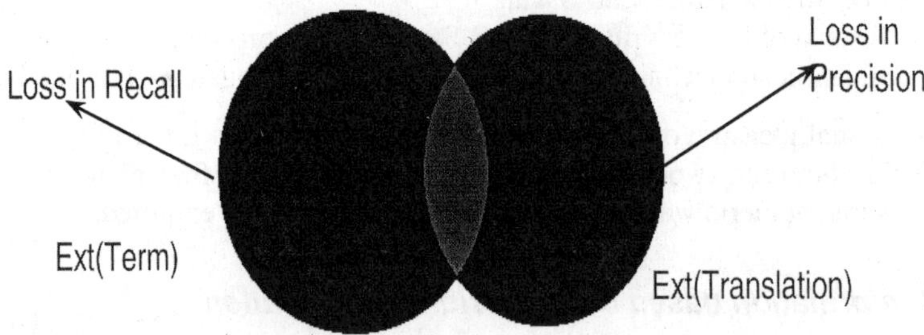

Fig. 9. Symmetric difference

The loss of information may be given as:

$$Loss = \frac{|\mathrm{Re}\,levantSet\,\Delta\,\mathrm{Re}\,trievedSet|}{|\mathrm{Re}\,levantSet| + |\mathrm{Re}\,trievedSet|}$$

$$Loss = 1 - \frac{1}{\dfrac{1}{2}\left(\dfrac{1}{\mathrm{Pr}\,ecision}\right) + \dfrac{1}{2}\left(\dfrac{1}{\mathrm{Re}\,call}\right)}$$

Semantic adaptation of precision and recall

Higher priority needs to be given to semantic relationships than those suggested by the underlying extensions. The critical step is to estimate the extension of Translation based on the extensions of terms in the target ontology. Precision and recall are adapted as follows:

- **Precision and recall measures for the case where a term subsumes its translation**. Semantically, we do not provide an answer irrelevant to the term, as Ext(Translation) $\subseteq$ Ext(Term) (by definition of subsumption). Thus, Ext(Term) $\cap$ Ext(Translation) = Ext(Translation). Therefore:

$$\Pr ecision = 1,$$

$$\text{Re} \, call = \frac{|\, Ext(Term) \cap Ext(Translation)\, |}{|\, Ext(Term)\, |} = \frac{|\, Ext(Translation)\, |}{|\, Ext(Term)\, |}$$

$$\text{Re} \, call.low = \frac{|\, Ext(Translation)\, |.low}{|\, Ext(Term)\, |},$$

$$\text{Re} \, call.high = \frac{|\, Ext(Translation)\, |.high}{|\, Ext(Term)\, |}$$

- **Precision and recall measures for the case where a term is subsumed by its translation.** Semantically, all elements of the term extension are returned, as Ext(Term) $\subseteq$ Ext(Translation) (by definition of subsumption). Thus, Ext(Term) $\cap$ Ext(Translation) = Ext(Term). Therefore:

$$\text{Re} \, call = 1,$$

$$\Pr ecision = \frac{|\, Ext(Term) \cap Ext(Translation)\, |}{|\, Ext(Translation)\, |} = \frac{|\, Ext(Term)\, |}{|\, Ext(Translation)\, |}$$

$$\Pr ecision.low = \frac{|\, Ext(Term)\, |}{|\, Ext(Translation)\, |.high},$$

$$\Pr ecision.high = \frac{|\, Ext(Term)\, |}{|\, Ext(Translation)\, |.low}$$

- Term and Expression are not related by any subsumption relationship. The general case is applied directly since intersection cannot be simplified. In this case the interval describing the possible loss will be wider as Term and Translation are not related semantically.

$$\Pr ecision.low = 0,$$

$$\Pr ecision.high = \max \left[\frac{\min\left[|Ext(Term)|, |Ext(Translation)|.high\right]}{|Ext(Translation)|.high}, \frac{\min\left[|Ext(Term)|, |Ext(Translation)|.low\right]}{|Ext(Translation).low|} \right]$$

$$\text{Re} \, call.low = 0, \quad \text{Re} \, call.high = \frac{\min\left[|\, Ext(Term)\, |, |\, Ext(Translation)\, |.low\right]}{|\, Ext(Term)\, |}$$

The various measures defined above are applied to the 4 translations and the loss of information intervals are computed. The values are illustrated in Table 1. For a detailed account of the computations involved, the reader may look at [Mena00].

TRANSLATION	LOSS OF INFORMATION
(**AND** document (**FILLS** doc-author-name "Carl Sagan"))	91.571% ≤ Loss ≤ 91.755%
(**AND** periodical-publication (**FILLS** doc-author-name "Carl Sagan"))	94.03% ≤ Loss ≤ 100%
(**AND** journal (**FILLS** doc-author-name "Carl Sagan"))	98.56% ≤ Loss ≤ 100%
(**AND** (**FILLS** doc-author-name "Carl Sagan") UNION(book, proceedings, thesis, misc-publication, technical report))	0 ≤ Loss ≤ 7.22%

Table 1: Various Translations and the respective loss of Information

7. Conclusions

Ontologies provide the semantic underpinning, while relationships are the back-bone for semantics in the Semantic Web or any approach to achieving semantic interoperability. For more semantic solutions, attention needs to shift from documents (e.g., searching for relevant documents) to integrated approach of exploiting data (content, documents) with knowledge (including domain ontologies). Relationships, their modeling, specification or representation, identification, validation or their use in query or information request evaluation are then the fundamental aspects of study. In this chapter, we have provided an initial taxonomy for studying various aspects of semantic relationships. To exemplify the some points in the broad scope of studying semantic relationships, we discussed four examples of our own research efforts during the past decade. Neither the taxonomy nor our empirical exemplification through four examples is a complete study. We hope it would be extended with study of extensive research reported in the literature by other researchers.

Acknowledgements

Ideas presented in this chapter have benefited from team members at the LSDIS Lab (projects: InfoHarmess, VisualHarness, VideoAnywhere, InfoQuilt, and Semantic Association Identification), and Semagix. Special acknowledgements to Eduardo Mena (for his contributions to the OBSERVER project), Kemafor Anyanwu, and Aleman Boanerges, (for their work on Semantic Associations), Brian Hammond, Clemens Bertram, Sena Arpinar, and David Avant (for their work on relevant parts of SCORE discussed here), and Krys Kochut (for discussions on semantic index and his work on SCORE).

References

[Anyanwu02]   K. Anyanwu and A. Sheth, "The ρ Operator: Computing and Ranking Semantic Associations in the Semantic Web", SIGMOD Record, December 2002.

[Arumugam02]   M. Arumugam, A. Sheth, and I. B. Arpinar, "Towards Peer-to-Peer Semantic Web: A Distributed Environment for Sharing Semantic Knowledge on the Web", Intl. Workshop on Real World RDF and Semantic Web Applications 2002, Hawaii, May 2002.

[Bailin01]   S. C. Bailin, and W. Truszkowski, "Ontology Negotiation Between Agents Supporting Intelligent Information Management", Workshop On Ontologies In Agent Systems, 2001.

[Berners-Lee01]   T. Berners-Lee, J. Hendler, and O. Lassila, "The Semantic Web, A new form of Web content that is meaningful to computers will unleash a revolution of new possibilities", Scientific American, May 2001.

[Boll98]   S. Boll, W. Klas and A. Sheth, "Overview on Using Metadata to Manage Multimedia Data", in Multimedia Data Management: Using Metadata to Integrate and Apply Digital Media, A. Sheth and W. Klas, Eds., McGraw-Hill Publishers, March 1998.

[Brezillon02]   P. Brezillon, and J.-C. Pomerol, "Reasoning with Contextual Graphs", European Journal of Operational Research, 136(2): 290–298, 2002.

[Brezillon01]   P. Brezillon, and J.-C. Pomerol, "Is Context a Kind of Collective Tacit Knowledge?", European CSCW 2001 Workshop on Managing Tacit Knowledge. Bonn, Germany. M. Jacovi and A. Ribak (Eds.), pp. 23–29, 2001.

[Brezillon99a]   P. Brezillon, "Context in Problem Solving: A Survey", The Knowledge Engineering Review, 14(1): 1–34, 1999.

[Brezillon99b]   P. Brezillon, "Context in Artificial Intelligence: I. A Survey of the Literature", Computer & Artificial Intelligence, 18(4): 321–340, 1999.

[Brezillon99c]   P. Brezillon, "Context in Artificial Intelligence: II. Key Elements of Contexts", Computer & Artificial Intelligence, 18(5): 425–446, 1999.

[Buneman00]   P. Buneman, S. Khanna, and W.-C. Tan, "Data Provenance: Some Basic Issues", Foundations of Software Technology and Theoretical Computer Science (2000).

[Buneman02a]   P. Buneman, S. Khanna, K. Tajima, and W.-C. Tan, "Archiving Scientific Data", Proceedings of ACM SIGMOD International Conference on Management of Data (2002).

[Chen99]   Y. Chen, Y. Peng, T. Finin, Y. Labrou, and S. Cost, "Negotiating Agents for Supply Chain Management", AAAI Workshop on Artificial Intelligence for Electronic Commerce, AAAI, Orlando, June 1999.

[Constan-   P. Constantopoulos, and M. Doerr, "The Semantic Index System - A topoulos93]   brief presentation", Institute of Computer Science Technical Report. FORTH-Hellas, GR71110 Heraklion, Crete, 1993.

[Cost02]   R. S. Cost, T. Finin, A. Joshi, Y. Peng, et. Al., "ITTALKS: A Case Study in DAML and the Semantic Web", IEEE Intelligent Systems Special Issue, 2002.

[Finnin88a]   T. Finin, "Default Reasoning and Stereotypes in User Modeling", International Journal of Expert Systems, Volume 1, Number 2, Pp. 131–158, 1988.

[Finin92]   T. Finin, R. Fritzson, and D. McKay, "A Knowledge Query and Manipulation Language for Intelligent Agent Interoperability", Fourth National Symposium on Concurrent Engineering, CE & CALS Conference, Washington, DC June 1–4, 1992.

[Heflin02]   J. Heflin, R. Volz. J. Dale, Eds., Requirements for a Web Ontology Language, March 07, 2002. http://www.w3.org/TR/webont-req/

[Hendler01]   J. Hendler, "Agents and the Semantic Web", IEEE Intelligent Systems, 16(2), March/April, 2001.

[Heuer99]   R. J. Heuer, Jr., "Psychology of Intelligence Analysis", Center for the Study of Intelligence, Central Intelligence Agency, 1999.

[Joshi00]   A. Joshi, and R. Krishnapuram, "On Mining Web Acceess Logs", Proc. SIGMOD 2000 Workshop on Research Issues in Data Mining and Knowledge Discovery, pp 63–69, Dallas, 2000.

[Joshi02]   K. Joshi, A. Joshi, Y. Yesha, "On Using a Warehouse to Analyze Web Logs", accepted for publication in Distributed and Parallel Databases, 2002.

[Kagal01a]   L. Kagal, T. Finin, and A. Joshi, "Trust-Based Security For Pervasive Computing Environments", IEEE Communications, December 2001.

[Kagal01b]   L. Kagal, T. Finin, and Y. Peng, "A Delegation Based Model for Distributed Trust Management", In Proceedings of IJCAI-01 Workshop on Autonomy, Delegation, and Control, August 2001.

[Kagal01c]   L. Kagal, S. Cost, T. Finin, and Y. Peng, "A Framework for Distributed Trust Management", In Proceedings of Second Workshop on Norms and Institutions in MAS, Autonomous Agents, May 2001.

[Krishnapuram01]   R. Krishnapuram, A. Joshi, O. Nasraoui, and L. Yi, "Low Complexity Fuzzy Relational Clustering Algorithms for Web Mining", IEEE Trans. Fuzzy Systems, 9:4, pp 595–607, 2001.

[Kashyap96]   V. Kashyap, and A. Sheth, "Semantic Heterogeneity in Global Information Systems: The Role of Metadata, Context, and Ontologies, in Cooperative Information Systems: Current Trends and Directions", M Papazoglou and G. Sclageter (eds), 1996.

[Kashyap95]   V. Kashyap, and A. Sheth, "Metadata for building the Multimedia Patch Quilt," "Multimedia Database Systems: Issues and Research Directions, S. Jajodia and V. S. Subrahmanium, Eds., Springer-Verlag, p. 297–323, 1995.

[Kashyap00]   V. Kashyap and A. Sheth, "Information Brokering Across Heteroge-

	neous Digital Data", Kluwer Academic Publishers, August 2000, 248 pages.
[Kass90]	R. Kass, and T. Finin, "General User Modeling: A Facility to Support Intelligent Interaction", in J. Sullivan and S. Tyler (eds.) Architectures for Intelligent Interfaces: Elements and Prototypes, ACM Frontier Series, Addison-Wesley, 1990.
[Kirzen99]	L. Kirzen, "Intelligence Essentials for Everyone, Occasional Paper Number Six", Joint Military Intelligence College, Washington, D.C., June 1999.
[Liere97]	R. Liere, and P. Tadepelli, "Active Learning with Committees for Text Categorization", Proc. 14th Conf. Am. Assoc. Artificial Intelligence, AAAI Press, Menlo Park, Calif., 1997, pp. 591–596.
[Mena00]	E. Mena, A. Illarramendi, V. Kashyap and A. Sheth, "OBSERVER: An Approach for Query Processing in Global Information Systems based on Interoperation across Pre-existing Ontologies", Distributed and Parallel Databases (DAPD), Vol. 8, No. 2, April 2000, pp. 223–271.
[Nonaka95]	I. Nonaka, and H. Takeuchi, "The Knowledge-Creating Company", Oxford University Press, New York, NY, 1995.
[Sebastiani02]	F. Sebastiani, "Machine Learning in Automated Text Categorization," *ACM Computing Surveys*, vol. 34, no. 1, 2002, pp. 1–47.
[Shah97]	K. Shah, A. Sheth, and S. Mudumbai, "Black Box approach to Visual Image Manipulation used by Visual Information Retrieval Engines", Proceedings of 2nd IEEE Metadata Conference, September 1997.
[Shah98]	K. Shah and A. Sheth, Logical Information Modeling of Web-accessible Heterogeneous Digital Assets, Proc. of the Forum on Research and Technology Advances in Digital Libraries," (ADL'98), Santa Barbara, CA. April 1998, pp. 266–275.
[Shah99]	K. Shah and A. Sheth, "InfoHarness: An Information Integration Platform for Managing Distributed, Heterogeneous Information," IEEE Internet Computing, November-December 1999, p. 18–28.
[Shah02]	U. Shah, T. Finin, A. Joshi, R. S. Cost, and J. Mayfield, "Information Retrieval on the Semantic Web", submitted to the 10th International Conference on Information and Knowledge Management, November 2002.
[Sheth96]	A. Sheth and V. Kashyap, „Media-independent correlation of Information: What? How?" Proceedings of the First IEEE Metadata Conference, April 1996. http://www.computer.org/conferences/meta96/sheth/
[Sheth90]	A. Sheth and J. Larson, "Federated Databases: Architectures and Issues," ACM Computing Surveys, 22 (3), September 1990, pp. 183–236.
[Sheth98]	A. Sheth, "Changing Focus on Interoperability in Information Systems: From System, Syntax, Structure to Semantics in Interoperating Geographic Information Systems", M. F. Goodchild, M. J. Egenhofer, R. Fegeas, and C. A. Kottman (eds.), Kluwer,

1998.

[Sheth02b]   A. Sheth, S. Thacker and S. Patel, "Complex Relationship and Knowledge Discovery Support in the InfoQuilt System", VLDB Journal, 2002 .

[Sheth02a]   A. Sheth, C. Bertram, D. Avant, B. Hammond, K. Kochut, and Y. Warke, "Semantic Content Management for Enterprises and the Web", IEEE Internet Computing, July/August 2002.

[Srinivasan02]   N. Srinivasan, and T. Finin, "Enabling Peer to Peer SDP in Agents", Proceedings of the 1st International Workshop on "Challenges in Open Agent Systems, July 2002, University of Bologna, held in conjunction with the 2002 Conference on Autonomous Agents and Multiagent Systems.

[Tolia02a]   S. Tolia, D. Khushraj, and T. Finin, "ITTalks: Event Notification Service: An illustrative case for services in the Agent cities Network", Proceedings of the 1st International Workshop on Challenges in Open Agent Systems, July 2002.

[Wiederhold92]   G. Wiederhold, "Mediators in the Architecture of Future Information Systems", IEEE Computer 25(3): 38–49, 1992.

[Zadeh65]   L.A. Zadeh. *Fuzzy sets*. In Information and Control, pages 338–353, 1965.

Subjective Enhancement and Measurement of Web Search Quality

M. M. Sufyan Beg and Nesar Ahmad

Department of Electrical Engineering, Indian Institute of Technology, New Delhi, 110 016, India, {mmsbeg,nahmad}@ee.iitd.ac.in

Abstract

The web search efforts started with the application of classical content-based information retrieval techniques to the web. But due to the commercial pressures, *spamming* became the major threat to the success of the content-based techniques. Attention was then diverted to exploit the hypertextual structure of the web. But, both the content based approach as well as the hyperlink based approach are objective ones, which are totally dependent on the effectiveness of their "feature extraction" mechanisms, with no apparent consideration to the preference of the searcher. In this chapter, a "user satisfaction" guided web search procedure is proposed. We calculate the importance weight of each document viewed by the user based on the feedback vector obtained from his actions. This document weight is then used to update the index database in such a way that the documents being consistently preferred go up the ranking, while the ones being neglected go down. Our simulation results show a steady rise in the satisfaction levels of the modeled users as more and more learning goes into our system. We also propose a couple of novel additions to the web search querying techniques. The user feedback obtained on the search results is also utilized for evaluating the performance of the search engine that supplied the results to the user. We also include a brief overview of the existing web search techniques.

Introduction

The recent past has seen a paradigm shift in web searching (Beg 2002) from the conventional *content based* searching (Meadow et al 2000, Rijsbergen 1979, Salton and McGill 1983, Srivastava et al 2001) to the more crisp *connectivity based* searching (Henzinger 2001, Beg and Ravikumar 2000, Beg 2001, Beg and Ahmad 2001). A few years ago, the query term frequency was the single main heuristic in ranking the web pages. But the emergence of novel search engines like *Google* (Google 2002) has marked the beginning of the era of *connectivity based* or *cita-*

tion based, or more commonly known as *hyperlink based* (or simply *link based*) web searching. It all started with the revolutionary work done by Kleinberg (Kleinberg 1998), and consolidated by Chakrabarti et al (Chakrabarti et al 1999a, Chakrabarti et al 1999b) and Brin & Page (Brin and Page 1998), culminating in what is now famous as the *Google* (Google 2002) search engine for the web. These works have primarily been concentrating on harnessing the additional information hidden in the hypertext structure of the web pages. But, whether the content based approach or the link based approach, both are totally dependent on the effectiveness of their "feature extraction" mechanisms, with no apparent consideration to the preference of the searcher. *DirectHit* (DirectHit 2000a) is a search engine that claims to harness the collective intelligence of the millions of the daily Internet searchers to improve the web search results. For this, it monitors attributes like which documents are selected by the user from the search result presented to him, how much time is spent at those sites, etc. (DirectHit 2000b). But our recent study (Beg and Ravikumar 2002) shows that *DirectHit* falls below the satisfaction of an average user.

In this chapter, we have tried to improve upon the approach of *DirectHit* by learning from the feedback obtained from the user. While doing so, we have also proposed a few novel additions to the web search querying techniques. A procedure for evaluating the performance of public web search engines is also discussed. We begin with a brief overview of the content based and the connectivity based techniques of web searching.

Content-Based Web Search

Going by the instinct, the contents of a page are the best descriptors of that page. But to get the index terms of a given document is a non-too-trivial task. It is a well-known fact that for a good amount of time, *Yahoo!* relied on the human intelligence to get this indexing done. Equivalently, we may have to employ the complex *Natural Language Processing* techniques to automate the job of indexing (Chakrabarti et al 2002). However, some easier techniques have been researched upon by the IR community for this task of keyword selection (Meadow et al 2000, Rijsbergen 1979, Salton and McGill 1983). But before we proceed, we must have a look at the text pre-processing operations, which are a pre-requisite for the application of any of the content-based techniques.

Text Pre-Processing

First of all, we need to remove the *stop-words*, the words that have very little semantic content. They are frequent homogeneously across the whole collection of documents. They are generally the prepositions or articles, like *the, an, for*, etc. Over a period of time, people have come up with a list of stop-words pertaining to a general domain. However, it may be argued that a stop-word is very much context dependant. A word like *web* may be treated as a stop word in a collection of web-related articles, but not so in a set of literary documents.

The text pre-processing also includes an important step of word *stemming*, wherein all the words with the same root are reduced to a single form. This is achieved by stripping each word of its suffix, prefix or infix, if any. That is to say that all the words are reduced to their canonical form. For instance, the words like *borrow*, *borrower*, *borrowed* and *borrowing*, all would be reduced to the stem word *borrow*. This way the burden of being very specific while forming the query, is taken off from the shoulders of the user. A well-known algorithm for carrying out word stemming is the *Porter Stemmer* algorithm (Porter 1980). It may be noted that the stemming techniques are very much dependent on the language of the document. For instance, just the removal of suffixes may usually suffice as the stemming technique in the case of English language, but not necessarily so with other languages.

Term Frequency

The frequency of a word in a given document is taken as a good measure of the importance of that word in the given document. This, of course, holds true only after the text pre-processing has been carried out.

An improved version of the term frequency (TF) method is the *term frequency-inverse document frequency* (TF-IDF) method. The intuition is to gauge the uniqueness of a keyword in a document, and hence the relevance. We decide upon a keyword after looking at the fact that this word should be frequent in the current document but not so frequent in the rest of the documents of the given collection. The TF-IDF value for a given keyword w is given as:

$$f_{TF-IDF} = \frac{f_w}{f_{w_{max}}} \log \frac{\rho}{\rho_w}$$

where, f_w is the frequency of the keyword w in the document, $f_{w_{max}}$ is the maximum frequency of any word in the document, ρ_w is the number of documents in which this keyword occurs, and ρ is the total number of documents in the collection. An alternative formulation is given as:

$$f_{TF-IDF} = \frac{\left[0.5 + 0.5 \dfrac{f_w}{f_{w_{max}}} \right] . \log \dfrac{\rho}{\rho_w}}{\sqrt{\sum_{all\ docs} \left[0.5 + 0.5 \dfrac{f_w}{f_{w_{max}}} \right]^2 \left[\log \dfrac{\rho}{\rho_w} \right]^2}}$$

Some miscellaneous heuristics may include having some enhanced weightage for the terms occurring in the title, abstract, meta-data or section headings etc. An alternative motivation is that the authors generally introduce the keywords earlier in a document. So, if a word w_k has m occurrences as the $l_1{}^{th}$, $l_2{}^{th}$, ..., $l_m{}^{th}$ word in the given document, the weight associated with it may be found as:

$$weight(w_k) = \frac{\sum_{j=1}^{m} \frac{1}{2^{l_j}}}{\sum_{i=1}^{num\_words} \frac{1}{2^i}}$$

where, *num_words* indicates the total number of words in the given document. This way the words occurring in the title, abstract or the introduction would get a higher weight, and the frequency of the word would also be taken into account. Now sorting the words on the basis of their weights and taking some top N words as the keywords is expected to give good result.

Vector Space Model

An n-dimensional vector space is taken with one dimension for each possible word or term. So n would be the number of words in a natural language. Each document or query is represented as a *term vector* in this vector space. Each component of the *term vector* is either 0 or 1, depending on whether the term corresponding to that axis is absent from or present in the text. Alternately, each component of the term vector is taken as a function that increases with the frequency of the term (corresponding to that axis) within the document and decreases with the number of documents in the collection that contain this term. This is to take into account the TF-IDF factor. In such a case, the term vectors of documents must be normalized to one so as to compensate for different document lengths.

Once the term vectors are obtained, the similarity between a document and a query may be obtained by computing the dot product of their term vectors. The larger the dot product, the greater would be the similarity. We can also calculate the angles between the term vectors of the query and the documents. Then, in response to the given query, the documents may be returned in an increasing order of the angles of their respective term vectors with that of the query.

Boolean Similarity Measures

In literature we also find the *Boolean Similarity Measure* (Li and Danzig 1997) for computing the similarities of one document to another and documents to queries for automatic classification, clustering and indexing. Such well-known similarity measures are *Dice's Coefficient, Jaccard's Coefficient, Cosine Coefficient,* and *Overlap Coefficient. Radecki* proposed two similarity measures, S and S^*, based on *Jaccard's Coefficient.* We denote $\varphi(Q)$ and $\varphi(R)$ as the sets of documents in the response to query Q and in the cluster of documents represented by R. The similarity value S between Q and R is defined as the ratio of the number of common documents to the total number of documents in $\varphi(Q)$ and $\varphi(R)$.

$$S(Q,R) = \frac{|\varphi(Q) \cap \varphi(R)|}{|\varphi(Q) \cup \varphi(R)|}$$

As $\varphi(Q) \subseteq \varphi(R)$, therefore, $S(Q,R) = \dfrac{|\varphi(Q)|}{|\varphi(R)|}$.

Since this measure requires the actual results to the query, it is mainly useful as an index of comparison.

In another measure S^*, Boolean expression Q is transformed into its *reduced disjunctive normal form* (RDNF), denoted as $\tilde{Q}$, which is the disjunction of a list of *reduced atomic descriptors*. If set T is the union of all the descriptors that appear in the to-be-compared Boolean pair, then the reduced atomic descriptor is defined as the conjunction of all the elements in T in either their original or negated forms. Let T_Q and T_R be the sets of the descriptors that appear in Q and R respectively. Suppose $T_Q \cup T_R = \{t_1, t_2, ..., t_k\}$, where k is the set size of $T_Q \cup T_R$. Then the RDNFs of Q and R are

$$\left(\tilde{Q}\right)_{T_Q \cup T_R} = \left(\tilde{q}_{1,1} \wedge \tilde{q}_{1,2} \wedge ... \wedge \tilde{q}_{1,k}\right) \vee \left(\tilde{q}_{2,1} \wedge \tilde{q}_{2,2} \wedge ... \wedge \tilde{q}_{2,k}\right) \vee ... \vee \left(\tilde{q}_{m,1} \wedge \tilde{q}_{m,2} \wedge ... \wedge \tilde{q}_{m,k}\right)$$

$$\left(\tilde{R}\right)_{T_Q \cup T_R} = \left(\tilde{r}_{1,1} \wedge \tilde{r}_{1,2} \wedge ... \wedge \tilde{r}_{1,k}\right) \vee \left(\tilde{r}_{2,1} \wedge \tilde{r}_{2,2} \wedge ... \wedge \tilde{r}_{2,k}\right) \vee ... \vee \left(\tilde{r}_{n,1} \wedge \tilde{r}_{n,2} \wedge ... \wedge \tilde{r}_{n,k}\right)$$

where m and n are the numbers of reduced atomic descriptors in $\left(\tilde{Q}\right)_{T_Q \cup T_R}$ and $\left(\tilde{R}\right)_{T_Q \cup T_R}$ respectively, and

$$\tilde{q}_{i,j} = \begin{cases} t_j ......original \\ \neg t_j ...negated \end{cases} .....1 \le i \le m, 1 \le j \le k$$

$$\tilde{q}_{i,j} = \begin{cases} t_j ......original \\ \neg t_j ...negated \end{cases} .....1 \le i \le m, 1 \le j \le k$$

where $\neg$ is the *not* operator. The similarity value S^* between two Boolean expressions (Q and R) is defined as the ratio of the number of common reduced atomic descriptors in $\tilde{Q}$ and $\tilde{R}$ to the total number of reduced atomic descriptors in them;

$$S^*(Q,R) = \frac{\left| \left(\tilde{Q}\right)_{T_Q \cup T_R} \cap \left(\tilde{R}\right)_{T_Q \cup T_R} \right|}{\left| \left(\tilde{Q}\right)_{T_Q \cup T_R} \cup \left(\tilde{R}\right)_{T_Q \cup T_R} \right|}$$

Another similarity measure $S^\oplus$, with lesser computation, is proposed, by transforming Boolean expression Q to its *compact disjunctive normal form* (CDNF), denoted as $\hat{Q}$, which is the disjunction of a list of *compact atomic descriptors*. Each compact atomic descriptor is the conjunction of a subset of descriptors that appear in its own Boolean expression. The CDNFs of Q and R are

$$\hat{Q} = \left(\hat{q}_{1,1} \wedge \hat{q}_{1,2} \wedge ... \wedge \hat{q}_{1,x_1}\right) \vee \left(\hat{q}_{2,1} \wedge \hat{q}_{2,2} \wedge ... \wedge \hat{q}_{2,x_2}\right) \vee ..... \vee \left(\hat{q}_{m,1} \wedge \hat{q}_{m,2} \wedge \hat{q}_{m,x_m}\right),$$

$$\hat{R} = \left(\hat{r}_{1,1} \wedge \hat{r}_{1,2} \wedge ... \wedge \hat{r}_{1,y_1}\right) \vee \left(\hat{r}_{2,1} \wedge \hat{r}_{2,2} \wedge ... \wedge \hat{r}_{2,y_2}\right) \vee ..... \vee \left(\hat{r}_{n,1} \wedge \hat{r}_{n,2} \wedge \hat{r}_{n,y_n}\right),$$

where m and n are the numbers of compact atomic descriptors in $\hat{Q}$ and $\hat{R}$, x_i is the number of descriptors in the ith *(1 ≤ i ≤ m)* compact atomic descriptor of $\hat{Q}$, and

y_j is the number of descriptors in the jth *(1≤ j ≤n)* compact atomic descriptor of $\hat{R}$. Each $\hat{q}_{i,u}$ and $\hat{r}_{j,v}$ in the CDNFs represents a descriptor in T_Q and T_R respectively. Specifically;

$$\hat{q}_{i,u} \ \varepsilon \ T_Q, \text{ where } (i,u) = (1...m,1...x_i)$$

$$\hat{r}_{j,v} \ \varepsilon \ T_R, \text{ where } (j,v) = (1...n,1...y_j)$$

The individual similarity measure is defined as

$$s^{\oplus}\left(\hat{Q}^i,\hat{R}^j\right)= \begin{cases} 0 \dotsfill if.T_Q^i \cap T_R^j = 0.or.\exists t \in T_Q^i, \neg t \in T_R^j \\[2ex] \dfrac{1}{2^{\left|T_R^j - T_Q^i\right|} + 2^{\left|T_Q^i - T_R^j\right|} - 1} .....otherwise \end{cases}$$

where $\hat{Q}^i$ indicates the ith compact atomic descriptor of CDNF $\hat{Q}$, $\hat{R}^j$ indicates the jth compact atomic descriptor of CDNF $\hat{R}$. T_Q^i and T_R^j are the sets of descriptors in $\hat{Q}^i$ and $\hat{R}^j$ respectively. We define the similarity of two Boolean expressions as the average value of the individual similarity measures ($s^{\oplus}$) between each compact atomic descriptor as

$$S^{\oplus}(Q,R) = \frac{\displaystyle\sum_{i=1}^{\left|\hat{Q}\right|}\sum_{j=1}^{\left|\hat{R}\right|} s^{\oplus}\left(\hat{Q}^i,\hat{R}^j\right)}{\left|\hat{Q}\right| \times \left|\hat{R}\right|}$$

This later method reduces time and space complexity from exponential to polynomial in the number of Boolean terms.

Problems with Content Based Web Search

The main problem of the content-based web searching stems from what is called as "*search engine persuasion*" or "*keyword spamming*" or simply "*spamming*". The commercial interest of many web page authors to have their pages ranking high for certain queries forces them to do some illegitimate manipulations. These manipulations may go as far as adding text in an invisible ink. For instance, if the word "car" is repeated some thousand times in an invisible link, to the web surfer it would not appear awkward, but the search engine would rate it very high for the queries on "car". The power of hyperlink analysis, as we discuss in the subsequent section, comes from the fact that it uses the content of other pages to rank the current page. This gives an unbiased factor to the ranking.

There may also be cases where the keyword we are looking for may not at all appear in a page, which, otherwise, is very much relevant to the query. For instance, the home page of IBM did not contain the word "computer". That would mean that if we rely solely on the content-based techniques, the pointer to the home page of IBM would not at all be returned in response to a query on "computer". This is very clearly an undesirable situation. We shall see how these problems are taken care of in the next section.

Connectivity-Based Web Search

The web is modeled as a directed graph, with the web pages being the nodes and the hyperlinks between them being the edges of the model graph (Broder et al 2000). One of the earliest efforts for harnessing the hyperlink structure of the web is reported in (Kleinberg 1998). Therein, we come across, what is called as the *Hyperlink-Induced Topic Search* (HITS) algorithm. Beginning with a set of seed pages obtained from a term-based search engine (such as *AltaVista*) pertaining to the topic, a base sub-graph is obtained by including in the set all those pages that are pointed to by the seed pages, as also all those which point to the seed pages. This sub-graph is then subjected to a weight propagation algorithm, so as to determine the numerical estimates of hub and authority weights by an iterative procedure. HITS, then returns as hubs and authorities for the search topic those pages that have the highest weights.

Clever (Chakrabarti et al 1999b) realizes the concepts of *Authorities* (pages which provide the best source of information on a given topic) and *Hubs* (pages which provide collection of links to authorities). We associate a non-negative authority weight x_p and a non-negative hub weight y_p with each page. We update the value of x_p, for a page p, to be the sum of y_q over all pages q that link to p:

$$x_p = \sum_{q \ such \ that \ q \to p} y_q$$

where the notation $q \to p$ indicates that q links to p. In a strictly dual fashion, if a page points to many good authorities, we increase its hub weight via

$$y_p = \sum_{q \ such \ that \ p \to q} x_q$$

Let us number the pages $\{1,2,\ldots,n\}$ and define their adjacency matrix A. Let us also write the set of all x values as a vector $x=(x_1,x_2,\ldots,x_n)$, and similarly define $y=(y_1,y_2,\ldots,y_n)$.

Then the update rule for x can be written as $x \leftarrow A^T y$,

and the update rule for y can be written as $y \leftarrow A x$.

Unwinding these one step further, we have

$$x \leftarrow A^T y \leftarrow A^T A x \leftarrow \left(A^T A\right)x$$

and $\quad y \leftarrow A x \leftarrow A A^T y \leftarrow \left(A A^T\right)y$

Linear Algebra tells us that, when normalized, this sequence of iterates converges to the principal eigenvector of $A^T A$ for x and $A A^T$ for vector y.

Google adopts the concept of *PageRank* (Brin and Page 1998, Page et al 1998). The quality measure of a page is its indegree. Suppose there are T total pages on the web. We choose a parameter d such that $0 < d < 1$, a typical value of d might lie in the range $0.1 \le d \le 0.15$. Let pages $p_1, p_2, \ldots, p_m$ link to page p_j, and let $C(p_i)$ be the number of links out of p_i. Then the *PageRank* $R(p_j)$ of the page p_j is defined to satisfy

$$R(p_j) = \frac{d}{T} + (1-d)\sum_{i=1}^{m}\frac{R(p_i)}{C(p_i)}$$

Note that the *PageRanks* form a probability distribution over web pages, so that the sum of all web pages' *PageRank* will be one. *PageRank* can be calculated using a simple iterative algorithm, and corresponds to the principal eigenvector of the normalized link matrix of the web. *PageRank* can be thought of as a model of user behavior. We assume there is a "random surfer" who is given a web page at random and keeps clicking on links, never hitting "back" but eventually gets bored and starts on another random page. The probability that the "random surfer" visits a page is its *PageRank*. The *d* damping factor is the probability at each page that the "random surfer" will get bored and request another page. A page can have a high *PageRank* if there are many pages that point to it, or alternatively, if there are some pages with high *PageRanks* that point to it.

A note is in order here. Apparently it seems that *Clever* is more versatile than *Google*, but still *Google* is found to give reasonably good results. The prime reason behind this is that a good hub would become a good authority in due course of time. The analogy may be taken from the referencing of research papers. An original research paper is a good authority and is often cited by other papers, while a good survey paper is a good hub as it contains pointers to some good original research works. But we see that in due course of time, a survey paper also gets cited very frequently and so qualifies to become as good an authority as any other research paper. This is the reason why *Google* gets away with just the notion of *PageRank*, without separating the concepts of authorities and hubs.

Some improvements in calculating the hub and authority weights are suggested in (Borodin et al 2001). Instead of summing up the authority weights to get the hub weight of a page that links to all those authorities, it is reasoned out to do the averaging. This way, a hub would be better only if it links to good authorities, rather than linking to both good and bad authorities. Moreover, a hub-threshold as well as an authority-threshold is also suggested. While computing the authority weight, only those hubs should be counted whose hub weight lies above the current average hub weight. This means that a site should not be considered a good authority simply because a lot of poor hubs point to it. Similarly, while computing the hub weight, only some top K authorities should be counted. This amounts to saying that to be considered a good hub, the site must point to some of the best authorities.

A machine learning perspective of connectivity based web search is given in (Chakrabarti 2000). The application of data mining and machine learning procedures for analyzing the hypertextual structure of the web is surveyed therein. The algorithms from all three domains, namely, supervised, semi-supervised and unsupervised, are discussed vis-à-vis hyperlinks.

Integrating Content Based and Connectivity Based Techniques

Of late, there have been efforts to merge the content based and the connectivity based techniques for better effects. For instance, an exclusive study has been made in (Li et al 2002) to combine four different content based techniques with the HITS-based algorithms. It has been argued that since the pure connectivity based techniques are affected by the Tight Knit Community (TKC) Effect (Lempel and Moran 2000), they may give poor results. This problem is specifically attributed to the cases where the number of in-links to a page is smaller as compared to the number of out links. It is claimed that this problem cannot be solved but by the introduction of content based techniques along with the connectivity based HITS algorithm.

In another work (Haveliwala 2002), the *PageRank* algorithm is shown to be improved substantially with the introduction of what is called as topic sensitivity. In the conventional *PageRank* algorithm, the *PageRank* vector is pre-computed independent of any particular search query. For topic sensitive *PageRank* algorithm, a set of *PageRank* scores is pre-computed for each page, each with respect to one of various topics. At query time, these *PageRank* scores are combined based on the topics of the query. The topic of the query may be picked up from either the history of queries, or the user's bookmarks and browsing history, or the document itself in the case of the query term highlighted by the user.

Efforts have also been in made in (Dwork et al 2001), (Ahmad and Beg 2002) and (Beg and Ahmad 2003) to carry out Rank Aggregation of the results obtained from different public web search engines. This is how we can get the combined advantage of different search techniques being employed by the participating search engines.

Web Characterization

Web searchers are eager to know how do the various public search engines compare. What is their web coverage in terms of their index size? Broader questions may include the ones like how many web pages are there in all, what is the percentage of .com pages, what is the average size of the web pages, and many more. To satisfy these questions with precise answers is not quite possible. We may, however, provide an estimated answer for these types of question. But for that, we need to have a uniform sample of the web. Researchers have resorted to Random Walks on the web to get this uniform sample. In this section, we take up these quantification measures.

An attempt has been made to quantify the user's satisfaction to the search results in (Zhang and Dong 2000). If r_1, r_2,..., r_n are the resources in a decreasing order of their value according to the user's judgement, then the "reverse order number" of the web resource r_i is defined under the rank function as

$$\psi(r_i) = \left| \{ r_k | (r_k \in R) \wedge (1 \le k < i) \wedge (rank(r_k) < rank(r_i)) \} \right|$$

where R is the set of the web resources. Now the user's satisfaction function for a query q is given as

$$sat(q, rank) = \frac{\sum_{i=1}^{n} \psi(r_i)}{(n-1)(n-2)/2}$$

For the query set $Q = \{q_1, q_2, ..., q_m\}$, the average satisfaction function would be

$$sat(Q, rank) = \frac{\sum_{j=1}^{m} sat(q_j, rank)}{m}$$

But then the main question with this approach remains as how to obtain from the user, the resources $r_1, r_2, ..., r_n$ in a decreasing order of their value according to the user's judgement. With a subjective measure explained later in this chapter, we have devised a method to do so, in a subtle and effective way.

To the best of our knowledge, no other attempt has been made to quantify the user's satisfaction to the search results. However, by other means, efforts have been made to compare the performance of various search engines. The underlying principle is to collect a uniform sample of the web pages by carrying out random walks on the web. This uniform sample is then used to measure the size of indices of the various search engines. This index size is an indirect means to estimate the performance of a search engine. Larger the index size, more is the web coverage and so more likely is the emergence of good search results. Some of the efforts in this direction are given in (Henzinger et al 1999, Henzinger et al 2000, Bar-Yossef et al 2000).

Using the concept of *PageRank*, we can estimate the quality of indices in an index-based search engine using *Random Walks* as follows (Henzinger et al 1999). Suppose each page p of the web is given a weight $w(p)$. We define the quality of the index as $w(S) = \sum_{p \in S} w(p)$. Since w is scaled, we always have $0 \le w(S) \le 1$. In fact, the *PageRank* can be thought of as a probability distribution over pages and hence a weight function. If a surfer visits page p, the random walk is in state p. The pages with high *PageRank* are more likely to be visited than pages with low *PageRank*. We can approximate the quality of a search engine index S according to the weight w by independently selecting pages p_1, p_2, ...,p_n, and testing whether each page is in S. Let $I[p_i \varepsilon S]$ be 1 if page p_i is in S, and 0 otherwise. Our estimate for $w(S)$ is $\overline{w}(S) = \frac{1}{n} \sum_{i=i}^{n} I[p_i \varepsilon S]$. So we merely need to measure the fraction of pages in our sample sequence that are in the index S in order to approximate the quality of S.

The above approach is improved upon in (Henzinger et al 2000). In order to remove the bias towards the popular pages, the pages have been sampled from the walk based on the inverse of an estimate for the *PageRank* of each page. The estimate of the *PageRank* of a page p has been made to be the visit ratio, i.e. the fraction of number of times the page p was visited during the walk to the length of the walk. In (Bar-Yossef et al 2000), we see the adoption of the conventional technique of carrying out random walk on a graph for that on the web. The random walk on a graph requires that the graph must be connected, undirected and regular. The natural graph model of the web, however, does not satisfy any on these three conditions. It is, therefore, assumed that the random walk can follow the links either forward or backwards, so as to make the web graph appear undirected. The non-indexible and hence inaccessible portion of the web is anyway out of bound for our study. What remains as the accessible web is thus connected as well. Now to make the graph regular, a trick has been used. Each web page has been assumed to be carrying different number of self-loops so as to make the degree of each web page the same. With these manipulations, the graph model of the web is now suitable for the conventional random walk algorithms for a graph.

The relative size and overlap of search engines has also been found in (Bharat and Broder 1998), but this time, instead of random walks, random queries have been used. These random queries are generated from a lexicon of about 400,000 words, built from a broad crawl of roughly 300,000 documents in the *Yahoo* hierarchy. The authors, however, admit that this technique is subjected to various biases. For instance, the longer pages are more likely to be selected by their query approach than the short ones. In (Lawrence and Giles 1998), the same goal, viz. comparing the search engines, has been achieved using a standard query log like that of NEC Research Institute. But this leaves questions as to the statistical appropriateness and the repeatability of the test by other researchers. Further work (Lawrence and Giles 1999b) used an approach based on random generation of IP addresses. One may have a genuine doubt about the scalability of this approach for the future 128 bit IP addresses. In another instance (Hawking 1999), a test data set has been made available for evaluation of web search systems and techniques by freezing a 18.5 million page snapshot of a part of the web.

One of the recent works in search engine evaluation appears in (Li and Shang 2000). For two different sets of ad-hoc queries, the results from *AltaVista*, *Google* and *InfoSeek* are obtained. These results are automatically evaluated for relevance on the basis of vector space model. These results are found to agree with the manual evaluation of relevance based on precision. Precision scores are given as 0, 1 or 2. But then this precision evaluation is similar to the form-filling exercise, which may turn out to be too demanding for the users. Most of the users may not be willing to waste time in filling up such a form. Even if we provide some kind of an incentive for filling up these forms, the users might provide any careless or even intentionally wrong feedback, just to bag that incentive, if at all he wishes to. The vector space model is a content-based technique with no consideration to the satisfaction of the user. Our work of evaluating web search quality, discussed later in this chapter, is driven by the ultimate goal - the satisfaction of the users.

An automated procedure for evaluating the systems that find related pages is outlined in (Haveliwala et al 2002). For this, a *familial distance* is defined between two documents as the distance between the most specific classes of the two documents in the class hierarchies on the web, such as Open Directory. Based on the familial distance between the sample document and each of the result documents, an ordering of the result documents may be carried out. This ordering of the documents is compared with the ordering returned by the system of finding related pages.

With a critical overview of the existing web search systems, we are all set for discussing the user feedback based web search system in the next section.

Search Engine Model

Our search engine maintains a database of document URLs, which are indexed by the keywords in the documents. Let R_{dk} be a variable in the interval [0,1] indicating the relevance of document d for the keyword k, and vice versa. Let $R=[R_{dk}]$ be a matrix of relevance indices, for all the documents in the database and for all the keywords extracted from the documents. Figure 1 depicts the architecture of our search engine.

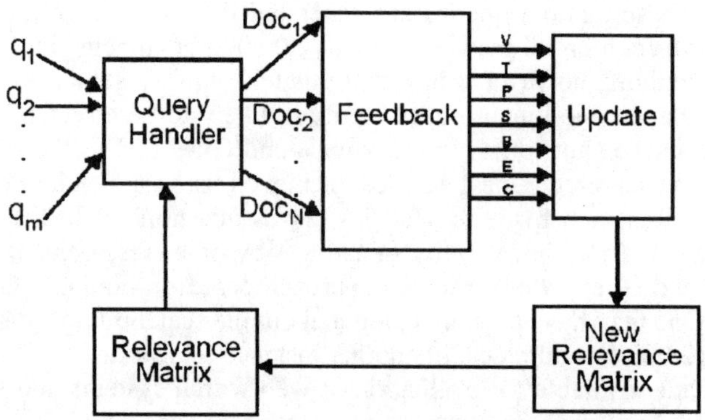

Figure 1: Search Engine Architecture

As shown in Figure 1, the user gives the query in the form of a combination of a few keywords. With the inputs from the relevance matrix R, the query is processed by a query handler, and the results are presented before the user. The feedback from the user is collected as explained later in this chapter. The feedback vector thus obtained is used to evaluate a new relevance matrix, which in turn, is used for subsequent queries.

The overall technique of user feedback based searching procedure can be described using the following pseudocode.

```
R = R₀
for every query Q do
        D = SortDocuments(Q, R);
        DisplayDocuments(D);
        F = FeedbackVector(V, T, P, S, B, E, C);
        R = UpdateMatrix(R, F)
end
```

We begin with some initial valued relevance matrix R. Based on the query Q coming from the user, we process the relevance matrix R to get the documents list D as the result to the query Q. The result is displayed before the user and the feedback to it obtained in the form of the feedback vector. This feedback vector is then utilized for updating the relevance matrix R in such a way that it gets closer to the satisfaction of an average user. These steps are further explained in subsequent sections.

Query Handler

A query Q may contain a number of keywords. These keywords may be combined conventionally using logical operators such as AND, OR and NOT. We may represent a general query Q as follows.

$$Q = \left(q_{1,1} \wedge q_{1,2} \wedge \ldots \wedge q_{1,p_1}\right) \vee \left(q_{2,1} \wedge q_{2,2} \wedge \ldots \wedge q_{2,p_2}\right) \vee \ldots \ldots \vee \left(q_{m,1} \wedge q_{m,2} \wedge \ldots \wedge q_{m,p_m}\right)$$

(1)

$$\text{where, } \quad q_{i,j} = \begin{cases} t_j & original \\ \\ \neg t_j & negated \end{cases} , \quad 1 \le i \le m, \ 1 \le j \le p_i$$

The relevance matrix R is used by the search engine to select the best documents for the query Q. This is done by identifying the columns corresponding to the keywords in the query and combining them correspondingly through "fuzzy-AND", "fuzzy-OR" and "fuzzy-NOT" operations to result in a single column of real values that represents the query-relevance vector QR. The element QR_d of this vector indicates the relevance of the document d for the given query. A sorted list of documents corresponding to the descending values in the vector QR would form the search result to the given query Q.

Example 1 Let the relevance matrix R with document set $\{d_1, d_2, d_3, d_4\}$ and the set of keywords $\{t_1, t_2, t_3, t_4, t_5\}$ be given as

$$R = \begin{array}{c} d_1 \\ d_2 \\ d_3 \\ d_4 \end{array} \begin{bmatrix} 0.5 & 0.2 & 0.4 & 0.3 & 0.7 \\ 0.3 & 0.6 & 0.7 & 0.5 & 0.3 \\ 0.0 & 0.3 & 0.3 & 0.2 & 0.5 \\ 0.7 & 0.0 & 0.4 & 0.0 & 0.4 \end{bmatrix}$$

For the query $Q = t_2 \vee \left(\neg t_4 \wedge t_5\right)$, the query-relevance vector QR is found as follows.

$$QR = fuzzy\_OR_{i=1}^{4}\left((R_{i,2}),\left(fuzzy\_AND_{j=1}^{4}\left(\left(fuzzy\_NOT_{k=1}^{4}(R_{k,4})\right),(R_{j,5})\right)\right)\right)$$

$$= \max_{i=1}^{4}\left((R_{i,2}),\left(\min_{j=1}^{4}\left((1.0 - R_{j,4}),(R_{j,5})\right)\right)\right)$$

$$= \max\left([0.2 \quad 0.6 \quad 0.3 \quad 0.0]',\left(\min\left([0.7 \quad 0.5 \quad 0.8 \quad 1.0]',[0.7 \quad 0.3 \quad 0.5 \quad 0.4]'\right)\right)\right)$$

$$= \max\left([0.2 \quad 0.6 \quad 0.3 \quad 0.0]',[0.7 \quad 0.3 \quad 0.5 \quad 0.4]'\right)$$

$$= [0.7 \quad 0.6 \quad 0.5 \quad 0.4]'$$

A descending order sort on the elements of the query-relevance vector QR gives the result of the query Q as $d_1 \succ d_2 \succ d_3 \succ d_4$.

Alternatively, we have expanded our query language by making provision for combining the keywords in the query using either

(a) linguistic operators such as "most", "as many as possible", etc., or

(b) weighted terms combined with logical operators AND, OR and NOT.

Query Formulation using Linguistic Operators

There are the cases where the logic of classifying a document as acceptable or unacceptable is too complicated to encode in any simple query form. These are the situations where the user knows a list of keywords that collectively describe the topic, but doesn't know precisely whether to use AND operators or the OR operators to combine these keywords. It may be noted that the AND operator requiring *all* the keywords to be present in the document would return too few documents, while the OR operator requiring *any* of the keywords in the document would return too many documents. It becomes preferable, therefore, to resort in such situations to the OWA operator, which is explained below.

Ordered Weighted Averaging (OWA) operator

The Ordered Weighted Averaging (OWA) operators were originally introduced in (Yager 1988) to provide a means of aggregation, which unifies in one operator the conjunctive and disjunctive behavior. The OWA operators, in fact, provide a parameterized family of aggregation operators including many of the well-known operators like maximum, minimum, k-order statistics, median and arithmetic mean. For some n different scores as x_1, $x_2,\ldots,x_n$, the aggregation of these scores may be done using the OWA operator as follows.

$$OWA(x_1, x_2, \ldots, x_n) = \sum_{i=1}^{n} w_i y_i$$

where y_i is the ith largest score from amongst x_1, $x_2,\ldots,x_n$. The weights are all non-negative $(\forall i, w_i \geq 0)$, and their sum equals one $\left(\sum_{i=1}^{n} w_i = 1\right)$. We note that the

arithmetic mean function may be obtained using the OWA operator, if $\forall i, w_i = \dfrac{1}{n}$.

Similarly, the OWA operator would yield the minimum function with $w_1=1$ and $w_i = 0$ if $i \neq 1$. The maximum function may be obtained from the OWA operator when $w_n=1$ and $w_i = 0$ if $i \neq n$.

In fact, it has been shown in (Yager 1988) that the aggregation done by the OWA operator is always between the maximum and the minimum. However, it remains to be seen what procedure should be adopted to find the values of the weights w_i. For this, we need to make use of the linguistic quantifiers, explained as follows.

Relative Fuzzy Linguistic Quantifier

A relative quantifier, $Q:[0,1]\rightarrow[0,1]$, satisfies:
$$Q(0)=0,$$
$$\exists r \in [0,1] \text{ such that } Q(r)=1.$$
In addition, it is non-decreasing if it has the following property:
$$\forall a,b \in [0,1], \text{ if } a>b, \text{ then } Q(a)\geq Q(b).$$
The membership function of a relative quantifier can be represented as:

$$Q(r)=\begin{cases} 0 & \text{if } r<a \\ \dfrac{r-a}{b-a} & \text{if } b\leq r\leq a \\ 1 & \text{if } r>b \end{cases}$$

where $a,b,r \in [0,1]$.

Some examples of relative quantifiers are shown in Figure 2, where the parameters are (0.3,0.8), (0,0.5) and (0.5,1), respectively. We may define many more relative quantifiers with customized values of the parameters a and b.

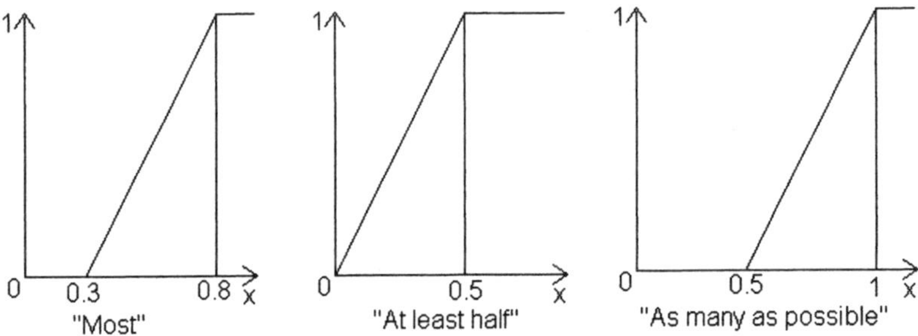

Figure 2: Relative Quantifiers

Yager (Yager 1988) computes the weights w_i of the OWA aggregation from the function Q describing the quantifier. In the case of a relative quantifier, with m criteria,

$$w_i = Q(i/m)-Q((i-1)/m), i=1,2,\ldots,m, \text{ with } Q(0)=0.$$

Example 2 For the number of criteria $(m)=7$, the fuzzy quantifier "most", with the pair $(a=0.3, b=0.8)$, the corresponding OWA operator would have the weighting vector as $w_1=0$, $w_2=0$, $w_3=0.257143$, $w_4=0.285714$, $w_5=0.285714$, $w_6=0.171429$, $w_7=0$.

OWA Operator Based Queries

We propose to prompt the user to issue linguistic operators like "most", "at least half", "as many as possible", etc., along with the keywords of the desired topic. The linguistic operator issued by the user would correspondingly mean if the user wishes to have documents that contain "most" of the keywords, or "at least half" of the keywords, or "as many as possible" keywords, etc. Let the general form of the query Q be given as

$$Q = "linguistic\_operator"\left(t_{i_1}, t_{i_2}, ..., t_{i_n}\right).$$

The kth element of the query-relevance vector QR may then be obtained as follows.

$$QR_k = \sum_{j=1}^{n} w_j . x_j$$

where x_j is the jth largest element in the set of elements $\left\{R_{k,i_1}, R_{k,i_2}, ..., R_{k,i_n}\right\}$, and $w's$ are the weights of the OWA operation, found as explained already.

Example 3 Let the relevance matrix R with document set $\{d_1, d_2, d_3, d_4\}$ and the set of keywords $\{t_1, t_2, t_3, t_4, t_5, t_6, t_7\}$ be given as

$$R = \begin{array}{c} d_1 \\ d_2 \\ d_3 \\ d_4 \end{array} \begin{bmatrix} 0.2 & 0.6 & 0.0 & 0.6 & 0.4 & 0.2 & 0.4 \\ 0.5 & 0.4 & 0.1 & 0.4 & 0.7 & 0.6 & 0.0 \\ 0.3 & 0.2 & 0.5 & 0.0 & 0.0 & 0.3 & 0.6 \\ 0.0 & 0.5 & 0.3 & 0.2 & 0.5 & 0.4 & 0.1 \end{bmatrix}$$

For the query $Q = $ "most" $(t_2, t_3, t_5, t_6, t_7)$, we would have the weights for the OWA aggregation as $w_1=0.0$, $w_2=0.2$, $w_3=0.4$, $w_4=0.4$, $w_5=0.0$ (see Example 2). The query-relevance vector QR is then given as follows.

$$QR = \begin{bmatrix} 0.0 \times 0.6 + 0.2 \times 0.4 + 0.4 \times 0.4 + 0.4 \times 0.2 + 0.0 \times 0.0 \\ 0.0 \times 0.7 + 0.2 \times 0.6 + 0.4 \times 0.4 + 0.4 \times 0.1 + 0.0 \times 0.0 \\ 0.0 \times 0.6 + 0.2 \times 0.5 + 0.4 \times 0.3 + 0.4 \times 0.2 + 0.0 \times 0.0 \\ 0.0 \times 0.5 + 0.2 \times 0.5 + 0.4 \times 0.4 + 0.4 \times 0.3 + 0.0 \times 0.1 \end{bmatrix} = \begin{bmatrix} 0.32 \\ 0.32 \\ 0.30 \\ 0.38 \end{bmatrix}$$

(c) A descending order sort on the elements of the query-relevance vector QR gives the result of the query Q as $d_4 \succ d_1 \approx d_2 \succ d_3$.

Query Formulation using Weighted Logical Operators AND, OR and NOT

Consider a case in which the user wishes to search for *analog computing systems*. Very clearly, the emphasis has to be on the term *analog* more than the other two terms. If this is treated as an AND query, only those documents would be returned which necessarily contain all the three terms. In the case of OR query, all those documents would be returned which contain any of these terms. We may not even give the query as (*analog* $\neg$*computing* $\neg$*systems*), because that would try to exclude the two terms altogether. The use of linguistic operator, say "most", would return even those documents that contain just the terms *computing* and *systems*, and not necessarily *analog*. All the above mentioned situations are not up to the user's expectations. We, therefore, feel it appropriate to provide the flexibility to the user to assign some importance weight to each of the terms in the query. For instance, the query may be given as (*analog*,0.8) $\wedge$ {(*computing*,0.4) $\vee$

(*sytems*,0.3)}. This would mean that the user wishes to give an importance weight of 8/15 to the term *analog*, 4/15 to the term *computing*, and 3/15 to the term *systems*. The general form of the weighted query may be given as follows.

$$Q = \left((q_{1,1}, f_{1,1}) \wedge (q_{1,2}, f_{1,2}) \wedge \dots \wedge (q_{1,p_1}, f_{1,p_1})\right) \vee \left((q_{2,1}, f_{2,1}) \wedge (q_{2,2}, f_{2,2}) \wedge \dots \wedge (q_{2,p_2}, f_{2,p_2})\right) \vee \dots$$

$$\dots \vee \left((q_{m,1}, f_{m,1}) \wedge (q_{m,2}, f_{m,2}) \wedge \dots \wedge (q_{m,p_m}, f_{m,p_m})\right)$$

$$\text{where, } q_{i,j} = \begin{cases} t_j & original \\ \neg t_j & negated \end{cases}, \quad 1 \le i \le m, \ 1 \le j \le p_i,$$

and $f_{i,j} \in \left]0,1\right]$ indicates the importance weight associated with the corresponding term $q_{i,j}$ of the query. In case, the user does not specify the value of $f_{i,j}$, it is taken as $f_{i,j} = 1$, by default.

The kth element of the query-relevance vector QR shall be given as

$$QR_k = \max_{i=1}^{m} \left(\min_{j=1}^{p_i} (R_{k,j} \times f_{i,j}) \right)$$

Example 4 Let the relevance matrix R with document set $\{d_1, d_2, d_3, d_4\}$ and the set of keywords $\{t_1, t_2, t_3, t_4, t_5\}$ be given as

$$R = \begin{array}{c} d_1 \\ d_2 \\ d_3 \\ d_4 \end{array} \begin{bmatrix} 0.5 & 0.2 & 0.4 & 0.3 & 0.7 \\ 0.3 & 0.6 & 0.7 & 0.5 & 0.3 \\ 0.0 & 0.3 & 0.3 & 0.2 & 0.5 \\ 0.7 & 0.0 & 0.4 & 0.0 & 0.4 \end{bmatrix}$$

For the query $Q = (t_2, 0.3) \vee (\neg(t_4, 0.6) \wedge (t_5, 0.2))$, the query-relevance vector QR is found as follows.

$$QR = \max_{i=1}^{4} \left((R_{i,2} \times 0.3), \left(\min_{j=1}^{4} \left((1.0 - (R_{j,4} \times 0.6)), (R_{j,5} \times 0.2) \right) \right) \right)$$

$$= \max \left([0.06 \ 0.18 \ 0.09 \ 0.0]', \left(\min([0.82 \ 0.7 \ 0.88 \ 1.0]', [0.14 \ 0.06 \ 0.1 \ 0.08]') \right) \right)$$

$$= \max \left([0.2 \ 0.6 \ 0.3 \ 0.0]', [0.14 \ 0.06 \ 0.1 \ 0.08]' \right)$$

$$= [0.2 \ 0.6 \ 0.3 \ 0.08]'$$

A descending order sort on the elements of the query-relevance vector QR gives the result of the query Q as $d_2 \succ d_3 \succ d_1 \succ d_4$.

Updating Relevance Matrix from the User Feedback

Once the results are obtained from the relevance matrix R in response to the user query Q using any of the techniques described earlier, the results are displayed before the user in the form of documents listing. The feedback of the user on these results is then collected as explained below.

User Feedback Vector

We characterize the feedback of the user by a vector (V, T, P, S, B, E, C), which consists of the following.

(a) The sequence V in which the user visits the documents, $V=(v_1, v_2, ...v_N)$. If document i is the kth document visited by the user, then we set $v_i = k$. If a document i is not visited by the user at all before the next query is submitted, the corresponding value of v_i is set to -1.

(b) The time t_i that a user spends examining the document i. We denote the vector $(t_1, t_2, ..., t_N)$ by T. For a document that is not visited, the corresponding entry in the array T is 0.

(c) Whether or not the user prints the document i. This is denoted by the Boolean p_i. We shall denote the vector $(p_1, p_2, ..., p_N)$ by P.

(d) Whether or not the user saves the document i. This is denoted by the Boolean s_i. We shall denote the vector $(s_1, s_2, ..., s_N)$ by S.

(e) Whether or not the user book-marked the document i. This is denoted by the Boolean b_i. We shall denote the vector $(b_1, b_2, ..., b_N)$ by B.

(f) Whether or not the user e-mailed the document v to someone. This is denoted by the Boolean e_i. We shall denote the vector $(e_1, e_2, ..., e_N)$ by E.

(g) The number of words that the user copied and pasted elsewhere. We denote the vector $(c_1, c_2, ..., c_N)$ by C.

The motivation behind collecting this feedback is the belief that a well-educated user is likely to select the more appropriate documents early in the resource discovery process. Similarly, the time that a user spends examining a document, and whether or not he prints, saves, bookmarks, e-mails it to someone else or copies & pastes a portion of the document, indicate the level of importance that document holds for the specified query.

When feedback recovery is complete, we compute the following weighted sum σ_j for each document j selected by the user.

$$\sigma_j = \left(w_V \frac{1}{2^{(v_j - 1)}} + w_T \frac{t_j}{t_{j_{max}}} + w_P p_j + w_S s_j + w_B b_j + w_E e_j + w_C \frac{c_j}{c_{j_{total}}} \right) \tag{2}$$

where $t_{j_{max}}$ represents the maximum time a user is expected to spend in examining the document j, and $c_{j_{total}}$ is the total number of words in the document j. Here, w_V,

w_T, w_P, w_S, w_B, w_E and w_C, all lying between 0 and 1, give the respective weightages we want to give to each of the seven components of the feedback vector. The sum σ_j represents the importance of document j.

The intuition behind this formulation is as follows. The importance of the document should decrease monotonically with the postponement being afforded by the user in picking it up. More the time spent by the user in glancing through the document, more important that must be for him. If the user is printing the document, or saving it, or book-marking it, or e-mailing it to someone else, or copying and pasting a portion of the document, it must be having some importance in the eyes of the user. A combination of the above seven factors by simply taking their weighted sum gives the overall importance the document holds in the eyes of the user. We may include more metrics like this, if so desired.

After the importance weight σ_j of each document j picked up by the user is calculated, the updating of the relevance matrix in accordance with the user feedback then proceeds as follows.

Dealing with Simple One-Keyword Query

One of our main concerns in this chapter is to modify the relevance matrix R so that the matrix gets tuned to the satisfaction of a (average) user. For this, we need to boost up the weights of the documents being consistently picked up by the users and penalize those, which are being mostly ignored by the users. To simplify the presentation of our enhancement procedure, let us assume initially that the query Q is constituted by a single keyword, say k. Let σ_d be the document importance factor calculated using equation (2) corresponding to the document d. We can, then, update the element R_{dk} of the matrix R as follows.

$$R_{dk}^{i+1} \leftarrow \frac{R_{dk}^i + \mu.\sigma_d}{1 + \mu.\sigma_d} \tag{3}$$

Here, R_{dk}^i denotes the value of the element R_{dk} of the matrix R in the ith iteration and μ is a suitable learning rate. Before we proceed further, we must show that while using (3) for index updating, R_{dk} always remains bounded by one, and also that R_{dk}^{i+1} is an improvement over R_{dk}^i. We prove these two points in the following two Lemmas, respectively.

Lemma 1: R_{dk} always remains bounded by one

i.e. $\forall i, \quad R_{dk}^{i+1} < 1$, given $R_{dk}^i < 1$.

Proof. Given that $R_{dk}^i < 1$

$$\Rightarrow \quad R_{dk}^i + \mu.\sigma_d < 1 + \mu.\sigma_d$$

$$\Rightarrow \quad \frac{R_{dk}^i + \mu.\sigma_d}{1 + \mu.\sigma_d} < 1$$

$$\Rightarrow \quad R_{dk}^{i+1} < 1. \qquad\qquad \{\text{from (3)}\}$$

Hence the result.

<u>Lemma 2:</u> R_{dk}^{i+1} is an improvement over R_{dk}^i

$$\text{i.e. } \forall i, \quad R_{dk}^{i+1} > R_{dk}^i, \text{ given } R_{dk}^i < 1.$$

<u>Proof.</u> Given that $R_{dk}^i < 1$

$$\Rightarrow \quad R_{dk}^i.\mu.\sigma_d < \mu.\sigma_d$$

$$\Rightarrow \quad R_{dk}^i + \mu.\sigma_d > R_{dk}^i + R_{dk}^i.\mu.\sigma_d$$

$$\Rightarrow \quad \frac{R_{dk}^i + \mu.\sigma_d}{1 + \mu.\sigma_d} > R_{dk}^i$$

$$\Rightarrow \quad R_{dk}^{i+1} > R_{dk}^i. \qquad\qquad \{\text{from (3)}\}$$

Hence the result.

Now, having proved Lemmas 1 and 2, we must explain how do we modify (3) so as to compensate for the multiple keywords in the query Q connected by logical connectives AND, OR and NOT.

Dealing with Complex Multiple Keywords Query

As given in equation (1), we may represent a general query Q as follows.

$$Q = \left(q_{1,1} \wedge q_{1,2} \wedge ... \wedge q_{1,p_1}\right) \vee \left(q_{2,1} \wedge q_{2,2} \wedge ... \wedge q_{2,p_2}\right) \vee ...... \vee \left(q_{m,1} \wedge q_{m,2} \wedge ... \wedge q_{m,p_m}\right)$$

$$\text{where, } q_{i,j} = \begin{cases} k_j & original \\ \\ \neg k_j & negated \end{cases}, \quad 1 \le i \le m, \ \ 1 \le j \le p_i$$

For such a general scenario, we need to distribute appropriately the document importance factor σ_d into all the constituent positive keywords of the query Q. This is done as follows.

$$R_{dk}^{i+1} \leftarrow \frac{R_{dk}^i + \mu.\dfrac{\sigma_d}{m}}{1 + \mu.\dfrac{\sigma_d}{m}} \qquad\qquad (4)$$

where m is the number of OR terms in the query Q. The intuition behind (4) stems from the thought that fuzzy-AND is taken as the minimum operation, while fuzzy-OR is taken to be the maximum operation. So, heuristically, we decided to distribute the document importance factor σ_d evenly between the OR terms and then equally between the positive AND terms within each OR term.

Example 5 Let us reconsider Example 1 with the query $Q = t_2 \vee \left(t_4 \wedge t_5\right)$. Let the document importance factors obtained from the user feedback vector be

$\sigma_{d_1} = 0.2$, $\sigma_{d_2} = 0.4$, $\sigma_{d_3} = 0.3$, $\sigma_{d_4} = 0.1$. Let the learning rate be $\mu = 0.1$. We would then have R updated from the user feedback as follows.

$$R_{d_1,t_2} = \frac{0.2 + \left(0.1 \times \dfrac{0.2}{2}\right)}{1 + \left(0.1 \times \dfrac{0.2}{2}\right)} = 0.208,$$

$$R_{d_1,t_4} = \frac{0.3 + \left(0.1 \times \dfrac{0.2}{2}\right)}{1 + \left(0.1 \times \dfrac{0.2}{2}\right)} = 0.307,$$

$$R_{d_1,t_5} = \frac{0.7 + \left(0.1 \times \dfrac{0.2}{2}\right)}{1 + \left(0.1 \times \dfrac{0.2}{2}\right)} = 0.703.$$

Proceeding this way, we get the updated relevance matrix as

$$R = \begin{array}{c} d_1 \\ d_2 \\ d_3 \\ d_4 \end{array}\begin{bmatrix} 0.5 & 0.208 & 0.4 & 0.307 & 0.703 \\ 0.3 & 0.608 & 0.7 & 0.51 & 0.314 \\ 0.0 & 0.31 & 0.3 & 0.212 & 0.507 \\ 0.7 & 0.005 & 0.4 & 0.005 & 0.403 \end{bmatrix}$$

Dealing with NOT Terms in the Query

For any negative (i.e. NOT) term k_i in the query Q, suppose a document d_j is picked up by the user from the search results presented before him. This means that the document d_j is not relevant to the keyword k_i. Hence, we can subtract a penalty factor $\delta \in [0,1[$ from R_{d_j,k_i}. The subtraction, however, is made to remain saturated to zero when the value of R_{d_j,k_i} reaches zero.

Example 6 Let us reconsider Example 5 with the query $Q = t_2 \vee (\neg t_4 \wedge t_5)$. Let the penalty factor for the NOT term t_4 be $\delta = 0.05$. We would then have R updated from the user feedback as follows.

$$R = \begin{array}{c} d_1 \\ d_2 \\ d_3 \\ d_4 \end{array}\begin{bmatrix} 0.5 & 0.208 & 0.4 & 0.25 & 0.703 \\ 0.3 & 0.608 & 0.7 & 0.45 & 0.314 \\ 0.0 & 0.31 & 0.3 & 0.15 & 0.507 \\ 0.7 & 0.005 & 0.4 & 0.0 & 0.403 \end{bmatrix}$$

It may be noted from Example 6 that even though $R_{d_4,t_4} = 0.0 - 0.05 = -0.05$, but it is made to saturate at 0.0.

Dealing with Linguistic and Weighted Queries

The linguistic queries and weighted queries have been introduced earlier in this chapter. For updating the relevance matrix R in the case of such queries, we propose to distribute the document importance factor σ_d in proportion to the weights associated with each term in the query. We know from our earlier discussion that the weight for each term is specified directly by the user in the case of weighted queries, while the weight corresponding to each term in the linguistic queries is calculated as the weights of the OWA operation.

Example 7 Let us reconsider Example 3. Let the document importance factors obtained from the user feedback vector be $\sigma_{d_1} = 0.2$, $\sigma_{d_2} = 0.4$, $\sigma_{d_3} = 0.3$, $\sigma_{d_4} = 0.1$. Let the learning rate be $\mu = 0.1$. We would then have R updated from the user feedback as follows.

$$R_{d_1,t_2} = \frac{0.6 + (0.1 \times 0.0 \times 0.2)}{1 + (0.1 \times 0.0 \times 0.2)} = 0.6,$$

$$R_{d_1,t_5} = \frac{0.4 + (0.1 \times 0.2 \times 0.2)}{1 + (0.1 \times 0.2 \times 0.2)} = 0.402,$$

$$R_{d_1,t_7} = \frac{0.4 + (0.1 \times 0.4 \times 0.2)}{1 + (0.1 \times 0.4 \times 0.2)} = 0.405,$$

$$R_{d_1,t_6} = \frac{0.2 + (0.1 \times 0.4 \times 0.2)}{1 + (0.1 \times 0.4 \times 0.2)} = 0.206,$$

$$R_{d_1,t_3} = \frac{0.0 + (0.1 \times 0.0 \times 0.2)}{1 + (0.1 \times 0.0 \times 0.2)} = 0.0.$$

Proceeding this way, we get the updated relevance matrix as

$$R = \begin{array}{c} d_1 \\ d_2 \\ d_3 \\ d_4 \end{array} \begin{bmatrix} 0.2 & 0.6 & 0.0 & 0.6 & 0.402 & 0.206 & 0.405 \\ 0.5 & 0.41 & 0.114 & 0.4 & 0.7 & 0.603 & 0.0 \\ 0.3 & 0.21 & 0.503 & 0.0 & 0.0 & 0.308 & 0.6 \\ 0.0 & 0.5 & 0.303 & 0.2 & 0.501 & 0.402 & 0.1 \end{bmatrix}$$

Example 8 Let us reconsider Example 4 with the query $Q = (t_2, 0.3) \vee ((t_4, 0.6) \wedge (t_5, 0.2))$. Let the document importance factors obtained from the user feedback vector be $\sigma_{d_1} = 0.2$, $\sigma_{d_2} = 0.4$, $\sigma_{d_3} = 0.3$, $\sigma_{d_4} = 0.1$. Let the learning rate be $\mu = 0.1$. We would then have R updated from the user feedback as follows.

$$R_{d_1,t_2} = \frac{0.2 + (0.1 \times 0.2 \times 0.3)}{1 + (0.1 \times 0.2 \times 0.3)} = 0.205,$$

$$R_{d_1,t_4} = \frac{0.3 + (0.1 \times 0.2 \times 0.6)}{1 + (0.1 \times 0.2 \times 0.6)} = 0.308,$$

$$R_{d_1,t_5} = \frac{0.7 + (0.1 \times 0.2 \times 0.2)}{1 + (0.1 \times 0.2 \times 0.2)} = 0.701.$$

Proceeding this way, we get the updated relevance matrix as

$$R = \begin{array}{c} d_1 \\ d_2 \\ d_3 \\ d_4 \end{array} \begin{bmatrix} 0.5 & 0.205 & 0.4 & 0.308 & 0.701 \\ 0.3 & 0.605 & 0.7 & 0.512 & 0.306 \\ 0.0 & 0.306 & 0.3 & 0.214 & 0.503 \\ 0.7 & 0.003 & 0.4 & 0.006 & 0.401 \end{bmatrix}$$

Exponential Penalization for Documents Left Untouched by the User

For the documents that have not at all been touched by the user from the search results presented before him, we employ an exponential penalization policy. Suppose a positive term k_i is in the query Q, and a document d_j is not touched by the user from the search results presented before him. R_{d_j,k_i} would then be penalized as follows.

$$R_{d_j,k_i}^{r+1} \leftarrow R_{d_j,k_i}^{r} - \frac{2^{\alpha}}{2^{\alpha_{max}} + 1} \tag{5}$$

where the age α is counted up each time the document d_j is neglected by the user, and is reset to zero whenever a user selects that document. α_{max} is the maximum count at which the age α is made to saturate. The intuition behind (5) is that a document being consistently ignored by the successive users must be penalized exceedingly heavily. Once again, the subtraction in (5) is made to saturate when the value of R_{d_j,k_i} reaches zero.

Subjective Measurement of Web Search Quality

The feedback collected from the users, as explained above, may also be utilized for search engines comparison. Before we proceed with the explanation of such a method, let us give some useful definitions.

Definition 1 Given a universe U and $T \subseteq U$, an *ordered list* (or simply, a *list*) l with respect to U is given as $l = [d_1, d_2, ..., d_{|T|}]$, with each $d_i \in T$, and $d_1 \succ d_2 \succ ... \succ d_{|T|}$, where "$\succ$" is some ordering relation on T. Also, for $i \in U \wedge i \in l$, let $l(i)$ denote the position or rank of i, with a higher rank having a lower numbered position in the list. We may assign a unique identifier to each element in U, and thus, without loss of generality, we may get $U=\{1,2,...,|U|\}$.

Definition 2 Full List: If a list l contains all the elements in U, then it is said to be a *full list*.

Example 9 A full list l given as $[c,d,b,a,e]$ has the ordering relation $c \succ d \succ b \succ a \succ e$. The universe U may be taken as $\{1,2,3,4,5\}$ with, say, $a \equiv 1$, $b \equiv 2$, $c \equiv 3$, $d \equiv 4$ and $e \equiv 5$. With such an assumption, we have $l = [3,4,2,1,5]$. Here $l(3)\equiv l(c)=1$, $l(4)\equiv l(d)=2$, $l(2)\equiv l(b)=3$, $l(1)\equiv l(a)=4$, $l(5)\equiv l(e)=5$.

Definition 3 Preference Ordering of the Alternatives: A full list, which is a permutation over the set $(1,2,\dots,N\}$, giving an ordered vector of alternatives, from best to worst.

Definition 4 Utility Function Ordering: Preferences on some $X = \{x_1,x_2,\dots,x_N\}$ are given as a set of N utility values, $\{u_i; i = 1,\dots,N\}$, $u_i \in [0,1]$, where u_i represents the utility evaluation given to the alternative x_i.

Definition 5 Fuzzy Preference Relation: Preferences on some $X = \{x_1,x_2,\dots,x_N\}$ are described by a fuzzy preference relation, $P_X \subset X \times X$, with membership function $\mu_{P_X} : X \times X \to [0,1]$, where $\mu_{P_X}(x_i,x_j)= p_{ij}$ denotes the preference degree of the alternative x_i over x_j. $p_{ij} = 1/2$ indicates indifference between x_i and x_j, $p_{ij} = 1$ indicates that x_i is unanimously preferred over x_j, $p_{ij} > 1/2$ indicates that x_i is preferred over x_j. It is usual to assume that $p_{ij} + p_{ji} = 1$ and $p_{ii} = 1/2$.

With the above definitions in mind, we can now begin with the actual procedure of finding the Fuzzy Search Quality Measure (FSQM).

Step I of FSQM: Transformation of Each Component of User Feedback Vector to a Corresponding Fuzzy Preference Relation

First, we convert the *preference ordering* given by the sequence vector $V=(v_1, v_2, \dots v_N)$ into a *fuzzy preference relation* R_V as (Herrera-Viedma et al 1999):

$$R_V(i,j) = \frac{1}{2}\left(1 + \frac{v_j - v_i}{N-1}\right)$$

Example 10 For $V = (2,1,3,6,4,5)$, we see that $N=6$. We may thus get:

$$R_V = \begin{bmatrix} 0.5 & 0.4 & 0.6 & 0.9 & 0.7 & 0.8 \\ 0.6 & 0.5 & 0.7 & 1 & 0.8 & 0.9 \\ 0.4 & 0.3 & 0.5 & 0.8 & 0.6 & 0.7 \\ 0.1 & 0 & 0.2 & 0.5 & 0.3 & 0.4 \\ 0.3 & 0.2 & 0.4 & 0.7 & 0.5 & 0.6 \\ 0.2 & 0.1 & 0.3 & 0.6 & 0.4 & 0.5 \end{bmatrix}$$

Next, we convert the utility values given by normalized time vector $T/T_{max} = \left(t_1/t_{1_{max}}, t_2/t_{2_{max}}, \dots, t_N/t_{N_{max}}\right)$, normalized cut-paste vector $C/C_{max} = \left(c_1/c_{1_{max}}, c_2/c_{2_{max}}, \dots, c_N/c_{N_{max}}\right)$, and all the remaining vectors, namely, $P = (p_1, p_2, \dots, p_N)$, $S = (s_1, s_2, \dots, s_N)$, $B = (b_1, b_2, \dots, b_N)$, $E = (e_1, e_2, \dots, e_N)$, into the corresponding *fuzzy preference relations* as (Herrera-Viedma et al 1999):

$$R_X(i,j) = \frac{\left(x_i^2\right)}{\left(x_i^2\right) + \left(x_j^2\right)}$$

where X may be substituted by T or P or S or B or E or C, as the case may be.

Example 11 For $T / T_{\max} = (0.3, 0.9, 0.4, 0.2, 0.7, 0.5)$, we see that $N=6$. So, we get:

$$R_T = \begin{bmatrix} 0.50 & 0.10 & 0.36 & 0.69 & 0.16 & 0.26 \\ 0.90 & 0.50 & 0.84 & 0.95 & 0.62 & 0.76 \\ 0.64 & 0.16 & 0.50 & 0.80 & 0.25 & 0.39 \\ 0.31 & 0.05 & 0.20 & 0.50 & 0.08 & 0.14 \\ 0.84 & 0.38 & 0.75 & 0.92 & 0.50 & 0.66 \\ 0.74 & 0.24 & 0.61 & 0.86 & 0.34 & 0.50 \end{bmatrix}$$

This way, we would have a set of seven fuzzy preference relations, with a one-to-one correspondence with the seven components of the user feedback vector. These fuzzy preference relations are then aggregated using the OWA operator.

Step II of FSQM: Aggregation of Individual Fuzzy Preference Relations

We must carry out the aggregation of the individual preference relations into a combined preference relation (R_C). This is achieved by what is called as Linguistic OWA (Ordered Weighted Averaging), or LOWA, in short (Herrera and Herrera-Viedma 1994). We can find the combined preference relation as:

$$R_A(i,j) = \sum_{k=1}^{7} w_k \cdot z_k \,,$$

where z_k is the kth largest element in the collection $\{R_V(i,j), R_T(i,j), R_P(i,j), R_S(i,j), R_B(i,j), R_E(i,j), R_C(i,j)\}$.

Example 12 Let us assume that $R_V(5,2) = 0.342$, $R_T(5,2) = 0.248$, $R_P(5,2) = 0.0$, $R_S(5,2) = 1.0$, $R_B(5,2) = 0.0$, $R_E(5,2) = 0.0$, $R_C(5,2) = 0.637$. Then, using the fuzzy majority criterion with the fuzzy quantifier "most" with the pair $(0.3, 0.8)$, and the corresponding LOWA operator with the weighting vector as in Example 11, for the combined preference relation, $R_A(5,2) = [w_1, w_2, w_3, w_4, w_5, w_6, w_7].[\text{Descending\_Sort}(R_V(5,2), R_T(5,2), R_P(5,2), R_S(5,2), R_B(5,2), R_E(5,2), R_C(5,2))]^T = 0.0 \times 1.0 + 0.0 \times 0.637 + 0.257143 \times 0.342 + 0.285714 \times 0.248 + 0.285714 \times 0.0 + 0.171429 \times 0.0 + 0.0 \times 0.0 = 0.1588$.

Step III of FSQM: From Combined Fuzzy Preference Relation to Combined Fuzzy Preference Ordering

After the combined preference relation R_A has been found using the LOWA operator and taking into account all the seven components of the user feedback vector, we need to convert R_A into the combined utility function ordering, Y_C. We can

do this using what is called as the *Quantifier Guided Dominance Degree* (*QGDD*)
(Herrera-Viedma et al 1999), as follows.

$$Y_c(i) = \sum_{k=1}^{N} w_k . z_k ,$$

where, w_k is the kth element of the weighting vector of the LOWA operator, and z_k
is the kth largest element of the row $R_C(i,j)$, $j = 1, 2,.., N$. If we use the linguistic
quantifer "as many as possible" for finding the weights of the LOWA operator in
QGDD, this would imply that we wish to combine "as many as possible" of the
elements of the the the ith row of $R_C(i,j)$, to get a quantification of the dominance that
the ith document has over all others in a fuzzy majority sense.

Example 13 Let $N = 6$. Then, for the fuzzy quantifier "as many as possible",
with the pair ($a = 0.5$, $b = 1$), the corresponding LOWA operator would have the
weighting vector as W=[0,0,0,1/3,1/3,1/3]. For, say, $R_A(4, j) = $ [0.332, 0.228,
0.260, 0.500, 0.603, 0.598], we would have $Y_C(4) = 0.0 \times 0.603 + 0.0 \times 0.598 +
0.0 \times 0.500 + 0.333 \times 0.332 + 0.333 \times 0.260 +0.333 \times 0.228 = 0.273$.

Now, on sorting the elements of Y_C in a descending order, we get the combined
preference ordering O_C.

Example 14 For $Y_C = $ (0.437,0.517,0.464,0.273,0.286,0.217), sorting in descending order would give us $O_C = $ (2,3,1,5,4,6).

Step IV of FSQM: Finally Getting FSQM

The combined preference ordering O_C, may be compared with ρ, the sequence
in which the documents were initially short-listed by the search engine. The
Spearman Rank Order Correlation Coefficient (r_s) is found between the preference orderings O_C and ρ. We repeat this procedure for a representative set of queries and take the average of r_s. The resulting average value of r_s is the required
fuzzy measure of the search quality (FSQM). The overall procedure of evaluating
FSQM is depicted in Figure 3.

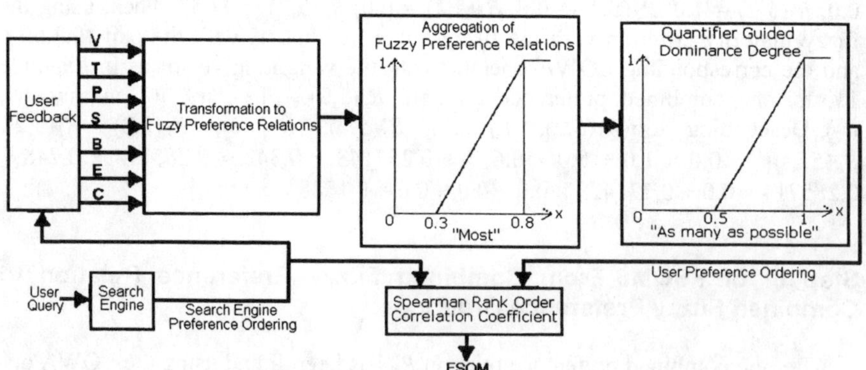

Figure 3: Evaluating Fuzzy Search Quality Measure (FSQM)

Experiments and Results

In a real application of user feedback based web search enhancement, the responses of actual users will be monitored. In a performance analysis experiment, it would be more convenient to use a probabilistic model to generate the user responses automatically. We begin with a randomly generated matrix $P=[P_{ij}]$, where P_{ij} is the probability of finding the term j in document i. This is taken to be proportional to the assumed importance of the term j in the ith document. In fact, this probability distribution captures the bias the cross section of users are expected to have while making their choices in selecting the documents from the results of a given query. So, this should not be mistaken for the index matrix R. The difference between the matrices R and P is the same as that between *fuzziness* (which describes the ambiguity of an event (Ross 1997)) and *randomness* (which describes the uncertainty in the occurrence of the event (Ross 1997)), respectively. We also have a query generator, that selects keywords randomly and then combine them with either AND, OR and NOT operators by the flip of coin, or the linguistic operators or the weighted operators randomly. Based on the matrix P and the generated query Q, we generate the feedback vector. We then learn from the user feedback vector thus obtained. After each learning step, we find out the Spearman Rank Order Correlation Coefficient (SROCC) between the result listing given by our search engine based on the index matrix R and the preferred listing given by the user model based on the probability matrix P.

As shown in Figures 4 and 5, the correlation coefficient (SROCC) between the listings given by our search engine and that preferred by the user model increases as the training set size (the number of feedback vectors used for learning) increases. This shows that after our search engine has learnt, it renders a better level of user satisfaction. Figures 4 and 5 differ in the rate of learning μ. With $\mu = 0.01$ in Figure 5, the learning is slower but finer than with $\mu = 0.1$, as in Figure 4. In both Figures 4 and 5, IIRV indicates that all the elements of the index matrix R are initialized with random values.

The simulation results show a steady rise in the satisfaction levels of the modeled users as more and more learning goes into our system. We need to ascertain this performance with the real queries. For this, we need to have a real search query log, such as the one from NEC Research Institute used in (Lawrence and Giles 1998, Lawrence and Giles 1999a). But unfortunately, we don't have access to one, at the moment. This is being taken as the future direction of research.

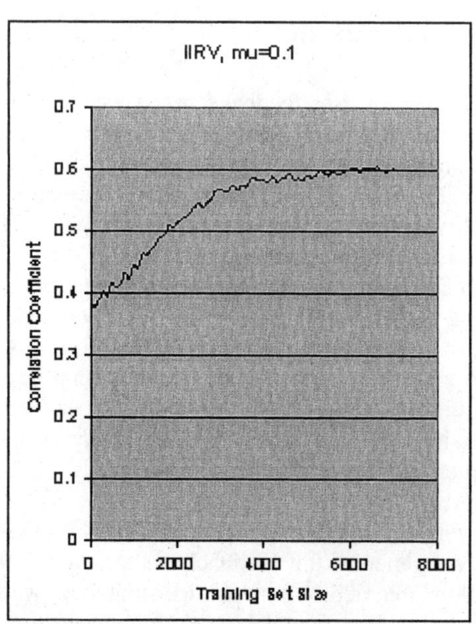

Figure 4: User Satisfaction versus the Number of User Feedback Vectors, with the learning rate $\mu = 0.1$

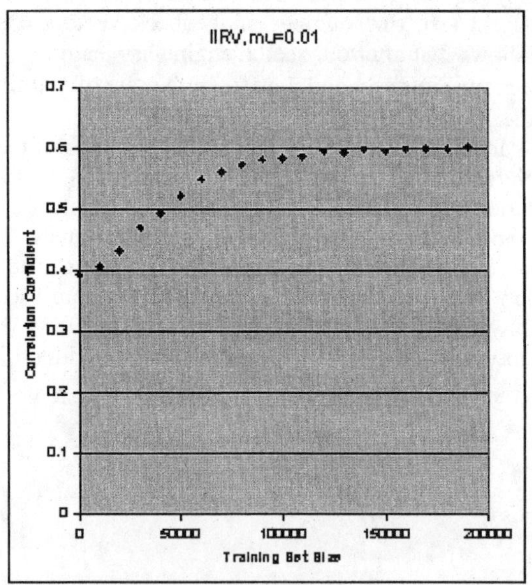

Figure 5: User Satisfaction versus the Number of User Feedback Vectors, with the learning rate $\mu = 0.01$

One of the advantages of our technique of user feedback based web search is that the performance of the search engine is no more dependent on the effectiveness of the heuristic used for characterization of the documents. Rather we go by "what the user wants" as our guideline. The second advantage is that our technique is equally valid for multimedia repositories. We know that the information available on the Internet has many diverse forms - text documents, images, audio and video data, formatted text documents, binary files, and so on. With our approach, we need not have separate techniques of "feature extraction" for different kinds of entities. The user feedback is handy for any type of web repository. To conclude, we can say that our technique combines improved quality with increased diversity.

Results Pertaining to the Fuzzy Measure of Web Search Quality

We experimented with a few queries on seven popular search engines, namely, *AltaVista, DirectHit, Excite, Google, HotBot, Lycos* and *Yahoo*. It may be noted here that our emphasis is more to demonstrate the procedure of quality measurement, rather than to carry out the actual performance measurement of these search engines. It is for this reason, as also to simplify our experiments, that we are taking into account just the three components V, T and P, out of the seven components of the user feedback vector. For instance, the observation corresponding to the query *document categorization query generation* is given in Table 1.

Table 1: Results for the Query *document categorization query generation*

Search Engine	Document Preference(V)		Fraction of time (T)	Printed (P)
	Sequence	Document No.		
AltaVista	1	2	0.0	0
	2	1	0.0011	0
DirectHit	1	10	0.00091	0
Excite	1	7	0.012	0
Google	1	1	0.092	0
	2	2	0.88	0
	3	3	0.0	0
	4	5	0.94	1
HotBot	1	1	0.092	0
	2	6	0.88	0
Lycos	1	1	0.0	0
	2	2	0.88	0
	3	7	0.92	0
Yahoo	1	1	0.092	0
	2	2	0.88	0
	3	3	0.0	0
	4	5	0.92	0
	5	9	0.47	0

Table 1 shows that from the results of *AltaVista*, the second document was picked up first by the user, but a zero time was spent on it, most probably because the documents must not be at this location anymore, and so no printout could be taken. The second pick up was made by the user on what was listed as the first document by *AltaVista*, the document was read by the user for just 0.11% of the time required to read it completely, and no printout was taken once again. This way the values corresponding to the rest of the search engines are shown in Table 1 for the query *document categorization query generation.*

We experimented with 15 queries in all. These queries are listed in Table 2 and their results are summarized in Table 3. This experiment has been carried out in August 2001. The value of FSQM is obtained using the relative quantifier "*most*" with the parameters ($a = 0.3$, $b = 0.8$) for aggregating the individual preference relations, and then "*as many as possible*" with the parameters ($a = 0.5$, $b = 1.0$) for getting the combined preference ordering. The results given in Table 3 are graphically represented in Fig. 6. We have taken these 15 queries for the sake of demonstrating our procedure. We may have taken many more. In fact, the actual measure would require a continuous evaluation by taking a running average over an appropriate window size of the successive queries being posed by the users.

Table 2: List of Test Queries

1.	*"measuring search quality"*
2.	*"mining access patterns from web logs"*
3.	*"pattern discovery from web transactions"*
4.	*"distributed association rule mining"*
5.	*"document categorization query generation"*
6.	*"term vector database"*
7.	*"client-directory-server model"*
8.	*"similarity measure for resource discovery"*
9.	*"hypertextual web search"*
10.	*"IP routing in satellite networks"*
11.	*"focussed web crawling"*
12.	*"concept based relevance feedback for information retrieval"*
13.	*"parallel sorting neural network"*
14.	*"spearman rank order correlation coefficient"*
15.	*"web search query benchmark"*

Table 3: Fuzzy Search Quality Measures for the Queries given in Table 2

Query	*AltaVista*	*DirectHit*	*Excite*	*Google*	*HotBot*	*Lycos*	*Yahoo*
1.	-0.539	0.030	-0.297	0.685	-0.242	0.248	0.770
2.	0.324	0.248	0.248	0.058	-0.079	0.358	0.330
3.	-0.297	0.576	0.576	0.770	0.685	-0.045	0.770

4.	0.030	0.030	0.030	0.103	0.248	0.721	0.685
5.	0.685	0.030	0.030	0.542	0.139	0.418	0.212
6.	0.030	0.030	0.030	0.248	0.576	0.673	0.685
7.	0.030	0.030	0.030	0.733	-0.297	0.139	0.576
8.	0.673	0.248	-0.261	0.285	0.685	0.212	0.285
9.	0.467	0.030	0.030	0.333	0.030	0.139	0.333
10.	0.030	0.576	0.030	0.139	0.333	0.430	0.091
11.	0.248	0.030	0.030	0.370	0.685	0.030	0.430
12.	0.248	0.576	-0.006	0.248	0.248	0.248	0.248
13.	0.248	0.673	0.248	0.285	0.030	0.212	-0.079
14.	0.139	-0.115	0.224	0.248	0.685	0.833	0.370
15.	0.248	0.030	0.139	0.576	0.685	0.576	0.576
Average	**0.171**	**0.202**	**0.072**	**0.375**	**0.294**	**0.346**	**0.419**

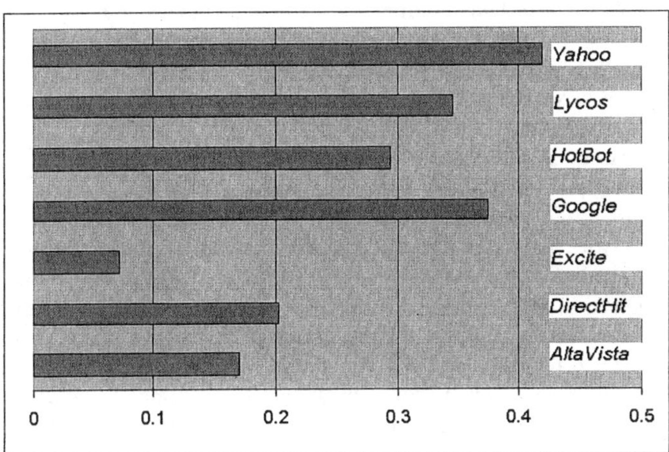

Figure 6: Performance of Web Search Engines based on three components of user feedback vector V, T, P.

From Figure 6, we see that *Yahoo* is the best, followed by *Google*, *Lycos*, *HotBot*, *DirectHit*, *AltaVista* and *Excite*, in that order. Let us, however, reiterate that these rankings of the search engines by us are just a pointer to what to expect, and should not be taken as the last word as yet. Our aim in this chapter is just to bring out a procedure for ranking search engines. We have just taken a few ad-hoc queries, 15 to be precise. For a more confident ranking of the search engines, we need to have a more comprehensive set of test-queries.

The strength of our search quality measure appears to be the fact that the ultimate intelligence of human beings is used for obtaining the "true" ranking (O_C) of the documents, which in turn, could be used as a standard to compare the ranking given by the search engines (ρ) against it. But for this, we are not at all bothering the user, rather inferring on our own the user feedback from his actions on the results. Moreover, our measure is absolute and not relative.

Summary

A "user satisfaction" guided web search procedure is presented in this chapter. The search index is made to improve from the users' feedback vectors. Improvement from the user feedback proceeds in such a way that the documents being consistently preferred by the users go up the ranking, while the ones being neglected go down. We also propose a couple of novel additions in the web search querying methods, so as to make the web search more convenient. In the performance analysis experiment, we have used a probabilistic model to generate the user responses automatically. We observe that the correlation coefficient between the listings given by our search engine and that preferred by the user model increases as the training set size (the number of feedback vectors used for learning) increases. This shows that after sufficient learning has gone into our search engine, it renders a better level of user satisfaction. With our technique of user feedback based web search, we are no more dependent on the effectiveness of the heuristic used for characterization of the documents. Hence, our technique is simply applicable for all types of multimedia data repositories. Testing our system with real queries is being left as the future direction of research, for reason mentioned.

We have also tried to quantify the search quality of a search system. More satisfied a user is with the search results presented to him in response to his query, higher is the quality of the search system. The *"user satisfaction"* is being gauged by the sequence in which he picks up the results, the time he spends at those documents and whether or not he prints, saves, bookmarks, e-mails to someone or copies-and-pastes a portion of that document. We have proposed a fuzzy technique based on relative quantifiers "most" and "as many as possible" for the proper combination of the metrics of the user feedback. Our proposition is tested on 7 public web search engines using 15 queries. With this limited set of queries used herein, it has been found that *Yahoo* gives the best performance followed by *Google*, *Lycos*, *HotBot*, *AltaVista*, *Excite* and *DirectHit*. To say it with more confidence, we need to have a better set of queries. Our aim has been to bring out a subjective procedure for ranking search engines.

References

Ahmad N, Beg MMS (2002) Fuzzy Logic Based Rank Aggregation Methods for the World Wide Web. In: Proc. International Conference on Artificial Intelligence in Engineering and Technology (ICAIET 2002), Kota Kinabalu, Malaysia, June 17-18, pp. 363-368.

Bar-Yossef Z, Berg A, Chien S, Fakcharoenphol J, Weitz D (2000) Approximating Aggregate Queries about Web Pages via Random Walks. In: Proc. 26th VLDB Conference, Cairo, Egypt.

Beg MMS (2001) Integrating the Notion of Hubs and Authorities into that of PageRank for the Web. In: Proc. 25th National Systems Conference (NSC 2001), Coimbatore, India, Dec. 13-15, pp. 332-336.

Beg MMS (2002) From Content-Based to Connectivity-Based Search on the World Wide Web: A Journey Trail. Journal of Scientific and Industrial Research, vol. 61, September, pp. 667-679.

Beg MMS, Ahmad N (2001) Harnessing the Hyperlink Structure of the Web. J. IETE Technical Review special issue on IT-Enabled Services, vol. 18, no. 4, July-August, pp. 337-342.

Beg MMS, Ahmad N (2003) Soft Computing Techniques for Rank Aggregation on the World Wide Web. World Wide Web - An International Journal, Kluwer Academic Publishers, vol. 6, issue 1, March, pp. 5-22.

Beg MMS, Ravikumar CP (2000) Distributed Resource Discovery from the Internet for e-Commerce Applications. 43rd Annual Technical Convention (ATC-2000) of IETE, New Delhi, India, September 30 - October 1.

Beg MMS, Ravikumar CP (2002) Measuring the Quality of Web Search Results. In: Proc. 6th International Conference on Computer Science and Informatics - a track at the 6th Joint Conference on Information Sciences (JCIS 2002), March 8-13, Durham, NC, USA, pp. 324-328.

Bharat K, Broder A (1998) A Technique for Measuring the Relative Size and Overlap of Public Web Search Engines. In: Proc. 7th International World Wide Web Conference (WWW9), April, pp. 379-388.

Borodin A, Roberts GO, Rosenthal JS, Tsaparas P (2001) Finding Authorities and Hubs from Link Structures on the World Wide Web. In: Proc. 10th World Wide Web Conference, May 2-5, Hong Kong.

Brin S, Page L (1998) The Anatomy of a Large-Scale Hypertextual Web Search Engine. In: Proc. 7th World Wide Web Conference, Amsterdam, Elsevier Science, pp. 107-117.

Broder A, Kumar R, Maghoul F, Raghavan P, Rajagopalan S, Stata R, Tomkins A, Wiener J (2000) Graph Structure in the Web. In: Proc. Ninth Int. World Wide Web Conference, May 15-19, Amsterdem, Netherlands, http://www9.org/ w9cdrom/160/160.html. Also appeared in Computer Networks, 33 (2000), pp. 309-320.

Chakrabarti S (2000) Data Mining for Hypertext: A tutorial Survey. ACM SIGKDD Explorations, 1(2), pp. 1-11.

Chakrabarti S, Dom BE, Gibson D, Kleinberg J, Ravikumar S, Raghavan P, Rajagopalan S, Tomkins A (1999) Hypersearching the Web. Feature Article, Scientific American, June.

Chakrabarti S, Dom BE, Ravikumar S, Raghavan P, Rajagopalan S, Tomkins A, Gibson D, Kleinberg J (1999) Mining the Web's Link Structure. IEEE Computer, August, pp. 60-67.

Chakrabarti S, Punera K, Subramanyam M (2002) Accelerated Focused Crawling through Online Relevance Feedback. In: Proc. World Wide Web Conference (WWW2002), Honolulu, Hawaii, USA, May 7-11.

DirectHit search engine (2000) http://www.directhit.com

DirectHit White Paper (2000) http://directhit.com/about/products/technology_whitepaper.html, September.

Dwork C, Kumar R, Naor M, Sivakumar D (2001) Rank Aggregation Methods for the Web. In: Proceedings of the Tenth World Wide Web Conference. Hong Kong.

Google search engine (2002) http://www.google.com

Haveliwala TH (2002) Topic-Sensitive PageRank. In: Proc. World Wide Web Conference (WWW2002), Honolulu, Hawaii, USA, May 7-11.

Haveliwala TH, Gionis A, Klein D, Indyk P (2002) Evaluating Strategies for Similarity Search on the Web. In: Proc. World Wide Web Conference (WWW2002), Honolulu, Hawaii, USA, May 7-11.

Hawking D, Craswell N, Thistlewaite P, Harman D (1999) Results and Challenges in Web Search Evaluation. Toronto '99, Elsevier Science, pp. 243-252.

Henzinger MR (2001) Hyperlink Analysis for the Web. IEEE Internet Computing, Jan.-Feb., pp. 45-50.

Henzinger MR, Heydon A, Mitzenmacher M, Najork M (1999) Measuring Index Quality Using Random Walks on the Web. Computer Networks, 31, pp. 1291-1303.

Henzinger MR, Heydon A, Mitzenmacher M, Najork M (2000) On Near Uniform URL Sampling. In: Proc. 9th International World Wide Web Conference (WWW9), Amsterdam, Netherlands, May.

Herrera F, Herrera-Viedma E (1997) Aggregation Operators for Linguistic Weighted Information. IEEE Trans. Systems, Man and Cybernetics - Part A: Systems and Humans, vol. 27, no. 5, September, pp. 646-656.

Herrera-Viedma E, Herrera F, Chiclana F (1999) A Consensus Model for Multiperson Decision Making with Different Preference Structures. Technical Report DECSAI-99106, Department of Computer Science and Artificial Intelligence, University of Granada, Spain, April.

Kleinberg J M (1998) Authoritative Sources in a Hyperlink Environment. Journal of the ACM, 46(5):604-632, 1999. A preliminary version appeared in Proc. 9th ACM-SIAM Symposium on Discrete Algorithms, ACM Press, New York and SIAM Press, Philadelphia, 1998, pp. 668-677.

Lawrence S, Giles CL (1998) Searching the World Wide Web. Science, 5360(280):98.

Lawrence S, Giles CL (1999) Accessibility of Information on the Web. Nature, vol. 400, pp.107-109.

Lawrence S, Giles CL (1999) Searching the Web: General and Scientific Information Access. IEEE Communications Magazine, Jan., pp. 116-122.

Lempel R, Moran S (2000) The Stochastic Approach for Link-Structure Analysis (SALSA) and the TKC Effect. In: Proc. 9th International World Wide Web Conference (WWW9), Amsterdam, Netherlands, May 15-19.

Li L, Shang Y (2000) A New Method for Automatic Performance Comparison of Search Engines. World Wide Web, Kluwer Academic, vol. 3, no. 4, December, pp. 241-247.

Li L, Shang Y, Zhang W (2002) Improvement of HITS-based Algorithms on Web Documents. In: Proc. World Wide Web Conference (WWW2002), Honolulu, Hawaii, USA, May 7-11.

Li SH, Danzig PB (1997) Boolean Similarity Measures for Resource Discovery. IEEE Trans. Knowledge and Data Engineering, 9(6), pp. 863-876.

Meadow CT, Boyce BR, Kraft DH (2000) Text Information Retrieval Systems. Second edition, Academic Press.

Page L, Brin S, Motwani R, Winograd T (1998) The PageRank Citation Ranking: Bringing Order to the Web (http://google.stanford.edu/~backrub/pageranksub .ps).

Porter M (1980) An Algorithm for Suffix Stripping. Program: Automated Library and Information Systems, 14(3).

Rijsbergen CJ van (1979) Information Retrieval. Butterworth & Co. (Publishers) Ltd., London, second edition, online documentation (http://www.dcs.gla.ac.uk/Keith /Preface.html).

Ross TJ (1997) Fuzzy Logic with Engineering Applications. Tata McGraw Hill.

Salton G, McGill MJ (1983) Introduction to Modern Information Retrieval. McGraw Hill.

Srivastava A, Ravikumar CP, Beg MMS (2001) Enhanced Similarity Measure for Client-Directory-Server Model. In: Proc. Seventh National Conference on Communication (NCC-2001), IIT Kanpur, India, Jan. 26-28, pp. 42-46.

Yager RR (1988) On Ordered Weighted Averaging Aggregation Operators in Multicriteria Decision Making. IEEE Trans. Systems, Man and Cybernetics, vol. 18, no. 1, January/February, pp 183-190.

Zhang D, Dong Y (2000) An Efficient Algorithm to Rank Web Resources. In: Proc. Ninth International World Wide Web Conference, Amsterdam, May 15-19, 2000. Also appeared in Computer Networks, 33, pp. 449-455.

INTERNET-ENABLED SOFT COMPUTING HOLARCHIES FOR e-HEALTH APPLICATIONS
-Soft Computing Enhancing the Internet and the Internet Enhancing Soft Computing-

Mihaela Ulieru
Electrical and Computer Engineering Department
The University of Calgary
2500 University Dr. NW, Calgary, Alberta T2N 1N4 CANADA
http://isg.enme.ucalgary.ca/People/Ulieru/Default.htm

ABSTRACT

This work builds on the synergetic triad Soft Computing – Internet – Multi Agent Systems in developing technologies for remote diagnosis, prediction and ubiquitous healthcare. Our approach extends the holonic enterprise paradigm to the medical domain. A medical holarchy is a system of collaborative medical entities (patients, physicians, medical devices, etc.) that work together to provide a needed medical service for the benefit of the patient. Representing holons as software agents enables the development of e-health environments as web-centric medical holarchies with a wide area of application in telemedicine. Our approach exploits the triad's synergy twofold. On one side we use soft computing to enhance to power of the Internet by a fuzzy-evolutionary approach enabling emergence of virtual communities in Cyberspace. On the other side we use the Internet to empower soft computing both as enabler for the powerful integration of several soft computing strategies into a unified diagnosis and prediction methodology and as enhancer of learning neuro-fuzzy diagnostic rules and knowledge base refinement through a remote database creation and exploitation mechanism. A case study in glaucoma progression monitoring illustrates the benefits of our approach.

Keywords: Soft Computing, Multi-Agent Systems, Evolutionary, Self-Organizing Cyberspace, Emergent Virtual Organizations, Neuro-Fuzzy Diagnosis and Prediction, Web-Centric Database Management, Medical Holarchy.

1. INTRODUCTION: EMERGENT HOLARCHIES IN CYBERSPACE

1.1. The Holonic Enterprise: A Model for Internet-Enabled Workflow Management in Virtual Organizations

Multi-agent Systems enable cloning of real-life systems into autonomous software entities with a 'life' of their own in the dynamic information environment offered by today's Cyberspace. The Holonic Enterprise (HE) has emerged as a business paradigm from the need for flexible open reconfigurable models able to emulate the market dynamics in the networked economy [1], which necessitates that strategies and relationships evolve over time, changing with the dynamic business environment. In today's open environment connected via the dynamic Web, the HE paradigm provides a framework for information and resource management in global virtual organizations by modeling enterprise entities as software agents linked through the internet [3]. Building on the triad: Internet-MAS-Soft Computing we recently developed a model for the HE that endows virtual communities/societies with proactive self-organizing properties [2].

A HE, Fig. 1 is a holarchy of collaborative entities (here generically coined as 'enterprises'), where each entity is regarded as a holon and is modeled by a software agent with *holonic* properties [4], (that is: the software agent may be composed of other agents that behave in a similar way but perform different functions at lower levels of resolution.)

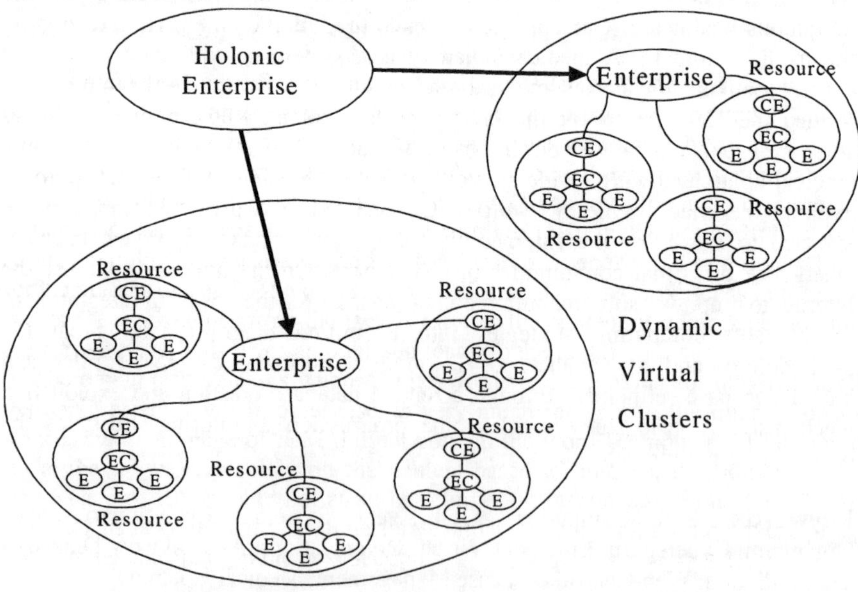

Fig. 1. Holonic Enterprise Model

The flow of information and matter across the HE defines three levels of granularity:

- **The inter-enterprise level.** At this level (Fig. 1), several holon-enterprises cluster into a collaborative holarchy to produce products or services. The clustering criteria support maximal synergy and efficiency. With each collaborative partner modeled as an agent that encapsulates those abstractions relevant to the particular cooperation, a dynamic virtual cluster emerges that can be configured on-line according to the collaborative goals.

- **The intra-enterprise level.** Once each enterprise has undertaken responsibility for the assigned part of the work, it has to organize in turn its own internal resources to deliver on time according to the coordination requirements of the collaborative cluster.

- **The physical resource level** - this level is concerned with the coordination of the distributed physical resources that actually perform the work. Each resource is cloned as an agent, which abstracts those parameters needed for the configuration of the holarchy at the physical level.

How does one build agents and groups of agents which fulfill the holonic philosophy? If a one-to-one mapping of holon to agent is performed, it is much more difficult to practically implement an agent (than it is to conceptualize a holon) which is itself a component of a higher level agent and which also contains lower level agents. Here, the concept of a *mediator agent* comes into play, Fig. 2 [3].

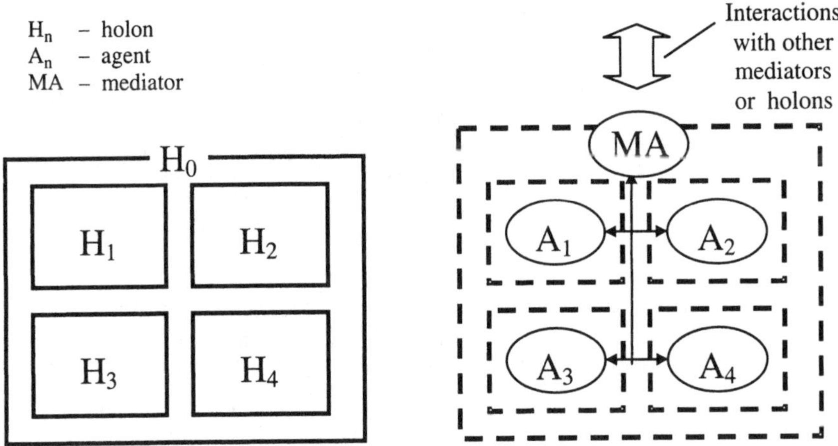

Fig. 2. Mapping Agents to Holons via Mediator

The mediator will fulfill two main functions. First, it acts as the interface between the agents in the holon and between the agents outside the holon (i.e. acts as a type of facilitator); conceptually, it can be thought of as the agent that represents the holon. Second, it may broker and/or supervise the interactions between the sub-holons of that holon; this also allows the system architect to implement (and later update) a variety of forms of interaction easily and effectively, thereby fulfilling the need for flexibility and reconfigurability. The mediator encapsulates the mechanism that clusters the holons into collaborative groups [5]. The architectural structure in such holarchies follows the design principles for metamorphic architectures, Fig. 3.

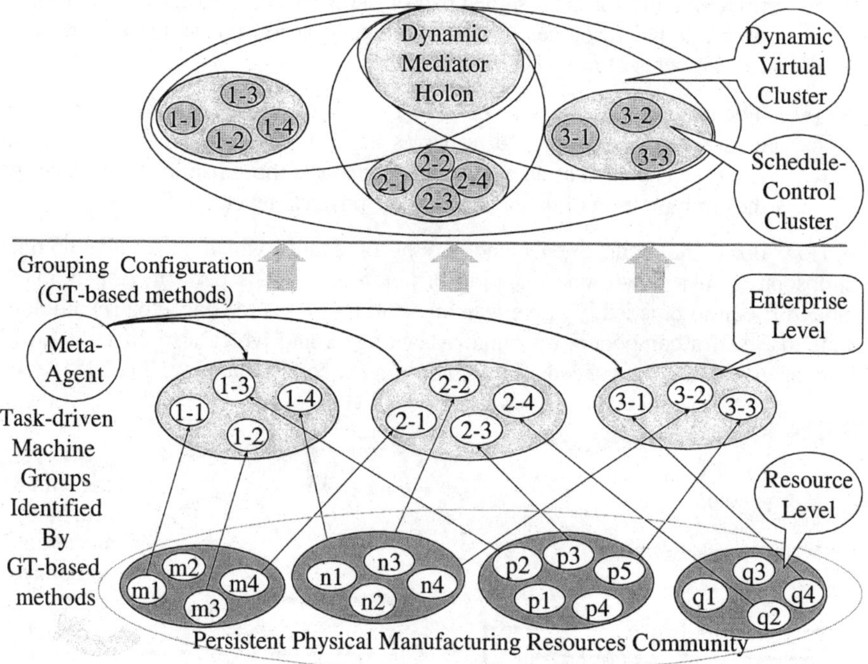

Fig. 3. Mediator-Centric Architecture

With each collaborative partner modeled as an agent that encapsulates those abstractions relevant to the particular cooperation, a *dynamic virtual cluster* emerges which can be configured on-line according to the collaborative goals. Backed by the recent advances in wireless and communications technologies such a dynamic collaborative holarchy can cope with unexpected disturbances (e.g. replace a collaborative partner who breaks commitments) through on-line reconfiguration of the open system it represents. It provides on-line task distribution across the available resources, Fig. 4. as well as deployment mechanisms that ensure real-time error reporting and on-demand workflow/information tracking (e.g. fault tracking in distributed discrete manufacturing, etc.).

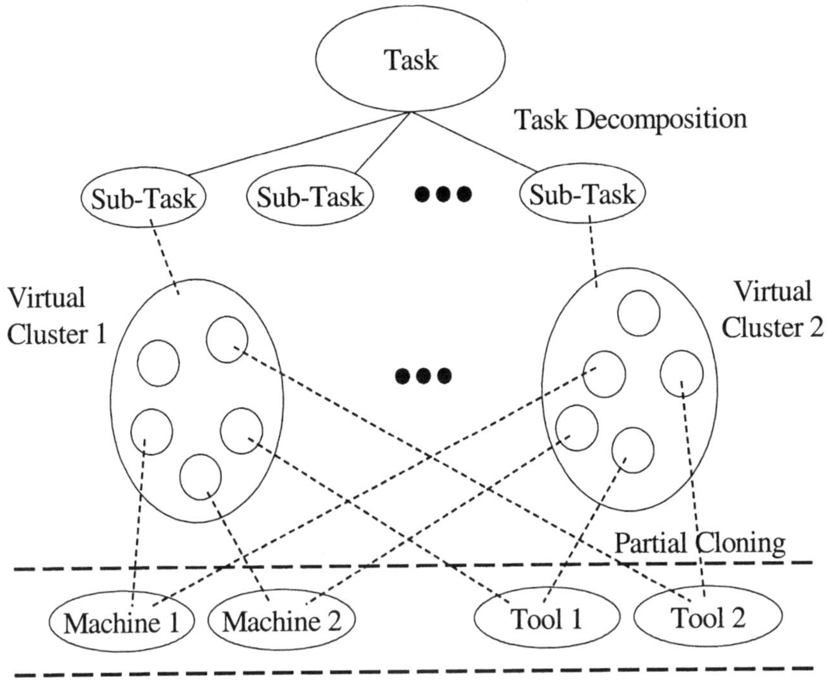

Fig. 4. Task Distribution Mechanism (Intra-Enterprise Level)

1.2. Soft Computing for the HE: Enabling the Evolutionary, Self-Organizing Cyberspace

As result of the process of evolution driven by the law of synergy, emergence endows the dynamics of composite systems with properties unidentifiable in their individual parts. The phenomenon of emergence involves:

- *self-organization* of the dynamical systems such that the synergetic effects can occur;
- interaction with other systems from which the synergetic properties can *evolve* in a new context.

The fuzzy-evolutionary approach introduced in [2] mimics emergence in Cyberspace as follows:

- it induces self-organizing properties by minimizing the entropy measuring the information spread across the virtual system/organization (holarchy) such that equilibrium is reached in an optimal interaction between the system's parts to reach the system's objectives most efficiently;

- it enables system's evolution into a better one by enabling interaction with external systems found via genetic search strategies (mimicking mating with most fit partners in natural evolution) such that the new system's optimal organizational structure (reached by minimizing the entropy) is better than the one before evolution.

We present the key points of this emergence mechanism in the next two Subsections.

1.2.1. *Self-Organization: Emergence of the Holonic Structure*

The essence of the approach stems from encapsulation into the mediator of a dynamic virtual clustering mechanism that optimizes the information and resource management across the HE to fulfill in the most efficient way the goal of the holarchy. The main idea is to minimize the entropy in the information spread across the VE (modeled as a multi-agent system) – such that each holon maximizes its knowledge of the task it is assigned, in order to best accomplish it. This naturally leads to the (self)organization of the VE in a holarchy (which defines a HE).

We regard a MAS (denoted here by $A_N = \{a_n\}_{n \in \overline{1,N}}$ the set of $N \geq 1$ agents) as a dynamical system in which agents exchange information and organize it through reasoning into knowledge about the assigned goal [6]. Once a goal is assigned for the MAS agents may cluster in various ways to work cooperatively towards the goal's accomplishment. We define a source-plan as a collection of $M_k \geq 1$ different clustering configurations in which the agents team-up to accomplish the holarchy goal: $P_k = \{P_{k,m}\}_{m \in \overline{1,M_k}}$. The only available information about P_k is the *degree of occurrence* associated to each of its clustering configurations (Fig. 3), $P_{k,m}$, which can be assigned as a possibility measure [7] $\alpha_{k,m} \in [0,1]$. Thus, the corresponding degrees of occurrence are members of a two-dimension family $\{\alpha_{k,m}\}_{k \in \overline{1,K}; m \in \overline{1,M_k}}$.

Optimal knowledge at the holarchy's highest level of resolution (inter-enterprise level) corresponds to an optimal level of information organization and distribution among the agents within all levels of the holarchy. We consider the entropy as a measure of the degree of order in the information spread across the multi-agent system modeling the holarchy. One can envision the agents in the MAS as being under the influence of an information "field" which drives the agent interactions towards achieving "equilibrium" with other agents with respect to this entropy[1]. This information is usually uncertain, requiring several ways of modeling to cope with the different aspects of the uncertainty. Fuzzy set theory offers an adequate framework for dealing with this uncertainty [8].

We model agent interactions through fuzzy relations considering that two agents are in relation if they exchange information. As two agents exchanging information are as well in the same cluster one can describe the clustering configurations by means of these fuzzy relations. The family of fuzzy relations, $\{\mathsf{R}_k\}_{k\in\overline{1,K}}$, modeling the clustering possibility over the MAS agents(A_N) is built using the possibility measures $\{\alpha_{k,m}\}_{k\in\overline{1,K};m\in\overline{1,M}_k}$ and the family of source-plans $\{\mathsf{P}_k\}_{k\in\overline{1,K}}$. This fuzzy relation of $\mathsf{A}_N\times\mathsf{A}_N$ is also uniquely associated to $P_{k,m}$. Thus a map associating a family of source plans $\mathsf{P}=\{\mathsf{P}_k\}_{k\in\overline{1,K}}$ to a fuzzy relation measuring their possibility of occurrence $\mathsf{R}=\{\mathsf{R}_k\}_{k\in\overline{1,K}}$, can be built:

$$T(\mathsf{P}_k)=\mathsf{R}_k\ ,\ \forall k\in\overline{1,K}.\tag{1}$$

We use the generalized fuzzy entropy [9] to measure the degree of order in the information spread across the holarchy:

$$S_\mu(\mathsf{R}_k)=-\sum_{i=1}^{N}\sum_{j=1}^{N}\mathsf{M}_k[i,j]\log_2\mathsf{M}_k[i,j]-\sum_{i=1}^{N}\sum_{j=1}^{N}]1-\mathsf{M}_k[i,j]]\log_2[1-\mathsf{M}_k[i,j]].\tag{2}$$

The generalized fuzzy entropy is the measure of the "potential" of this information field and *equilibrium* for the agents under this influence corresponds to an optimal organization of the information across the MAS with respect to the assigned goal's achievement [6]. When the circumstances change across the holarchy (due to unexpected events, such as need to change a partner that went out of business, machine break-down, raw materials unavailable, etc.) the equilibrium point changes as well inducing a new re-distribution of information among the agents with new emerging agent interactions.

---

[1] The information 'field' acts upon the agents much in the same manner as the gravitational and electromagnetic fields act upon physical and electrical entities respectively.

The optimal source plan:

$$P_{k_0} = T^{-1}(\arg\min_{k \in \overline{1,K}} S_\mu(P_k)) \text{ , where } k_o \in \overline{1,K} . \tag{3}$$

defines the most efficient clustering of resources across the holarchy relative to the goal achievement. P_{k_0} is the least fuzzy (minimally fuzzy), i.e. the least uncertain source-plan. The fuzzy relation encoding the agent clustering to fulfill the VE goal according to this optimal plan, R_{k_0} defines two types of source plans that can emerge:

- When R_{k_0} is a *similarity relation*, then clusters are associated in order to form new clusters, and a nested hierarchy emerges that organizes the MAS modeling the VE into a holarchy (that is a HE emerges from the VE[2]).

- When R_{k_0} is only a *proximity* relation, tolerance (compatibility) classes can be constructed as collections of eventually overlapping clusters (covers). This time, the fact that clusters could be overlapping (i.e. one or more agents can belong to different clusters simultaneously) reveals the capacity of some agents to play multiple roles by being involved in several tasks at the same time while the holonic properties of the organization are still preserved.

This procedure endows the VE with self-organizing properties that ensure emergence of the optimal holonic structure once the resources are known (that is once the distributed partners have committed to the common goal and allocated the resources they want to put in to accomplish it.) Thus the above procedure ensures emergence of an optimal HE from any virtual organization with predefined resources.

1.2.2. Evolution Towards the Best Structure

In the open environment created by the dynamic Web opportunities for improvement of an existing virtual organization arise continuously. New partners and customers alike come into the virtual game bidding their capabilities and money to get the best deal. Staying competitive in this high dynamics requires openness and ability to accommodate change rapidly through a flexible strategy

---

[2] A (unique) similarity relation can be constructed starting from the proximity relation R_{k_0}, by computing its *transitive closure*. Thus, the potential holonic structure of MAS can be revealed, even when it seems to evolve in a non-holonic manner.

enabling re-configuration of the organization to be able to respond to new market demands as well as to opportunities (e.g. in playing with a better partner when needed.) In response to this need we have designed an evolutionary search strategy that enables the virtual organization to continuously find better partners fitting the dynamics of its goals as they change according to the market dynamics.

We regard 'the living Web' as a genetic evolutionary system. The selection of the agents (partners) that best fit the holarchy's objective is done in a similar way to the natural selection by 'survival of the fittest' through which the agents/partners that best suit the HE with respect to the goal accomplishment are chosen from the offers available on the Web. In this search model the mutation and crossover operators (p_m and p_c) represent probabilities of finding 'keywords' (describing the attributes required from the new partners searched for) inside the search domain considered. Our construction is based on the observation that the search process on an agent domain[3] [10] containing information about a set of agents that 'live' on the Web (e.g. a directory 'look-up'-like table of 'yellow page' agents describing the services that the possible partners offer [11]) is analogous to the genetic selection of the most suitable ones in a population of agents meant to 'fit' the virtual organizations' goals.

The main idea is to express the fitness function (measuring how well the new agent fits the holarchy's goal) in terms of the fuzzy entropy (2):

$$F = S_\mu \tag{4}$$

With this, minimizing the entropy across the extended MAS (which includes the agents from the search domain) according to HE goal-reach optimization equates optimizing the fitness function which naturally selects the best agents fitting the optimal organizational structure of the HE. In the sequel we present the mathematical formalism for this evolutionary search.

According to (3) it results that minimizing S_μ leads to a fuzzy relation that encodes the best clustering configuration for the HE. This fuzzy relation being either a proximity or a similarity measure it is intuitive to consider it as a good measure for the relevancy **R** of the new agents to the HE goal-reach:

$$\mathbf{R} = R_{k_0} = \arg \min_{k \in \overline{1,K}} S_\mu(R_k) \tag{5}$$

Defining for example the fuzzy relation R_k in (2) as a preference relation encoding, e.g. the desire of agents to work cooperatively, gives a *relevancy measure* that perfectly fits the purpose of the search for better partners. That is –

---

[3] See the FIPA architecture standard at www.fipa.org

when agents are found for which the preference is higher then for the existing ones, they should replace the old ones. This increases the membership values of the preference relationship, which indicates that the relevance relative to our search is higher. The algorithm that searches for better partners in Cyberspace is presented in Fig. 5.

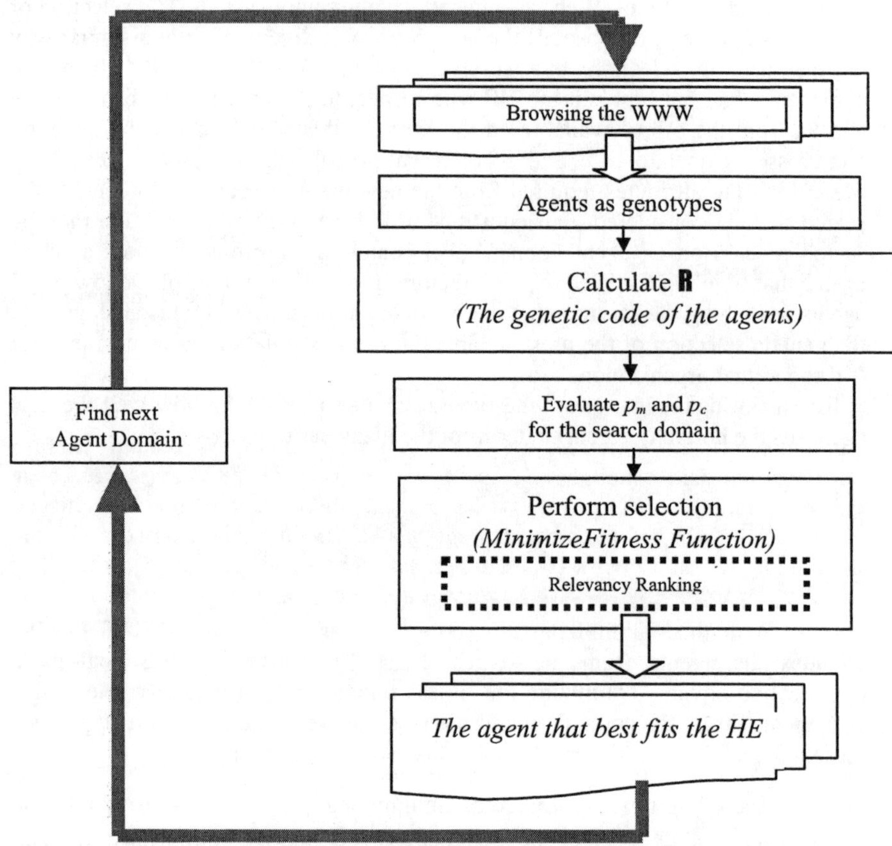

Fig. 5: Evolutionary Search in Cyberspace

We initialize the search process as follows:
- The initial population (*phenotype*) consists of the existing agents in the HE before the search.
- Calculate R for the phenotype
- Rank the preferences and determine the optimal source plan (3) by computing the corresponding α-cuts.
- The preferences for the optimal source plan (represented as binary strings) constitute the *genotype*. They encode all the relevant information needed to evolve the HE towards a better structure by selecting better agents while searching on an expanded domain.

140

- The phenotype evolves by reproduction according to how the probabilities of mutation and crossover (p_m and p_c) affect the genotype [12]. Each chromosome of the population (in the genotype) will be randomly affected.

The essence of this evolutionary search process stems from the recursive modification of the chromosomes in the genotype in each generation while monitoring the fitness function (4). At each iteration (that is whenever a new agent domain s searched) all members of the current generation (that is the existing agents in the holarchy and the new ones searched for) are compared with each other in terms of the preference measures. The ones with highest preferences are placed at the top and the worst are replaced with the new agents. The subsequent iteration resumes this process on the partially renewed population. In this way the openness to new opportunities for continuous improvement in the HE constituency is achieved and with this the emergence of an optimal structure for the holarchy. Embedding this strategy in the mediator (Fig. 3) endows the HE with the capability to continuously evolve towards a better and better structure by bringing to the table better and better partners as they are found.

The proposed emergence mechanism empowers the HE with self-adapting properties and moreover enables it to evolve like a social organism in Cyberspace, by mating its components with new partners as they are discovered in a continuous incremental improvement search process.

2. ON MEDICAL HOLARCHIES: PARTICULARITIES AND CHALLENGES

Applying this approach to the virtual societies 'living' on the dynamic Web endows them with behavioral properties characteristic to natural systems. In this parallel universe of information, enterprises enabled with the proposed emergence mechanism can evolve towards better and better structures while at the same time self-organizing their resources to optimally accomplish the desired objectives. Once a goal is set (by a customer) a HE emerges clustering available resources (modeled as software agents) to meet the need optimally (e.g. with minimal cost.)

2.1. Particularities of Medical Holarchies

A medical holarchy is a system of collaborative medical entities (patients, physicians, medical devices, etc.) that work together to provide a needed medical service for the benefit of the patient. Like any HE, medical holarchies are as well customer centric, the customer being (in most cases) – the patient.[4] The elements defining the levels of a medical HE are:

---

[4] In some medical holarchies the physician can be a customer (e.g. when the holarchy is used to retrieve information about a certain disease, or patient – for the physician's interests.)

- **Inter-Enterprise**: Hospitals, Pharmacies, Medical Clinics/Laboratories
- **Intra-Enterprise**: Sections/Units/ Departments of the each medical enterprise
- **Resource Level**: Machines for medical tests, medical monitoring devices, information processing resources (medical files, computers, databases, decision support systems), physicians, medical personnel (technicians, assistants, etc.)

When combined with the latest advances in communication technologies (supporting access to/from anywhere/anytime via wireless and other advanced mobile means, Fig. 6) medical holarchies ensure ubiquitous ad-hoc healthcare [13] by enabling:

- Emergence of a medical holarchy around patient's need. Search for the most suitable medical entities able to cooperate and organize their interaction to provide the desired services optimally
- Disease history tracking / disease progression monitoring through the integrated information model defining the emergent medical holarchy that enables workflow management throughout the levels [3]
- Patient-centric management of medical data (enabling system's configuration and re-organization for use to the patient's benefit)
- Cooperative medical decision-making and management of conflicting requirements in multiple disease condition [14]
- Patient condition monitoring with alerting the medical personnel in case of emergency [15]
- Medical emergency logistics with patient information retrieval and heterogeneous transaction workflow management throughout the medical holarchy.

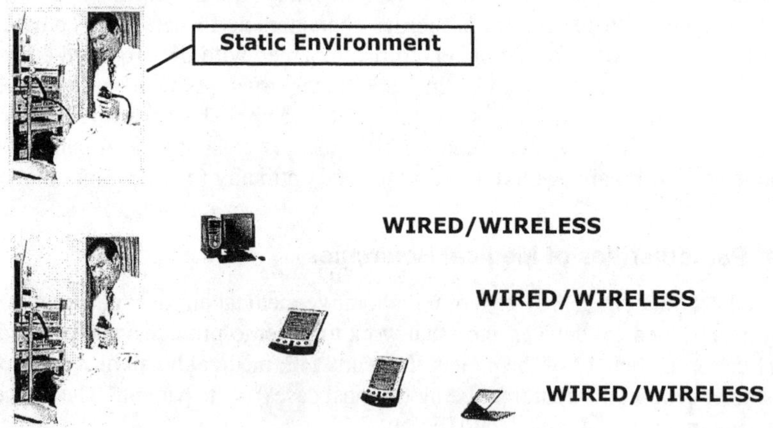

Fig. 6. IT enablers for emergent medical holarchies

From a software engineering perspective [16] the logical view on a medical holarchy (at the intra-enterprise level) involves the domain actors[5] (Fig. 7, Table 1) and the objects/agents – as domain entities[6] (Fig. 8, Table 2).

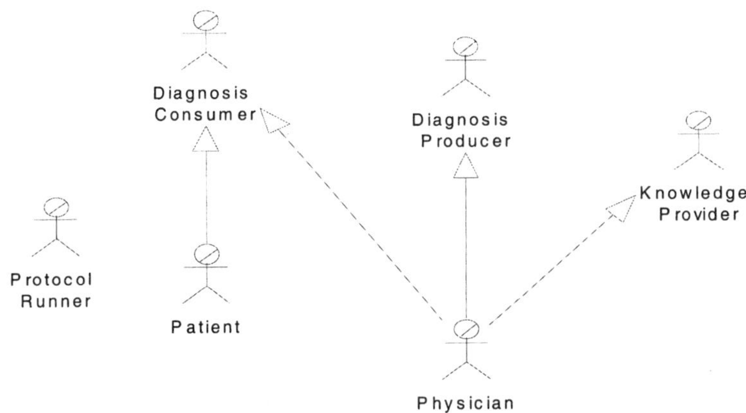

Fig. 7. Domain Actors in a medical holarchy

Table 1. Responsibilities of Domain Actors

ID	Actor	Description
1.	Diagnosis Producer	Performs Diagnoses.
2.	Diagnosis Consumer	Has an interest and a use for a Diagnosis
3.	Knowledge Provider	Contributes to the body of knowledge that supports diagnosis
4.	Physician	Primary medical contact and a Diagnosis Consumer, Diagnosis Producer, and Knowledge Provider.
5.	Patient	The person with the health care concern. The Patient is also a Diagnosis Consumer.
6.	Protocol Runner	Runs diagnostic tests in order to observe and measure relevant health care characteristics.

---

[5] An interesting observation about this abstraction of the roles in the diagnosis domain is that we have avoided making assumptions about the *number* of instances of these roles in the domain. An implementation of this model in any architecture could have, for example, many Diagnosis Producers, each one having acquired slightly different specializations

[6] a domain document (entity) is a characterization of the information that the domain actors exchange and act upon as they carry out their responsibilities

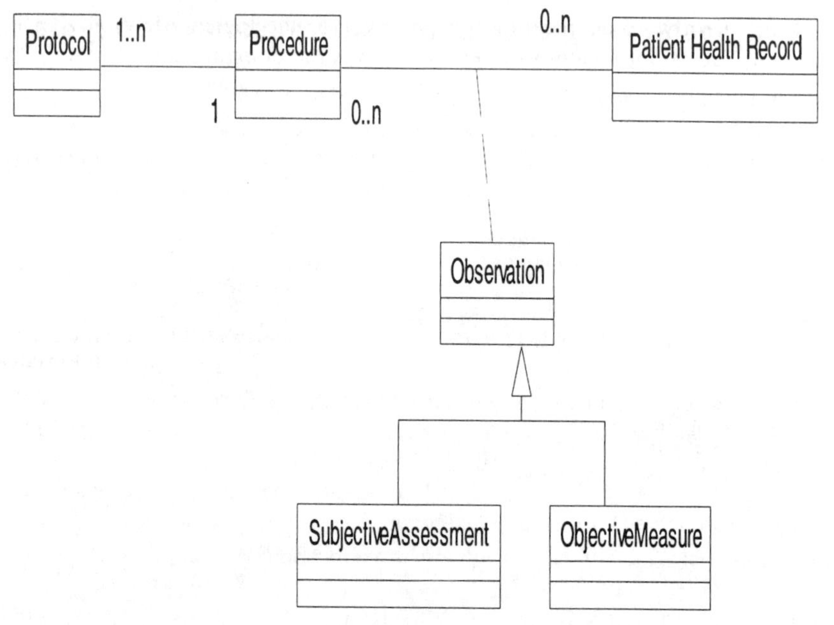

Fig. 8. Domain Entities (business documents)

Table 2. Description of Domain Entities

Domain Entity	Description
Protocol	Represents the set of steps that a Protocol Runner takes in order to collect the diagnostic measurements. This ensures that the data is collected in a controlled and measured fashion. The protocol has revisions in order to support continual improvement. The Diagnosis Producer will not make diagnoses if the diagnostic measures were not collected with an appropriate revision.
Procedure	Represents a diagnostic test or procedure.
Patient Health Record	Represents the sum of all health-related information about the Patient
Observation	Represents an observation that contributes to a Diagnosis Producer making a diagnosis.

Across the holarchy actors and domain entities collaborate to best serve the patient's need[7].

---

[7] It is interesting to observe that all the actors in the medical holarchy act at the physical resource level, except the patient – who can access the resource level and by this

Actors are cloned as software agents that interact with the domain entities and other agent-actors to achieve the goal of the holarchy. Each actor has defined a subgoal that it has to achieve to support the overall achievement of the holarchy's goal. The holarchy self-organizes in a way that ensures its goal's achievement optimally through the synergetic interaction of all actors. This ensures optimal goal achievement for all actors that collaborate towards the holarchic goal achievement. The individual goals of the domain actors in a medical holarchy are generically described in Table 3.

Table 3. Collaborations emerging around actors' goals in a medical holarchy

Actor	Goal Identifier	Goal Description	Collaborations defining Holons
Diagnosis Consumer	GS-001	To obtain a diagnosis	Physician, Patient, Pharmacist
Diagnosis Producer	GS-101	To assess the risk of a Patient having a designated condition	Physician, Expert Software, Patient
	GS-102	To predict the Patient's progress during treatment or recovery	
	GS-103	To assess the efficacy of treatment alternatives	
	GS-104	To interpret the measurements and observations that arise from diagnostic tests	
Knowledge Provider	GS-201	To map domain expertise to shareable rules	Physician, Expert Software
	GS-202	To share domain expertise	
	GS-203	To improve the protocols for conducting test procedures	
Protocol Runner	GS-301	To observe the clinical status of a Patient	Physician, Medical Assistant / Technician, Test Machines
	GS-302	To perform diagnostic tests and publish the measurements	
Patient	GS-401	To know the results of diagnostic procedures	Patient (basic holon)
	GS-402	To minimize the number of tests required	
Physician	GS-501	To maximize their capacity for treating Patients	Physician (basic holon)
	GS-502	To assist and guide the Patient to better health	

shortcutting throughout the holarchy. This is a particularity of HE in the medical domain. In other areas (e.g. manufacturing holarchies, having the customer the trigger of the supply chain [3]) customers do not have direct access to the resources at the physical level within the HE.

Usually the goal of the medical holarchy is set by a need (around which the holarchy actually emerges!). The holarchic goal is always to satisfy that particular need optimally. The interaction between different actors collaborating to achieve the holarchic (global) goal defines a dynamic network of subgoals (local goals/individual goals of each agent – Table 3) Fig. 4 from which the configuration of the holarchy emerges. Such collaborations define holons clustering the entities of the holarchy that work together to fulfill the goal. Thus the holarchy self-organizes in a way that ensures the most efficient interaction of holons at all levels – which in turn defines priorities for the goals of each individual holon.

A need around which a holarchy emerges is homonymous to a use case in software development [16]. The use case diagram in Fig. 9 identifies typical needs around which medical holarchies emerge[8].

The goals/needs can be identified with the various functions that the holarchy can undertake, functions that are distributed across the holons according to their responsibilities/roles/capabilities[9]. A description of such functions and the holons emerging from the collaborations needed to fulfill each function is presented in Table 4.

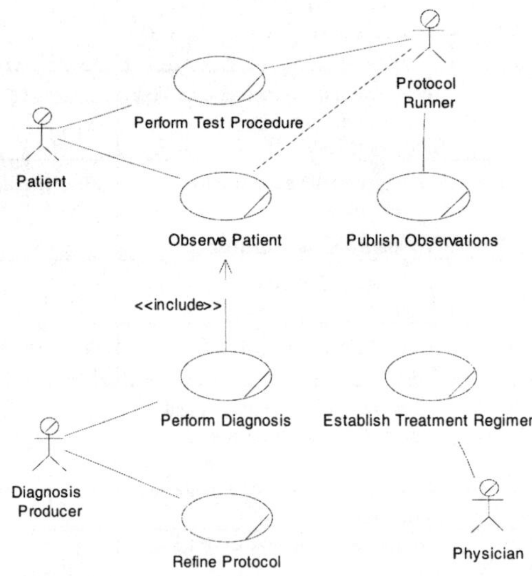

Fig. 9. Goals/Needs that a Medical Holarchy can achieve (a subset)

---

[8] Here we assume that an individual test or even treatment regimen is only a part of a long-term care plan that the Physician establishes and administers. This distinguishes the role of a Physician from that of a Diagnosis Producer, but recognizes that in many cases an individual person fulfills both of these roles.

[9] Some actors are higher level holons, consisting of clusters of other agents that collaborate to achieve the respective function. Other actors (such as #4 – Physician) are simply basic holons (singletons).

Table 4. Examples of Goal/Need/Function in a Medical Holarchy and the Emerging Collaborations

Actor (Holon)	Domain Function	Description	Collaborative entities
1. Diagnosis Producer	Perform Diagnosis	Assess risks, determine progress in treatment, or evaluate alternative treatments	Physician, Expert Software
4. Physician	Establish Treatment Regimen	Initiate long-term care plan for the Patient	Physician
6. Protocol Runner	Observe Patient	Make qualitative observations of the Patient's condition	Physician, Medical Assistant/Technician, Test Machines, Database (electronic files)
	Perform Test Procedure	Make quantitative observations of the Patient's condition	Physician, Medical Assistant/Technician, Test Machines,
	Publish Observations	Make the results of observations available	Physician, Medical Assistant/Technician, Test Machines, Database (electronic files)
	Refine Protocol	Adapt the test procedure protocol to meet new requirements	Physician, Medical Assistant/Technician, Test Machines, Database (electronic files), Expert Software

2.2. Agents vs. Web Services in Medical Holarchies Implementation

2.2.1. Web-Centric Implementation of Medical Holarchies

For distribution across the web, when implementing medical holarchies each domain actor maps to a web service with a specific set of functional responsibilities. The domain functions listed in Table 4 map to individual methods on the designated web services. The domain entities (Table 2) map to regular objects implemented in the language of choice. The result is a set of loosely coupled functional components that interact as required to carry out the requested diagnosis. The role of mediator agent (Fig. 3) in such a medical holarchy is played by a centralized workflow manager that coordinates the requests for diagnosis and gets the responses back to the originator, Fig. 10. Implementing the domain model in this fashion permits easy distribution across the web since implementers could deploy individual web services on different machines in different locations as required. In addition, the implementers could also deploy infrastructure services (such as database management) as web services.

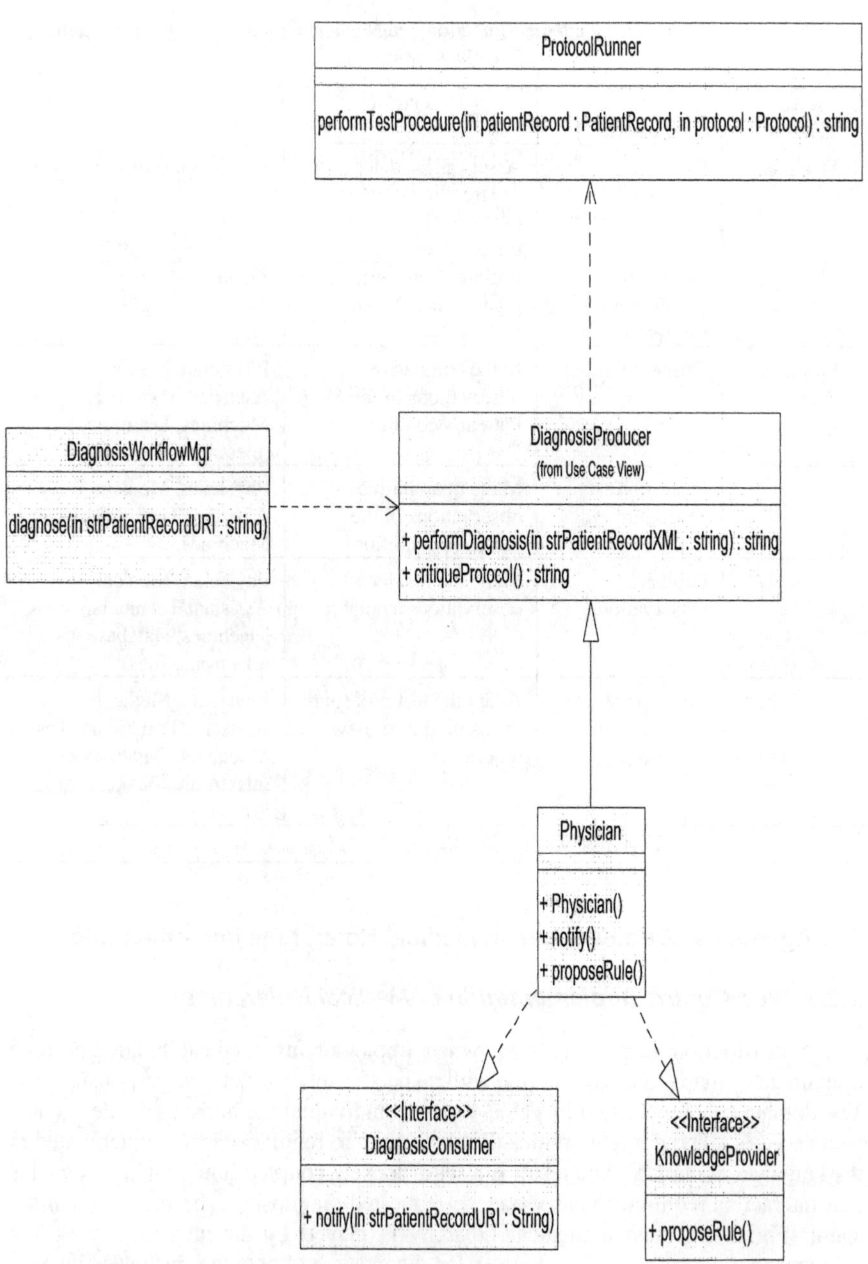

Fig. 10. Sample Medical Holarchy with Workflow Manager as Mediator (Web-Centric implementation)

The mappings of architectural elements that define the medical holarchy – into web services is presented in Table 5.

Table 5..Mapping of Holarchic Architectural Elements to Web Services

Generic Architecture Element	Mapping
Abstract Domain Actor	Interface
Concrete Domain Actor	Web Service
Goal	Web Service Method
Domain Entity	Class Module, XML Document

2.2.2. Multi-Agent System Implementation of Medical Holarchies

In Fig. 11, the domain actors map to agents and their functions/goals map to tasks (Table 6). The concrete domain actors (and the mediator) are implemented directly as agents. In the case of FIPA-OS[10], for example, this means that they would extend the FIPAOSAgent class[11].

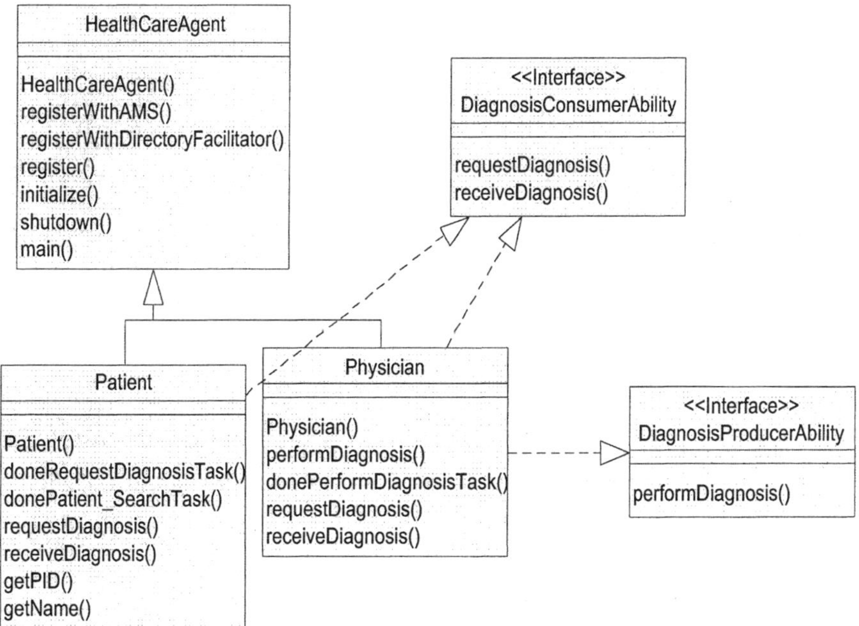

Fig. 11. Medical Holarchy (FIPA-OS implementation)

---

[10] See www.fipa.org

[11] In FIPA-OS, an agent can optionally have a listener class that receives the message and initiates the response.

Generic Architecture Element	Mapping
Abstract Domain Actor	Interface, Listener Task
Concrete Domain Actor	Agent
Goal	Task
Domain Entity	Class Module, XML Document

In order to provide the appropriate abstract behavior, the goals and functions/responsibilities (Tables 3 and 4) of the abstract domain actors are implemented as Java interfaces (e.g. DiagnosisProducerAbility and DiagnosisConsumerAbility in Fig. 11), but also as descendents of the Task class (Fig. 12). This duality is only necessary to coordinate the entities collaborating to achieve the goal. In FIPA-OS this is done via a mechanism allowing/requiring the agents to "listen" for incoming performatives from other agents.

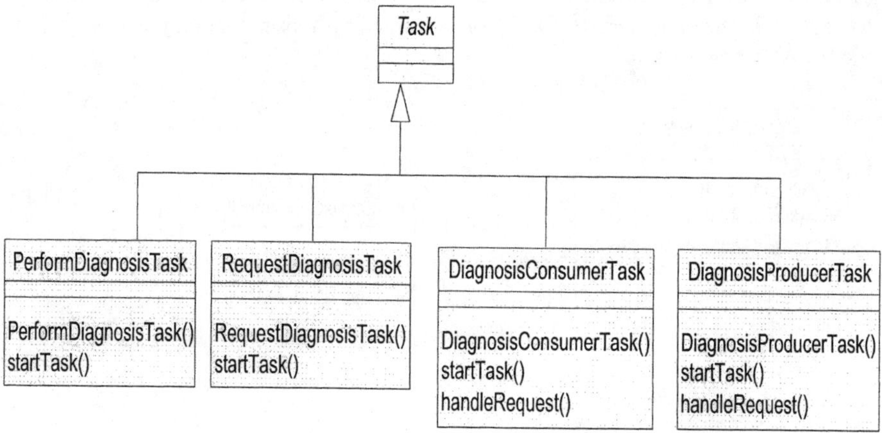

Fig. 12. Implementing goals as tasks in medical holarchy deployment

Both a web services perspective and an agent perspective require fundamentally the same domain analysis. The analysis of the roles (actors) and their goals/responsibilities provide an adequate basis for implementing the solution in either technique, although a hybrid solution where some domain actors are web services (the ones that don't interact other than to respond to requests) and other actors are agents (the ones that respond and initiate domain activity), may in some particular applications be more suitable. The greatest difference lies in the existence (or non-existence) of an object that handles the workflow vs. a mediator agent. In an agent-based system, the interoperability and interactions between agents and tasks is well-founded and easy to implement. In a web services solution, however, this interoperability must be moderated by an additional workflow manager type of web service. The need to program this service makes the solution more static and less dynamic and as well adds more code to the implementation. The industry tendency to promote web services as remote procedure calls, however, weakens their usefulness. The metaphor suggested here by to mapping a web service to a domain actor (Table 5) as opposed to a function library makes our approach more suitable to e-Health applications [16].

2.3. Challenges in Implementing Medical Holarchies

Besides the challenges involved in the development and deployment of any internet-enabled agent-based system medical holarchies imply particular challenges brought by the specifics of the healthcare system [17]. A major issue is reconciliation of the various standards of care across the continents. Especially security and privacy of electronic medical records is of major significance and has proven to be the major brake that slowed down the adoption of e-Health by major clinics around the world but especially in the North Americas [15]. A solution to both issues (that seems truly viable) is the electronic institution [18] - a normative framework which emulates regulatory mechanisms in real life social institutions. Such institutions define and police norms that guide individual agents collaborating in a medical holarchy. These norms set acceptable actions that each agent can perform in connection to the role(s) it plays and clearly specifies access restrictions on data according to these roles.

Another important issue pertains to the development, dissemination and utilization of common communication standards, vocabularies and ontologies [15]. The EU's CEN/TC 251 aim is to achieve compatibility and interoperability between independent systems, to support clinical and administrative procedures, technical methods to support interoperable systems as well as requirements regarding safety, security and quality. The US standardization bodies, the American Society for Testing and Materials' Committee on Healthcare Informatics (ASTM E31) [19] and Health Level Seven [20] are involved in similar work. ASTM E31 is developing standards related to the architecture, content, storage, security, confidentiality, functionality, and communication of information while HL7 is mainly concerned with protocol specifications for application level communications among health data acquisition, processing, and handling systems.

Bioinformatics and health care informatics are fields that already have active communities developing ontologies, yet the application of such ontologies as OpenGALEN [21], Unified Medical Language System (UMLS) [22], Systematized Nomenclature of Human and Veterinary Medicine (SNOMED) [23], has lagged behind their potential, despite the huge drive by health care professionals to bring bioinformatics and health care information into clinical workstations and onto the Internet.

The main reason appears to be that these existing ontologies are being developed to meet different needs, each with its own representation of the world, suitable to the purpose it has been developed for. There is as yet no common ontology. Of those that are being developed, OpenGALEN provides a common terminology that is currently of limited scope, while UMLS lacks a strong organizational structure, and SNOMED provides only diagnosis nomenclature and codification.

Besides the social acceptance of medical holarchies, professional acceptance – that is by the medical doctors is a major issue.

Health care professionals are quite reluctant to accept and use new technologies. In the first place, they usually have a very busy schedule, so they lack the time to be aware of the latest advances in technologies and how they could be used to reduce their workload [24]. They refuse to use new tools if they are not integrated smoothly into their daily workflow. They also often mention the lack of time and personnel to convert all the required medical data into an electronic format, so that it can be easily accessed and managed[12]. Some doctors also mention the "hype" built around Artificial Intelligence and, especially, expert systems, twenty years ago, which did not live up to their expectations, and they may reasonably argue that the "intelligent autonomous agent" paradigm, so fashionable today, may also fail to deliver real world results.

Despite the inherent challenges undertaken when developing a medical holarchy, the advantage of responding to the needs and requirements of today's healthcare system, especially to the need for ubiquitous access to healthcare services and ease of workflow management throughout the medical system, is worth the effort and associated risk. The major advantages that come with the adoption of Internet-enabled holarchies in medicine lay in the fulfillment of the following healthcare requirements [13]:

- ***Health service providers:***
— two-phase production of health services requiring the explicit (often: active) integration of the patient as so-called external factor of production,
— process management support,
— re-integration of small pieces of processes (process fractals) to complete processes
— activity and process coordination,
— identification, and elimination of institutional, organizational, media related breaks,
— thoroughly consistent, integrated, adequate and reliable information dissemination,
patient-centric knowledge processing

- ***Health service customers (patients):***
— health services are inherently individualized,
— demand for health services is inherently mobile (at work, at home, sports, leisure),
— mobility-related demand remains to be analyzed, quantified — and satisfied,
— single patients can easily be grouped by their diseases (or disease profiles) which makes it relatively easy to estimate the market volume.

---

[12] Medical records are usually hand written and distributed in different departments of a medical centre.

- ***Relevance of Internet-enabled technologies and mobile communication:***
— potential for high to very high numbers of users,
— in more or less all areas of private and work life,
— on all levels of public, and private health systems,
— relevant to all parts of healthcare value chains,
— key factor for customer contact / patient relationship management,
— extremely broad variety of mobile devices (which often require small bandwidths only), Fig. 6 provides for a lot of different mobile healthcare services, and
— significant relevance of B2B commerce practices for the whole healthcare sector.

In the sequel we will illustrate the impact that medical holarchies can have to improve diagnosis and prediction on a case study in glaucoma monitoring and diagnosis. In developing the system we worked closely with a renowned glaucoma specialist.

3. A CASE STUDY: REMOTE GLAUCOMA PROGRESSION MONITORING AND DIAGNOSIS

As a medical activity, diagnosis aims to determine if a patient suffers of a specific disease, and if the answer is yes, to provide a specific treatment [25]. In the particular case of glaucoma the main challenge for the ophthalmologist is not as much the diagnosis itself but rather the evaluation of the risk for its occurrence and the prediction of disease progression to establish a suitable follow up and treatment accordingly.

Glaucoma is a progressive eye disease that damages the optic nerve, usually associated with increased intraocular pressure (IOP). If left untreated, it can lead to blindness, being a leading cause of blindness. Glaucoma is affecting some 67 million people all over the world. In Canada there are about 200,000 glaucoma cases [3].

3.1. The Glaucoma Holarchy – Inter-enterprise level

A medical holarchy (inter-enterprise level) enabling progression monitoring of glaucoma patients is presented in Fig. 13. In the sequel we will illustrate on a simple example how soft computing can enhance the power of diagnosis and monitoring the progression of this fatal disease, working throughout the layers of the medical holarchy. At this highest level (Fig. 13) the mediator clusters all the resources involved in diagnosis, prediction and progression monitoring of the disease and manages the flow of information and interactions throughout the holarchy according to the particular need to be dealt with.

A typical scenario can be envisioned as follows. A patient complains of dizziness and headaches to the family doctor and is sent to an ophthalmologist (after the doctor determines that apparently there are no other causes that could explain the hoecake). The ophthalmologist diagnoses myopia and prescribes a set of glasses. After a few month the patient's vision further deteriorates considerably and the ophthalmologist (after administering a typical test) decides to send him to a glaucoma specialist. The specialist prescribes a certain treatment (sends the prescription to the pharmacy) and sets a follow up date after two month. When the patient returns (after having followed the treatment as prescribed) – the IOP test indicates high damage of the optic nerve and the specialist decides that surgery is the only way to stop progression of the disease. The patient enters the hospital (surgery department) immediately. In preparation for surgery a very high amount of information about the patient has to be gathered at short notice: medical history, related diseases and particular conditions (e.g. a history of strokes, etc.) that may influence the surgery outcome and/or require particular conditions for the surgery to be successfully performed. For this the patient's family doctor is contacted (who will provide the medical history) as well as the specialist monitoring the patient's special conditions.

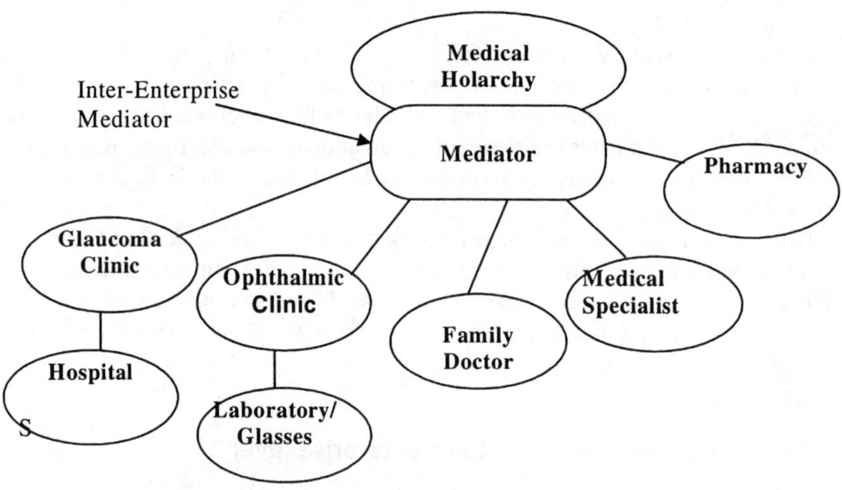

Fig. 13. Medical holarchy for progression monitoring of glaucoma patients

One can easily see from this scenario how all the medical holons in the holarchy (Fig. 13) collaborate to ensure a successful diagnosis and treatment. Even in the simplest case a lot of factors have to be considered when making a decision for surgery and the necessary information is difficult to find without strong informational support.

3.2. The Glaucoma Clinic Holarchy (Intra-enterprise level)

Zooming into the glaucoma clinic holon (intra-enterprise level, Fig. 14) one finds a collaborative holarchy specialized in progression monitoring of glaucoma. Table 7 maps the elements of this holarchy to the entities of the generic medical holarchy presented in Fig. 7/Table 1. To accomplish its goal each actor has to collaborate with other entities of the holarchy. Such collaborations define holons clustering the elements fulfilling these goals, as already presented in Tables 3 and 4 for a generic medical holarchy. In Table 7 each holon in the glaucoma holarchy consists of a set of collaborative agents and is viewed as an actor.

Table 7 The glaucoma holarhy

Actor (from Fig. 7)	Holon (from Fig. 14)
1. Diagnosis Producer	Physician, Prediction Expert System, Test Machines
2. Diagnosis Consumer	Patient, Pharmacy, Physician
3. Knowledge Provider	Physician, Patient, Expert System, Patient Database, Test Machines
6. Protocol Runner	Medical Assistant, Test Machines, Expert System, Physician

The patient (eventually via his personal assistant – a software agent) registers with the medical clinic to schedule a visit (including the preliminary tests that have to be taken) while providing preliminary information for the patient's file. (This information can be taken by contacting the mediator at the inter-enterprise level who coordinates gathering of the necessary data from the various entities – Fig. 13). During visit the medical doctor assesses the tests and the patient's medical history and checks the eye's condition in attempting to make a decision regarding the risk of damage progression. For this the medical specialist relies on his experience.

In about 70% of the cases the diagnosis of glaucoma is pretty evident for ophthalmologists. There are some cases however (at least one a day) where a specialist cannot determine the patient's condition (if the patient has glaucoma or not, if the disease will progress in a fatal way or not, etc.). To support the medical doctor in assessing such difficult cases we have developed an integrated diagnosis and prediction methodology [27] that uses several soft computing techniques embedded into an expert machine (Prediction Expert System in Fig. 14) processing patient data and test machine measurements all gathered into a database (fig. 15) as it will be further explained.

Such a system relies on a strong electronic data infrastructure supporting patient's records and collection of information from the expert machines. The outline of this infrastructure is presented in Fig. 16 and will be detailed in the next section together with the automatic diagnostic and prediction methodology embedded in the expert system.

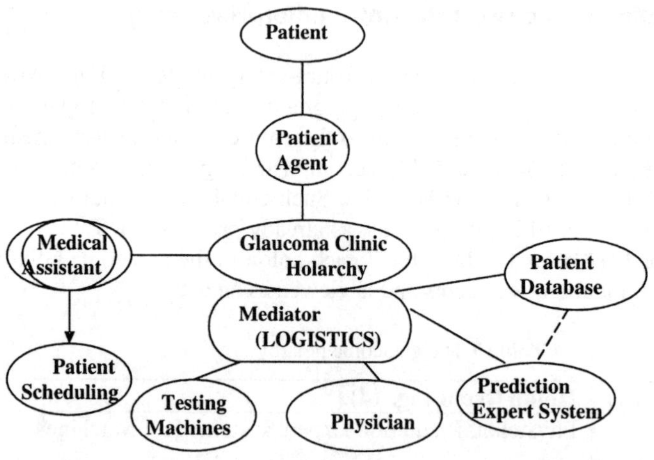

Fig. 14. Glaucoma Clinic Holarchy (intra-enterprise level).

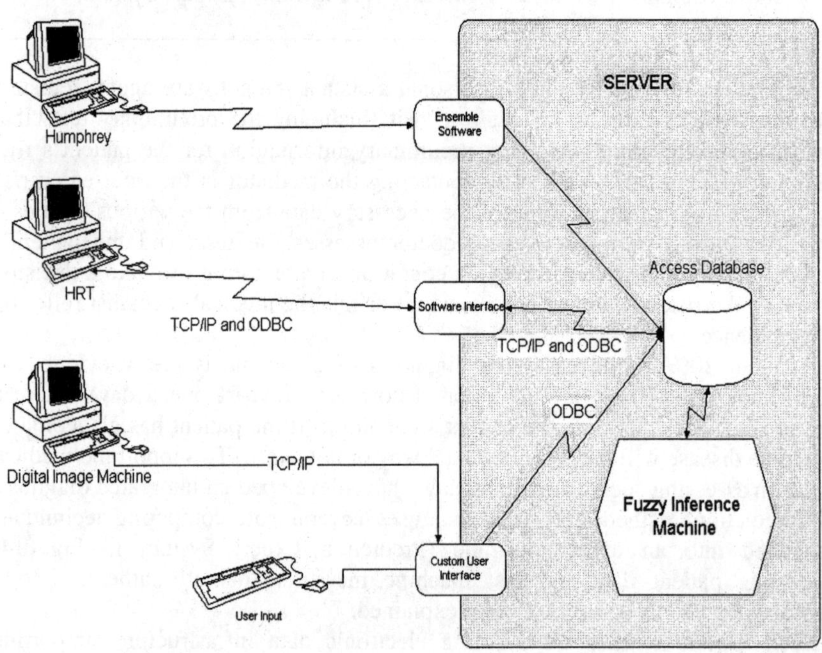

Fig. 15. Physical Resource (Machine) Level in the Glaucoma Clinic Holarchy

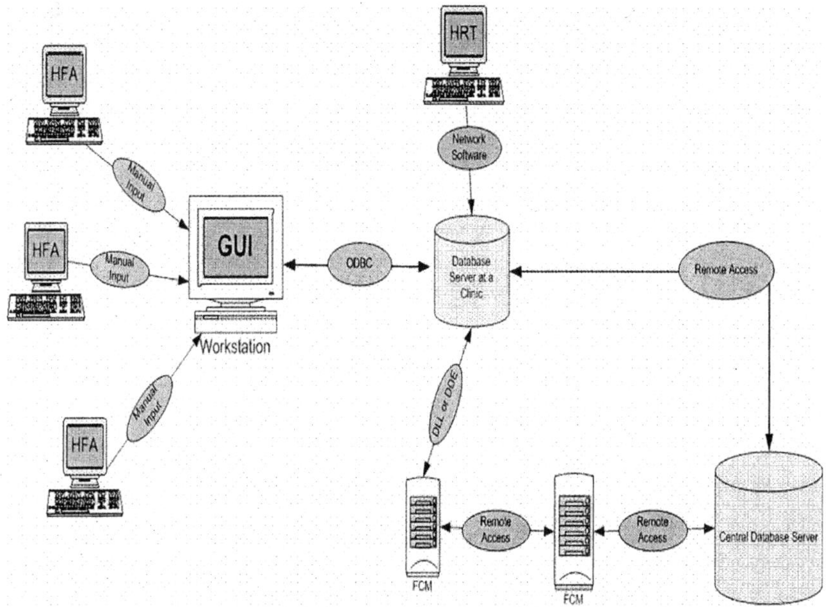

Fig. 16. Internet-enabled infrastructure supporting the soft computing diagnostic system

3.3. The Diagnosis Holarchy: Physical Resource Level

In this section we will zoom into the expert machine located at the physical resource level of the holarchy.

A standard investigation procedure [28] consists of several ophthalmic measurements done with various test machines measuring parameters relevant for the disease assessment, as presented in Fig. 15. The test data collected for each patient is stored in a database (on the same machine as the expert system, Fig. 15). Each new set of test parameters collected from the various ophthalmic devices) is run through the expert system, (eventually together with the medical expert opinion and subjective evaluations) and as result the 'assessment of the patent status' is obtained. The expert system has been designed around the software suite developed by Transfertech GmbH Germany [29], Fig. 17, by integrating several of their packages.

3.4. Internet-enabled soft computing methodology for glaucoma progression monitoring (Physical resource level)

The integrated diagnostic and prediction methodology (Fig. 17) aims on one side to emulate the assessment done by the expert physician, while at the same time it collects data relevant for predicting the disease progression.

To cover both these goals two fuzzy inference engines are run in parallel (Diagnosis Engine and respectively Prediction Engine in Fig. 17). The double function of the 'expert machine' is ensured by first running the diagnosis engine which emulates a standard diagnosis procedure made by the expert physician, and then running through the prediction engine the assessment obtained using the diagnosis engine together with the prescribed treatment and follow-up time interval determined by the medical doctor.

The output of the diagnostic engine evaluates the 'disease assessment' via the fuzzy variable *Risk* with three terms: **Low, Moderate** and **High**. The risk factors are mapped to the disease assessment via fuzzy if-then rules such as:

IF *Myopia* is High and *IOP* is High **THEN** *Risk* is High

IF *Hands/Feet Temp* is Cold and *Hypertension* is Present **THEN** *Risk* is Moderate

IF *Myopia* is High and *Age* is Old **THEN** *Risk* is High

IF *Hypertension* is Present and *Diabetes* is High **THEN** *Risk* is Moderate

For details regarding development of the fuzzy knowledge base see [24].

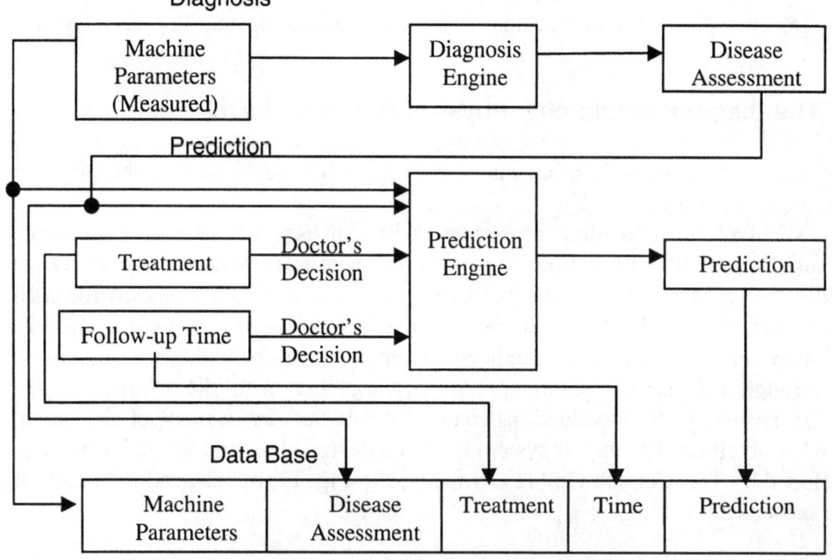

Fig. 17. Integrated Diagnostic and Prediction Methodology

For the prediction stage a continuous evaluation of the prediction rule base is done by assessing the accuracy of the previous prediction comparatively to the 'disease assessment' obtained at the follow-up (determined by running the prediction engine at the previous visit.) Based on the assumption that a database with sufficient patient information is already available[13] the Prediction Engine is developed in a three step learning process, Fig. 18.

1. Only once creation of CAM project

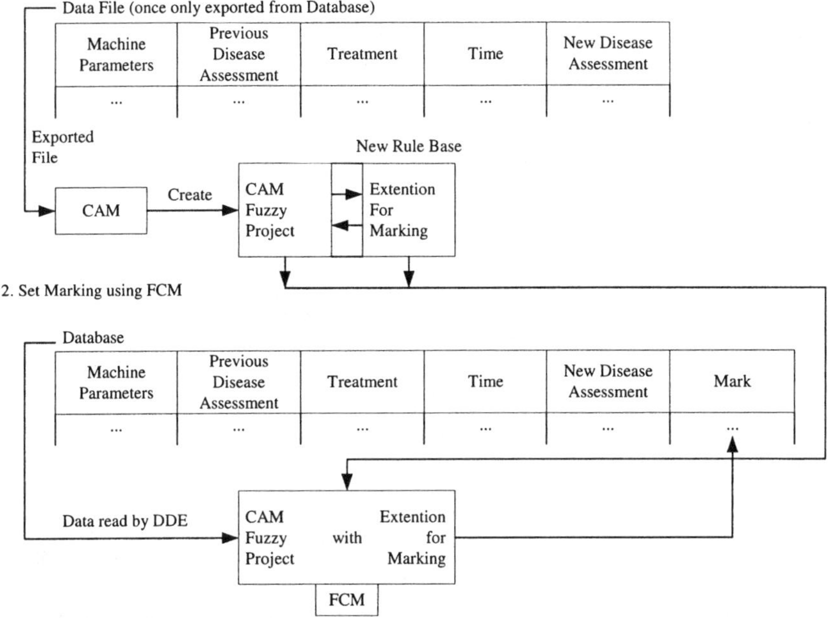

3. Learning Stage

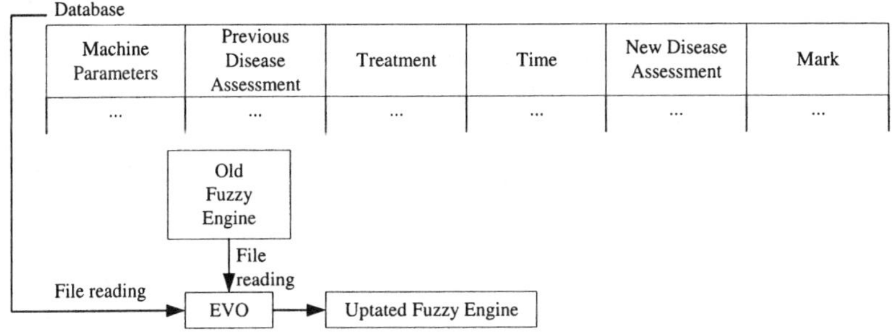

Fig. 18. Integrated Soft Computing Techniques for the Development of the Prediction Engine

---

[13] The design of the database itself was a challenging process, mainly because very few patient data was available in electronic form. We had to input most of the data from the patient files that were handwritten by the medical specialist. The database was organized by storing the measured parameters, the 'disease assessment' (made by the medical doctor at the investigation time during that particular patient visit) as well as the treatment and time interval decided by the medical expert for scheduling the next visit (patient's follow-up date) as well as the result of running the prediction engine [31].

The main idea is to compare the prediction made by the system at the previous visit with the assessment made at the current visit and depending on the result, that particular dataset is 'marked' in the database as either 'learning relevant' (if the predicted status was similar to the doctor evaluation), or as 'learning irrelevant' otherwise.

The marking of relevant data is made by using Clustering Analysis Manager (CAM) in combination with the Fuzzy Control Manager (FCM) developed by Transfertech, GmbH Germany [29], Fig. 18. If a 'basic' rule set is not available to support initial learning (by the Neuro-Com software[14]) for the diagnosis engine (which was our case in the beginning when all the information available was handwritten patient files) - then an initial step is required to create this basic fuzzy rule base.

For this (Step 1 in Fig. 18), first an initial database needs to be created in electronic form from patient files, containing information as previously explained. This data will be exported to the CAM[15] which will create based on it an initial fuzzy rule base (CAM-Fuzzy Project). Based on this initial fuzzy rule base a 'marking' rule base will be designed in the generic form:

IF {conditions} THEN {marking value}

where the {conditions} part contains the data fields to be evaluated for learning at Step 2 of the procedure in Fig. 18. Once this 'marking rule base' is determined the Fuzzy Control Manager can be run to determine the marking values for the database (Step 2 in Fig. 18)[16].

Now the database is prepared for the learning stage, which involves Transfertech's evolutionary optimizer (EVO), as illustrated in Step 3 of Fig. 18. The data marked as 'learning relevant' in the database will be processed by the EVO as follows:
- Set the optimization parameters (fitness function, etc.)
- Load the initial data (that is the rule base of the Prediction Engine in Fig. 17 – named 'old fuzzy engine' in Fig. 18, Step 3) and the learning data.
- The EVO (see footnote 14) will process this initial data and deliver a new fuzzy rule base containing the updated rules for the prediction (Updated Fuzzy Engine in Fig. 18).

---

[14] See: www.transfertech.de

[15] NOTE. An alternative to the use of the CAM to mark the learning data from the data base is the use of the Learning Data Generator (LDG) provided by Transfertech. This tool reads the input-output data set from a previous run of the diagnosis engine and provides a file containing the previous inputs together with the 'expected output' of the diagnosis engine. This would enable complete automation of the diagnosis procedure as there will be no more need for the 'doctor's decision' step in Fig. 17. However the disadvantage consists in this case in the requirement for more optimization rules to be 'crunched' by the EVO. The main advantage of using the LDG consists in that it enables evaluation of the accuracy of the prediction, giving very useful feedback for learning. We are in the process of experimenting with this tool.

[16] Due to nondisclosure agreements with Transfertech we cannot give more details regarding this procedure

As it results from Fig. 15 all the data involved in the diagnosis and prediction process is centralized in a database located on a server together with the fuzzy inference (diagnosis and respectively prediction) machines. This leads naturally to the idea of extending the learning capability of the diagnostic system by enabling data from several clinics to contribute to the knowledge refinement process, Fig. 16 [30].

The prediction system to be updated periodically using the EVO as explained in the previous Section is placed on a central server together with the central database (storing patient data from all the clinics involved in the diagnosis network). The central database will be updated periodically from each clinic. A copy of the diagnosis and prediction engines will function in each clinic and will be updated after the learning process is done on the central 'master' copy. This will be done at regular intervals, ranging between one week and one month, depending on the amount of new learning data marked from the patient examinations submitted by each clinic. To connect the local engines from each clinic to the 'master' engines located on the central server a secure and reliable connection has to be established, as patient data is sensitive.

It is remarkable how the Internet enables integration of several soft computing techniques and information systems for the refinement of the knowledge base of an expert system for diagnosis and prediction (Fig. 16). In the next Section we will emphasize on the other side how soft computing enables the power of the internet as infrastructure for the emergence of a medical holarchy that responds to an emergency case in glaucoma diagnosis.

3.5. Soft Computing for the Internet: Emergence of a Glaucoma Holarchy

In this Section we present (on a simple example) how a medical holarchy backed by the Internet in achieving the self-organizing, evolutionary properties presented in Section 1, can enable collaboration between the medical entities involved through a consistent workflow management across all levels, as well as by the remote ubiquitous access of the holarchy.

Suppose a glaucoma patient has an accident while on travel in a foreign country. From the PDA or cellular phone (Fig. 6) the intensive care doctor sets the goal for the holarchy via his personal assistant (interface) agent who initiates the request for creation of a medical holarchy serving the particular patient. The patient's personal assistant is immediately alerted and contacts the mediator associated with the patient's medical history (Fig. 13). All the hospitals/clinics associated with the patient are contacted and the workflow manager (mediator) structures the holarchy in a manner that enables access to the most relevant information related to patient's condition determined by the accident (provided by the personal assistant of the emergency doctor). Remote and immediate access to the relevant medical files and data is achieved by the workflow coordination and agent interactions throughout the holarchy. Suppose the medical specialist that usually treats the patient is on holidays. In this case the evolutionary search strategy embedded within the mediator will enable discovery of the next best specialist who will be contacted for advice by the mediator. The medical specialist will receive all the information (files, images, machine tests, etc.) needed to make a quick assessment of the situation. By taking into account the patient's multifactorial complex condition (available thanks to the enabled access to information within the holarchy) the specialist will be in a position to suggest immediate action and as well to receive feedback from the expert system (Fig. 14) and/or other specialists contacted by his personal assistant (agent) [14].

Now suppose that the patient is in stable condition but needs strict continuous monitoring to avoid any risk of recession. The holarchy (Fig. 14) will enable the patient to leave the hospital while being under continuous contact wit the specialist to report any abnormal condition that may develop. Pharmacies (Fig. 13) will as well be remotely contacted by the medical specialist so the prescription drugs can be provided on immediate need. The patient will be able to access the expert system to obtain on one side assessment of her/his condition as the symptoms will start to develop and on the other side a prognosis of the evolution of his/her condition that will enable a decision regarding possibility to travel back home or not.

4. CONCLUSIONS

Based on recently reached theoretical results inducing emergence in virtual organizations we proposed a methodology for the design and implementation of medical holarchies linking distributed medical entities into a collaborative environment. Despite the challenges in implementing and deploying medical holarchies the advantages of ubiquitous healthcare enabled by our holarchic model for the soft computing endowed Cyberspace are to be acknowledged not only in the particular case of glaucoma diagnosis presented here.

The strength of our approach as it relates to medicine consists in its dual character emerging from the synergetic interaction between soft computing techniques and the Internet. On one side soft computing is an enabler for the Internet by endowing it with self-organizing, evolutionary properties similar to those of living systems. From this perspective, r regarded as holonic enterprise medical holarchy emerges in Cyberspace around patents' need clustering the resources available to optimally fulfill it. We have illustrated on a glaucoma progression monitoring example how emergent medical holarchies enable patient information retrieval and (inter/intra)-hospital workflow management, disease progression tracking/monitoring and remote access to patient information for multifactorial decision making in case of emergency.

On the other side we have emphasized how the Internet itself can be an enabler for the power of soft computing by supporting integration of several soft computing techniques as well as the creation f databases that enable the soft computing technologies to work optimally.

Our current work is directed towards the extension of these results to other e-Health and ubiquitous healthcare applications (such as: tracking and monitoring of diabetes and/or cancer patients, assistance of elderly patients, etc.) as well as to technical problems such as fault tracking in circuit board manufacturing (jabil.com) and remote diagnosis of gas turbines.

ACKNOWLEDGEMENTS

The author gratefully acknowledges the contribution of the Glaucoma Team at the University of Calgary – the most inspirational example of collaborative holarchy supporting this work. I especially acknowledge the work of Adam Geras, MSc student – for the successful attempt to create a holonic model for telemedicine while implementing our crazy ideas into excellent code. The enormous efforts of Dr. Nicolae Varachiu, Glaucoma Project Manager, in holding together our holarchic Team through the glue of inspirational guidance, professional experience and priceless encouragement – are highly appreciated. Especially his successful attempt to build the essential link that connects the medical and software engineering perspectives within this challenging interdisciplinary environment deserves our highest gratitude. The excellent work done by Cynthia Karanicolas, MSc student – in building the diagnosis and prediction knowledge bases is highly praised. Her creativity, devotion to this work and endurance in facing insurmountable difficulties were the bricks from which each expert rule of our world class expert machine was built.

This work is funded by the National Science and Engineering Research Council of Canada, under NSERC – Collaborative Health Research Project Grant for which Dr. Ulieru is principal investigator.

REFERENCES

1. McHugh P, Wheeler W, Merli, G. (1995) Beyond Business Process Reengineering: Towards the Holonic Enterprise, John Wiley & Sons, Inc., Toronto.
2. Ulieru M (2002) Emergence of Holonic Enterprises from Multi-Agent Systems: A Fuzzy-Evolutionary Approach, Invited Chapter in: Loia V (Ed) Soft Computing Agents, IOS Press (in print).
3. Ulieru M, Brennan R, Walker S (2002), The Holonic Enterprise – A Model for Internet-Enabled Global Supply Chain and Workflow Management. In: International Journal of Integrated Manufacturing Systems, No 13/8, ISSN 0957-6061.
4. Koestler A (1967) The Ghost In The Machine, Arkana Press.
5. Maturana F, Norrie D H (1996), Multi-agent mediator architecture for distributed manufacturing. In: Journal of Intelligent Manufacturing, Vol 7, pp. 257-270.
6. Ulieru M, Ramakhrishnan S (1999) An Approach to the Modelling of Multi-Agent Systems as Fuzzy Dynamical Systems", Advances in Artificial Intelligence and

Engineering Cybernetics, Vol. V: Multi-Agent Systems/Space-Time Logic/Neural Networks (George Lasker, Ed.), IIAS-68-99, ISBN 0921836619

7. Dubois, D. and H. Prade (1988), Possibility Theory, Plenum Press NY
8. Klir G, Folger T (1988) Fuzzy sets, Uncertainty, and Information, Prentice Hall,
9. Zimmermann H-J (1991) Fuzzy Set Theory And Its Applications, Kluwer Academic
10. Ulieru M, Cobzaru M, Norrie D. (2001) A FIPA-OS Based Multi-Agent Architecture for Global Supply-Chain Applications. In: Proceedings of IPMM 2001 International Conference on Intelligent Processing and Manufacturing of Materials, July 29-August 3, 2001, Vancouver, BC.
11. Ulieru M, Norrie D, Kremer R, Weiming Shen (2000) A Multi-Resolution Collaborative Architecture for web-Centric Global Manufacturing. In: Information Science, volume 127, Journal no. 7669, ISSN # 0020-0255
12. Fogel D (1998) Evolutionary Computation: The Fossil Record, IEEE Press ISBN: 0780334817
13. Kirn, S (2002) Ubiquitous Healthcare: The OnkoNet Mobile Agents Architecture, In: Proceedings of Net-Object Days (NODe) 2002, October 7-10, 2002, Erfurt, Germany.
14. Heidin J, Bergquist M, Gater H, et al. (2002) Implementation of Feedback – an application of quality assurance, learning and e-Communication of diagnosis of Medical Images. In: Proceedings of SSGRR2002s – International Conference on Electronic Infrastructures for e-Business, e-Education, e-Science and e-Medicine, L'Acquila, Italy, July 29-August 4, 2002
15. Nealon J L, Moreno A (2002) The Application of Agent Technology to Healthcare. In: Proceedings of the First Agentcities Workshop: Challenges in Implementing Open Environments, AAMAS 2002, July 14-18, Bologna, Italy.
16. Geras A (2002) A generic Architecture for Medical Diagnosis Systems, Final Report for the CSCW Course Project (advisor Dr. Ulieru), Electrical and Computer Engineering Department, The University of Calgary
17. Shankararaman T P V, Ambrosiadou V, Robinson B (2000) Agents in health care. In: V. Shankararaman T P V (Ed) Workshop on Autonomous Agents in Health Care, pp 1–11
18. Cort´es U, L´opez-Navidad A, V´azquez-Salceda J, V´azquez A, Busquets D,. Nicol´as M, Lopes S, V´azquez F, Caballero F, Carrel (2000) An agent mediated institution for the exchange of human tissues among hospitals for transplantation. In: 3, Congr´es Catal`and'Intel.ligencia Artificial ACIA, pp 15–22
19. ASTM E31 - http://www.astm.org/COMMIT/COMMITTEE/E31.htm
20. HL7 (Health Level 7) - http://hl7.org
21. OpenGALEN: http://www.opengalen.org
22. UMLS: http://www.nlm.nih.gov/research/umls
23. SNOMED: http://www.snomed.org
24. Varachiu N, Karanicolas C, Ulieru M (2002) Computational Intelligence for Medical Knowledge Acquisition with Application to Glaucoma. In: Proceedings of the First IEEE Conference on Cognitive Informatics (ICCI'02), Calgary, Canada, August 17-19, 2002, IEEE Computer Society Order Number PR01724, ISBN 0-7695-1724-2, Library of Congress # 2002107061, pp 233-238,.
25. Marr D (1982) Vision, W. H. Freeman and Co., San Francisco
26. G. E. Trope (2001) Glaucoma: A Patient's Guide to the Disease, Univ. of Toronto Pr.

27. Ulieru M, Pogrzeba G, (2002) Integrated Soft Computing Methodology for Diagnosis and Prediction with Application to Glaucoma Risk Evaluation In: Proceedings of 6th

IASTED International Conference on Artificial Intelligence and Soft Computing, July 17-19, 2002, Banff, Canada

28. Kanski J J, McAllister J A (1989) Glaucoma: A Coulour Manual of Diagnosis and Treatment, (Butterworths, London, Boston, Singapore, Sydney, Toronto, Wellington,

29. Varachiu N (1999) A Fuzzy Shapes Characterization for Robotics. In: Reusch B (Ed) Lecture Notes in Computer Science, vol.1625, Computational Intelligence - Theory and Applications, Springer-Verlag, pp. 253-258

30. Ulieru M, Geras A (2002) Emergent Holarchies for e-Health Applications: A Case in Glaucoma Diagnosis. In: Proceedings of IECON 2002 – 28th Annual Conference of the IEEE Industrial Electronics Society, November 5-8, 2002, Sevilla, Spain (accepted).

31. Nagarajappa N, (2002) Uploading Data from the Local to the Centralized Database for Glaucoma Diagnosis and Prediction – Internal Report, Electrical and Computer Engineering, The University of Calgary

Searching and Smushing on the Semantic Web - Challenges for Soft Computing

T. P. Martin

University of Bristol, BS8 1TR, UK
Trevor.Martin@bristol.ac.uk

Abstract: The World Wide Web is an astonishing information repository, founded on the simple principles that any web resource can link to any other web resource and that as little as possible should be centrally regulated and imposed. In practice, access to the right information on the web is hampered by the sheer volume of data. Tim Berners-Lee, widely acknowledged as the "father of the web" has pointed out that this is mainly due to the data being machine-readable but not machine-understandable, and has proposed an extension to the current web, known as the semantic web. This allows relational knowledge to be encoded in web pages, enabling machines to use inference rules in retrieving and manipulating data. In turn, this will reduce the quantity of irrelevant data retrieved and increase the usefulness of the web. In order to facilitate the interface between human and machine understanding of data, there is a need to incorporate uncertainty into the knowledge representation and greater flexibility into matching and inference. This is in line with current practice - uncertainty is implicitly handled by most search engines that return a list of pages ranked by their "degree of relevance" to the query.

The need for web pages to include knowledge representation presents a tremendous opportunity for fuzzy researchers. In this paper we outline some knowledge representation issues involved and focus on the process of fuzzy matching within graph structures. In particular, we highlight a process known as "smushing" which allows syntactically different nodes to be combined, so that information drawn from multiple sources may be fused.

Introduction

Large quantities of data are stored in computer-based systems. Such data has arisen from the use of relational databases and from the explosive growth of the world wide web in recent years. Frequently, data is collected and then under-utilised - in itself, the data is useless unless it can be queried, retrieved and explored in order to realise its value. Questions of data storage and retrieval are fundamental to the information revolution - and the way in which information is stored has a large bearing on the ways in which it can be accessed.

The field of intelligent information involves

☐ sophisticated mechanisms for searching, extracting and presenting information
☐ metadata techniques in which a rich array of 'data about the data' is stored
☐ the ability to configure data to different users with different needs and
☐ the ability to make inferences from data when appropriate.

The semantic web [12, 15] offers a tremendous opportunity for fuzzy researchers to demonstrate the power of soft inference in making useful human-understandable deductions from the semi-structured and sometimes contradictory information available on the web. The semantic web aims to convert web content from a *machine-readable* form to a *machine-understandable* form without sacrificing human understandability. It is based on subject-relation-object style assertions, with additional layers to enforce semantic constraints and allow inference. The relational knowledge encoded in web pages can be based on any representation which gives the required expressive power, such as RDF, N3 or conceptual graphs [13, 14].

Conventional search techniques involve keyword-based approaches derived from information retrieval, with a second generation of search engines utilising additional information such as the link structure of the web to indicate page importance. Finding the "right" page is often a major problem - for example, searching for "orange" will retrieve pages concerned with Orange County, the mobile phone operator and defoliant chemicals. An additional problem with keyword-based approaches is the possibility of misleading information - particularly in web searching where pages can contain hidden text which is not displayed by the browser. This hidden text can be used to improve a page's rating in keyword searches, say by an unscrupulous advertiser.

It is difficult to combine multiple pages as different information sources are typically not homogeneous – that is, they may use different vocabularies or different interpretations of terms. Information integration systems attempt to provide "one point access" for querying, but must take account of the fact that semi-structured and unstructured data is largely free of formal semantics and requires human interpretation to make sense of it.

A further problem in finding the right information arises from lack of access to the "deep" (or "invisible") web, i.e. information that is stored in databases. This information can be extracted by query but is not explicit to a browser and is hence not readily indexed. Such systems may be updated frequently e.g. a news feed.

The semantic web allows machines to use ontologies and inference rules in retrieving and manipulating data. In particular, this enables information from a class hierarchy to be used in judging whether a page is relevant to a query.

For example, Google searches for "pub bristol lager" and "pub bristol beer" returned no common entries in the first twenty sites listed – by incorporating the knowledge that "lager is a type of beer" we can see that these queries should be virtually synonymous. Such knowledge is available online (see www.cs.umd.edu/projects/plus/DAML/onts/beer1.0. daml). Although there is promise in the semantic web approach, the lessons of AI over the past forty years show that the required level of knowledge representation and reasoning is far from trivial. The increasing use of XML for data interchange offers another path towards more intelligent search. XML is less expressive than RDF as a representation for metadata, but is more widespread and appears to be a *de facto* standard. It is possible for a human to deduce something about the nature of the data if the DTD is available – however, this relies on a great deal of background information triggered by the selection of appropriate tag names.

This paper outlines existing techniques for retrieval and representation, and points to some areas in which fuzzy methods could profitably be applied. We argue that there is a potential mis-match between the crisp categories required for logic and the imprecise, flexible terms used by humans. This has been noted by several researchers [16, 19, 20] and formed the basis for some preliminary ideas [31], which are further expanded here. We show how uncertainty in knowledge representation can be combined with conceptual graphs to implement the kind of flexible matching processes needed to bring intelligent reasoning to the semantic web.

Information Retrieval and Searching

Classical techniques for extracting data depend on the form of the stored data. At one extreme is the relational database which requires complete and well-known data, with a rigidly specified schema and a formal, highly structured querying mechanism. In return for this rigidity and adherence to structure, guarantees can be made about the semantics, integrity and consistency of the data (see for example [30, 37]). Data extraction, conversion, transformation, and integration are all well-understood database problems, with clear theoretical underpinnings. As soon as data moves away from the confines of this completely known and specified format, problems arise. For example, null values are necessary when data is unknown or irregular, leading to well-known problems of interpretation.

At the opposite extreme, we have completely unstructured data which is largely free of formal semantics and generally requires human interpretation to make sense of the data. The process of extracting and searching for data within free text documents has given rise to the field of information retrieval - see [1] or [38] for a good introduction. Searching is generally on the basis of weighted keywords [33], which can be generated automatically from a text corpus. Variations of this model include probabilistic retrieval, fuzzification of the vector model, neural nets, Bayesian nets, etc - see chapter 2 of [1] for a more complete discussion of classical information retrieval methods, and some recent extensions.

Falling in the middle of the spectrum, overlapping with both categories, we have semi-structured data such as classified directories, product catalogues, help systems, and much of the World Wide Web. In semi-structured data there are compulsory components, for example a business name and telephone number in a classified directory, and there are optional, less-structured components such as a description of products and services offered. Starting from the standard database approach, it is necessary to make substantial changes to accommodate semi-structured data. In order to allow flexibility in data, schema and querying, many of the rigid requirements must be relaxed, although the fundamental relational database operations are retained. An example of such an extended database system is LORE (Lightweight Object REpository), a DBMS designed specifically for semi-structured information [32] which has been extended to store XML-based data [24] .

More work has been carried out on extending information retrieval concepts to semi-structured data, particularly with the proliferation of the world wide web. Typically, the structured elements are ignored and keyword-based free-text search is used.

Extensions

Although many intranet search engines and document retrieval systems use the vector representation with the TF-IDF technique, additional refinement can be achieved by clustering documents in some way. The most obvious (and time consuming) approach to clustering is to use manual classification of documents. For example, Yahoo (www.yahoo.com) uses a hierarchical labelling to assign each web page to a particular category. Documents can then be retrieved by navigation through the category hierarchy as well as by keyword search.

Documents can be clustered automatically using the keyword vectors extended by synonyms, stemmed variations of the keywords and terms which are frequently close to the keyword in the text.

An interesting approach to clustering is provided by Kohonen [29] in which the vector representation of a document is based on trigram frequencies i.e. the number of occurrences of each triple of words in the document. The standard Kohonen net approach is used to reduce the dimensionality of the document vectors, yielding a 2 or 3 dimensional map. A query can be classified in a similar manner and retrieved documents should be close to it on the map.

This approach is computationally intensive, and difficult to update incrementally. However, it can be argued that Kohonen clusters are based on concepts rather than simple keywords, a claim that can also be made for some of the more advanced search-engine products such as Autonomy (www.kenjin.com)

An alternative to pure keyword search, exploited by the Google search engine, is to use the link structure of the web as a guide to the quality of pages [17]. Any web page which has a large number of links pointing to it is assumed to be important, and is allocated a high rating. This score is propagated through the link structure, as it is assumed that any "important" page (as indicated by a high rating) will contain links to other important pages; thus links to a particular page are weighted by the score calculated for the source of the link. By iterating this process, a *PageRank* is calculated for each page as the importance is propagated through the network of links. Formally, if we treat the web as a graph with pages as vertices and hyperlinks as edges, then the *PageRank* corresponds to the principal eigenvector of the normalised adjacency matrix.

Performance Evaluation

The relational database approach does not lend itself to any flexibility when evaluating query results - if the system does not return all of the correct answers (and only the correct answers) then it is logically inconsistent and needs to be modified. Information retrieval, on the other hand, recognises that in many cases the set of relevant answers is inherently ill-defined and that there is a degree of dependence on the query. Nevertheless, formal measures have been proposed to quantify how well a retrieval system is working. The most common measures are *precision* and *recall*, defined by partitioning the set of documents into

☐ those which are relevant to the query and those which are not relevant
☐ those which are retrieved and those which are not

In an ideal querying system these two partitions would coincide, i.e. all relevant documents would be retrieved, and no irrelevant documents would be retrieved. In practice, this is not the case.

Precision measures the proportion of retrieved documents that are relevant
Recall measures the proportion of relevant documents that are retrieved

These measures need to be viewed with some caution - we can obtain maximum recall by retrieving all documents, and retrieving no documents at all has infinite precision. In spite of such limitations, these measures are often used to rank the performance of different approaches to building queries, retrieving documents, document relevance, etc. Sets of known relevant documents are established, and average system performance against these sets can be assessed.(e.g. [39]). We note in passing that it is difficult to apply these measures reliably to the question of web retrieval, since the set of all "relevant" documents is frequently not known.

Metadata

There is a widely-recognised problem with keyword-based web searches which may return hundreds (or even thousands) of documents - on a good day the "relevant" documents may be in the first ten or twenty returned, but it is likely that a lot of relevant documents are missed. The scale of this problem is increasing daily as the number of web pages increases.

To save time and space, many search engines do not extract keywords from the entire page, but simply use HTML tags to focus on the important data. Searching for text marked up with <title> and heading <h1>, <h2>, ... tags is assumed to give sufficient information to index a page. In addition, the optional use of <meta> tags enables a list of keywords, author, creation date etc to be specified for use in indexing.

```
<p>
Department of Engineering Mathematics
<br>
  University of Bristol<br>
  Queens Building<br>
  University Walk<br>
  Bristol BS8 1TR <br>
  UK </p>
<p>Tel. +44 117 928 8200</p>
<p>Fax. +44 117 925 1154</p>
```

```
<ADDRESS>
 <DEPT> Department of Engineering Mathematics </DEPT>
 <ORG> University of Bristol </ORG>
 <BUILDING>Queens Building </BUILDING>
 <STREET> University Walk </STREET>
 <CITY> Bristol </CITY>
 <POSTCODE> BS8 1TR </POSTCODE>
 <COUNTRY> UK </COUNTRY>
 <PHONE> +44 117 928 8200 </PHONE >
 <FAX> +44 117 925 1154 </FAX>
</ADDRESS>
```

Fig. 1   HTML and XML markup of the same data

However, most of this structure is concerned with *how* to display the document rather than bearing any relation to its content. Although there are guideline on the use of <*meta*> tags, there is a danger that an unscrupulous web author could promote a site by improper use of such tags.

An additional problem, even when searching a limited number of sites such as a large corporate intranet, is the difficulty of extracting relevant information from a system which is being continuously updated - for example, a football fan might be interested only in news updates about one particular team.

A possible solution to this is the notion of "push" technologies, in which a simple user profile is used as a filter to screen out unwanted material and deliver only relevant information to the user. (see [18] or www.infogate.com for example). This requires mechanisms for determining the content of a page, representing the interests of a user, and judging how well the two match for any given page.

XML and RDF

In order to improve performance, we require knowledge about the content of a page, i.e. metadata. XML was created as a method of marking up documents and conveying the semantics - in the same way as HTML is used to specify the display properties of different pieces of text, XML can be used to convey a hierarchical relationship of data values and to indicate the significance of different components in the data. XML allows new tags to be defined, with a BNF grammar to determine the precise combinations of tags that are permitted. XML is becoming a de facto standard for electronic data interchange. An example is shown in Fig 1.

On its own, XML is not necessarily any more informative than HTML. The tags are arbitrary and can be defined by any creator of web pages, so that to a computer there is little difference between `<p>` ... `</p>` and `<address>` ... `</address>`. XML is an extendible syntax for document description, whereas RDF defines an extendible structure for expressing the semantics of document description [21, 40] by defining resources in terms of values of various properties. For example, the Dublin Core metadata standard defines a list of properties that can be used to describe web resources. These include obvious properties such as creator, title, format, and less well defined properties such as relation, description, rights.

The example in Fig. 2 () shows the use of some properties:

```
Title="Candle in the Wind"
Subject="Diana, Princess of Wales"
Date="1997"
Creator="John, Elton"
Type="sound"
Description="Tribute to a dead princess"
Relation="IsVersionOf Elton John's 1976
song Candle in the Wind"
```

Fig. 2 RDF example (taken from http:// dublincore.org / documents / 2001 / 04 / 12 / usageguide / generic.shtml)

Clearly it is impossible for a single, centrally defined resource to encompass all possible metadata schemas. Additionally, the very nature of the web - its ability to freely exchange information - militates against the imposition of standards other than those essential to ensure communication. Top-down standardisation would be unlikely to succeed especially when the diversity of domains, languages and cultures is considered. The extensible nature of RDF means that different metadata schemas can be established by different groups. As long as a definition is made available, the semantics of the data can be accessed by any interested party. This is in accordance with the task of the working group on RDF, i.e. "to specify semantics for data based on XML in a standardised interoperable manner". We can expect metadata from different sources to have some common elements and some unique elements; it is also possible that different sources may use different structures to refer to essentially the same information.

Additional difficulties may arise where the property values are provided in free text form by the author - one author's description of the content may not accurately reflect all information contained in the document.

To summarise, RDF is a model of metadata - corresponding to the semantics - whereas XML is the representation language, i.e. the syntactic level.

Inference and Ontologies

RDF is essentially a simple knowledge representation system, and can be used as the basis of reasoning. Much work has been done within the AI community in applying formal logic methods to inference for the semantic web, including first and higher order logics, description logics, etc (see www.semanticweb.org / inference.html). Many AI approaches assume a centralised definition for knowledge representation, with great care taken to make the language sufficiently expressive and unambiguous. Problems may occur when combining knowledge from separate sources, although much work has been carried out in portability of ontologies. (see [23] for example) The aim is summarised by Gruber [25]:

"To support the sharing and reuse of formally represented knowledge among AI systems, it is useful to define the common vocabulary in which shared knowledge is represented."

The purpose of an ontology is to capture the semantics of an information source, that is to define the terms and relationships between them. An ontology will normally contain named classes, their attributes and constraints and relations between them. Formally, an ontology is "a specification of a conceptualisation" which is expressed as a logical theory (see http://www-ksl.stanford.edu/kst/what-is-an-ontology.html). Translation between ontologies enables different information sources to be fused, i.e. ontologies can aid translation between representations and reasoning about the equivalence of values. An ontology essentially consists of a taxonomy of concepts, relations between concepts, and axioms (rules) which impose constraints and allow transformations of data. The idea of an ontology is central to the semantic web, although there can be a very high cost in ontology creation and maintenance. A number of environments have been proposed to partially automate this task [26, 36].

We can expect metadata from different sources to have some common elements and some unique elements; also, different sources may use different structure to refer to essentially the same information. Where these differences neatly follow set-theoretic relations (equivalence, subsethood), pure logic is adequate for reasoning. For example, one library schema might use the term "author", whereas another uses "creator", one supermarket schema might use "apples", "oranges", "bananas" etc where another uses "fruit".

In practice, when creating an ontology it is easy to become enmeshed in esoteric philosophical questions of exact definitions for concepts, precise limits on expressive power, etc. In part, this may arise from the insistence on crisp classification even when a very limited domain is consid-

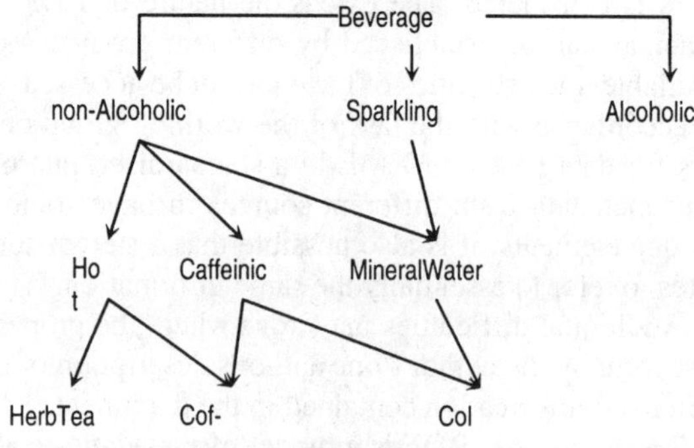

Fig. 3   fragment of a classification hierarchy

ered. For example, Fig 3 shows a fragment of a classification hierarchy taken from [35] (see also http:// users.bestweb.net /~sowa/ontology/toplevel.htm, also [22])

Clearly this should not be interpreted as a crisp hierarchy. Different blends of coffee contain caffeine to a greater or lesser degree; adding leaf tea to the hierarchy would introduce a category that contained caffeine but to a lower degree than coffee, etc.

The Fril++ approach, outlined below, allows us to build hierarchies in which partial memberships can be accommodated.

Opportunities for Soft Computing

At first sight, the semantic web is tailor made for the techniques of soft computing. The aim is to bridge the gap between machine-readable "hard logic" and human-understandable natural language (possibly in some restricted form). None of the usual logical requirements can be guaranteed - there is no centrally defined unique format for data, no guarantee of truth for assertions made, no guarantee of consistency, etc.

As is well known in the fuzzy community, natural language and crisp logic-based representations frequently do not mesh well because of the inherent vagueness used in many natural language terms. Creating a type hierarchy requires analysis of the problem domain into distinct classes, with a well-defined hierarchical structure. Each class should correspond to a set of objects in the "real world", and behave accordingly. Of course, the notion of real world is dependent on the application - the real world of a graphical user interface, for example, is the display on the screen. A window on the screen either belongs to the system, or to an application program, and it is either a document window, a modeless dialog, etc. However, in other applications, the decomposition into classes may not be so clear.

A class is a (crisp) set of objects which can be distinguished by particular properties from all other objects within a system. Clearly there is a potential mis-match between the crisp categories required for logic and the imprecise, flexible terms used by humans. Let us consider the classification of weather events. Although everyone understands the concept of "raining for a few minutes", very few could draw a precise boundary between this and (say) "raining for several minutes" or "drizzling for a few minutes". Online ontologies [http://opencyc.sourceforge.net/daml/cyc.daml and mnemosyne.umd.edu/~aelkiss/weather-ont.daml] provide crisp definitions as follows:

a few minutes duration	*2-10 minutes.*
rain (a LiquidPrecipitation Event)	*Precipitation, either in the form of drops larger than 0.02 inch (0.5 mm), or smaller drops which, in contrast to drizzle, are widely separated.*
drizzle(a LiquidPrecipi-	*Fairly uniform precipitation composed exclusively of*

The human understanding is less precise. Such divisions are obvious cases for a fuzzy approach - we have linguistic labels for categories (rain, drizzle, a few minutes) and instances (weather events) which belong to the categories to a greater or lesser extent. This form of uncertainty cannot be ignored if we are to fulfil the goal of making the knowledge as understandable as possible to humans.

However there are dangers in blindly proposing fuzzy logic as the solution to all problems on the semantic web. What is needed is not just a simple fuzzification of numerical terms - although this may well be a useful aid to certain search problems, given a commonly understood definition of membership. The fundamental difficulty is in providing an unambiguous (and preferably objective) definition of what is meant by the membership degrees. Without this, there is little prospect of interoperability (although it should be pointed out that search engines typically use numbers between 0 and 1, or 0-100, to rank retrieved pages without defining what is meant by the numbers, other than the upper end point corresponds to definitely relevant and the lower to definitely not). Too often, the interpretation of membership degree is side-stepped by appeal to the subjective opinion of a user, or as a context-dependent parameter whose interpretation is best decided by the system designer.

Baldwin's mass assignment theory [2] relates probabilistic and fuzzy uncertainty, giving a "voting model" interpretation which allows membership degrees to be obtained by reference to a population. By relating membership values to probabilistic and statistical concepts, the theory gives an unambiguous semantics to fuzzy sets.

Once the common semantics of fuzzy sets is adopted, the real need is for fuzzification and partial matching of concepts. It is relatively easy to search within a database for a restaurant which is "near" to one's current location, or which has a menu on which "most" items are "reasonably priced". It is much more difficult to search for a restaurant with a "high quality wine list" or which has "a good range of spicy vegetarian main courses".

Additionally, we must bear in mind the available computing power and the need for practical implementations of these systems. In order to prove the usefulness of fuzzy logic, examples are needed which either can't be done at all without fuzzy, or which can be done much more efficiently. It is essential that fuzzy extensions can demonstrate performance improvements over "classical" techniques.

Knowledge Representation using Conceptual Graphs and Fril++

Conceptual graphs [34] are a powerful and general form of graphical knowledge representation. Their similarity to RDF has been noted by several researchers, including Berners-Lee [13].

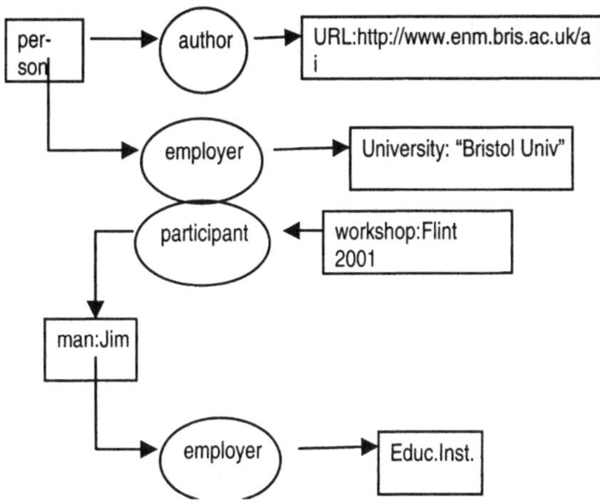

Fig 4   examples of conceptual graphs

A conceptual graph is a finite, connected bipartite graph representing a proposition. Two examples are shown in Fig 4, corresponding to the propositions "a person employed by the University of Bristol is author of the web page http://www.enm.bris.ac.uk/ai" and "a man called Jim, employed by an educational institute, is a participant at the FLINT2001 workshop".

Rectangular nodes are *concept* nodes, and consist of a *type* e.g. person, man, university.... and an optional referent, labelling an individual conforming to that type. A concept node may optionally contain a *referent* such as Jim, Flint 2001, etc in the example above, identifying a particular example of the type. The set of possible referents is derived from the universe U.

Several different forms are possible for the referent:

Kind of referent	Notation	Example
Generic (existential)	[DOG]	a dog
Universal	[DOG:]	all dogs
Individual	[DOG #12]	the dog
Named individual	[DOG:Jip]	Jip the dog
Measure	[HEIGHT:@6ft]	a height of 6ft
Generic Set	[DOG:{*}]	dogs
Set of individuals	[PERSON:{Jim,John}]	Jim and John
Set of Quantity	[DOG:{*}@5]	5 dogs
Partially specified	[PERSON:{Jim,*}]	Jim and others
Disjunctive set	[PERSON:{Jim|John}]	Jim or John

Types correspond to RDF classes. Elliptical nodes denote *relations* between concepts. Types are partially ordered, using a type hierarchy which enables subtypes to inherit properties from their supertypes and also allows generalisation and specialisation of graphs. Types can be defined using conceptual graphs.

New graphs can be constructed by formation rules:

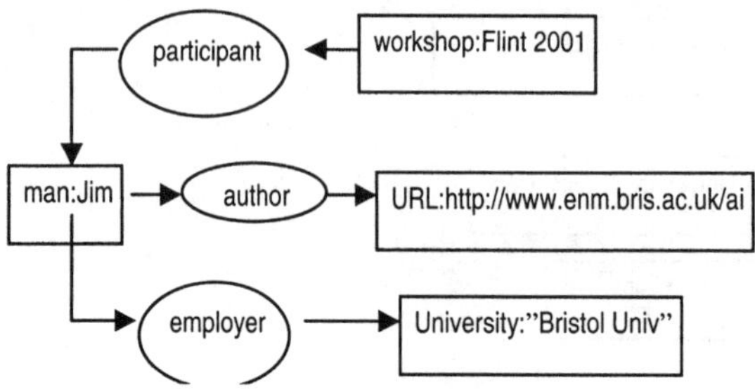

Fig 5   join of the conceptual graphs shown in Fig 4

☐ copy : a copy of the graph is taken.

☐ restriction : For any concept c in a graph u, type(c) may be replaced by a sub-type: if c is generic, its referent may be changed to an individual marker, only if referent(c) conforms to type(c) before and after the change.

☐ join : If a concept c in graph u is identical to a concept d in graph v, the join is the graph obtained by deleting d and linking to c all arcs of conceptual relations that had been linked to d.

☐ simplify : If conceptual relations r and s in the graph u are duplicates, then one of them may be deleted from u together with all its arcs.

Consider the two graphs defined in Fig 4. We restrict the concept [person] to [man], since *man* is a subtype of *person,* then restrict the referent to *Jim,* which conforms to the type *man.* Similarly, we can restrict *EducInst* to *University:Bristol,* and join the graph on these nodes, simplifying to the graph in Fig 5.

These operations preserve the property of meaningfulness rather than truth; however, starting from false graphs we can only derive false graphs.

In addition to concept and relation nodes, a third category - actors - implements computed relations. We also note that a node of type *Proposition* can contain a conceptual graph as referent.

Uncertainty in Conceptual Graphs

Baldwin and Morton [11] extended the conceptual graph framework to allow fuzzy referents. They partitioned the type lattice into mutually exclusive subtypes : entity, attribute, information, event, and state, and defined functions on the first three enabling fuzzy information to be represented. For example, an entity exhibits perceptual fuzziness, i.e. a compatibility between a referent and a type. Linguistic fuzziness arises in an attribute such as height or colour, which could have referents *tall* and *pale-red.* Finally, propositional fuzziness allows a fuzzy truth value to be associated with a graph which expresses a proposition. These require modified graph operations such as join, project. [41] extended the framework to allow fuzzy relations. [28] has applied fuzzy conceptual graphs to machine learning.

Conceptual graphs also have very powerful mechanisms for handling incomplete and default information.

- [] A *schema* is a graph which represents an example of the usage of a given concept, i.e. that indicates a context in which the concept is applicable.
- [] A *prototype* is a graph giving the central tendencies of a concept with generic concepts restricted to default values.
- [] A *canonical graph* for a type states necessary constraints on the use of that type. It is a common generalisation of all graphs that use that concept type.
- [] A *type definition* contains the necessary and sufficient conditions for defining a type. Type definitions support decompositional semantics where a high level concept type is decomposed into a graph of primitive types.

A conceptual graph toolkit implementing these operations and properties is available in Fril from ftp://xian.enm.bris.ac.uk/pub/CGPackage

Fril++

Fril++ [3, 7] is an extended Fril [9, 10] which allows an uncertain class hierarchy to be defined. As discussed above, much analysis and design is founded on the assumption that a real-world problem can be modelled using crisp sets of objects. For example, an employee will definitely be a member of the class (set) *person*, and not a member of the class *job*. The decomposition into classes may not always be so cut-and-dried. Consider partitioning the class *job* into managerial, technical, manual, etc. sub-classes. These are not crisply defined categories, and (for example) a research team leader may fall into both managerial and technical classes. Such a division is an obvious case for a fuzzy approach, since we have a linguistic label for a category (managerial occupations) and a number of instances (elements) which belong to the category to a greater or lesser extent. There are other, more obvious fuzzifications of the object oriented model, namely to allow uncertainty in data values and the propagation of uncertainty through local and inter-object computations.

Fuzzy classes can be of considerable use in conceptual graph modelling. Memberships do not need to be numbers, nor even explicitly stored or evaluated for any particular instance. An instance may be a close match to one class when one subset of its properties is considered; it may be close to another when another subset is considered. The programmer has a set of properties in mind when deciding on the class hierarchy, but it is by no means obvious that every inheritance must take class memberships into account.

Uncertainty in the Type Hierarchy

Creating a type hierarchy requires analysis of the problem domain into distinct classes, with a well-defined hierarchical structure. Each class should correspond to a set of objects in the "real world", and behave accordingly (see the section above on "Inference and Ontologies").

Fuzzy classes [4, 8] can be of considerable use in the data handled by a knowledge-based program. An instance can have partial membership in more than one class. Memberships do not need to be numbers, nor even explicitly stored or evaluated for any particular instance. An instance may be a close match to one class when one subset of its properties is considered; it may be close to another when another subset is considered. The programmer has a set of properties in mind when deciding on the class hierarchy, but it is by no means obvious that every inheritance must take into account the class memberships.

Fusion of concepts - "smushing"

Almost inevitably, there are will be small inconsistencies, both within RDF data from one source and between RDF data derived from different sources. To take a

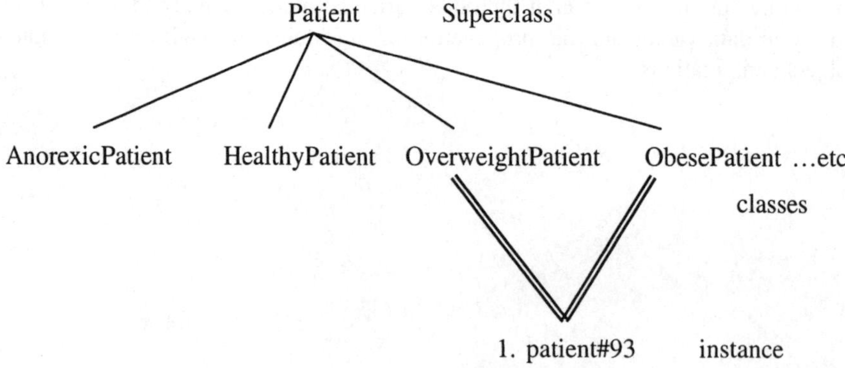

Fig. 6. an instance with partial membership in two classes

180

```
conference
      location = California
      start = 14 August 2001
      duration = 4 days
      title = Fuzzy Logic and the Internet
      participants = { ....etc }
```

```
workshop
     title = FLINT 2001 - 2001 BISC Int'l Workshop on Fuzzy Logic
        and the Internet
     url = bisc7.eecs.berkeley.edu/BISC/flint2001/
     location = University of California, Berkeley
     date = August 14-18, 2001
```

Fig.7.   different metadata representations of the same event

very simple example, we may have a document with author details
```
name="Martin, Trevor"
```
and another with
```
name="Trevor Martin"
```

It is highly likely that these represent the same person (particularly if other attributes such as institution, email address, etc are the same). A method of approximate matching is needed, enabling data from the different sources to be combined (or "inconsistencies" within data to be identified).

More complex problems may arise where slightly different representations have been used. For example, the two structures in Fig 7 clearly refer to the same event, but differ syntactically in almost all respects. The "smushing" problem (see [8] for the use of this term, also [27]) refers to the problem of combining disparate knowledge sources correctly. Damiani [19, 20] has made some initial inroads into this problem from an XML perspective using weighted directed graphs. IIis work assumes that the user has indicated the importance of edges as a weighted XML graph, and the degree of match is computed from these weights. In contrast, we focus on the matching of syntactically non-identical nodes. Clearly there are many different levels at which this graph matching problem can be tackled. At the simplest, we have essentially a syntactic match - the names "Martin, Trevor" and "Trevor Martin" require a simple re-ordering of words to become identical. However, syntactic closeness may be misleading in same cases. For example, the UK postcodes IP1 1AB and IG1 1AB differ only by 1 character but are over 50 miles apart, whereas IP1 1AB and IP5 2NP are syntactically very different (only two characters in common) but are within 5 miles of each other. Thus syntactic closeness is not necessarily a useful indicator for geographic closeness.

At a higher level, we may be faced with unifying and computing the match between two structures such as the *workshop* and *conference* examples above.

Such matching problems require context-dependent procedures attached to the graphs themselves. Fril++ with the conceptual graph toolkit represents a software system with all the required features to implement this

We take a simple example, for illustrative purposes only, to indicate the problem. Consider an electronic "yellow pages" style database of companies offering products and services. Each could have data as follows:

Field	sub-field	type
company name		string
address	street name	string
	number/building	number or string
	town/city	string
	postcode / zipcode	string / number
telephone number	area code	number
	subscriber	number
electronic contact details		string
opening hours		numerical time range
price list	item	string
	price	number
products/services offered		text
classification		code (numerical)

To search such a database we obviously need standard lookup techniques - "find the telephone number of the Restaurant at the End of the Universe, Bristol". More frequently, queries may be less precisely specified. To cater for this, we also need the following notions of approximate equality:

☐ string matching - to allow for mis-spelt items (e.g. in name, address, "search for a restaurant in Brsitol" - needs to be corrected to "Bristol")

☐ matching of continuous-valued numerical data - for example, "search for a restaurant in Bristol where a pizza costs less than £5-00"

☐ fuzzy matching of continuous-valued numerical data - "search for a restaurant in Bristol where a pizza is *cheap*" given some definition of the fuzzy term *cheap* in this context.

☐ matching of discrete numerical data e.g. "search for the nearest restaurant to map reference 51°29N, 2°39W", where the address in the database is converted to a map reference

☐ fuzzy matching of discrete numerical data e.g. "search for the restaurants within *easy walking distance* of map reference 51°29N, 2°39W", again with a suitable definition of the italicised fuzzy term

☐ matching using a concept hierarchy - "search for fish and chips in Bristol" using the hierarchical knowledge "fish and chips" < "takeaway food" < "fast food" where < represents a subtype relation. This must take account of partial membership in a class, where relevant.

We are not concerned with the (substantial) work in converting natural language queries to graphs, but focus on the problem of matching a query graph with the database. There is no universal method for calculating approximate equality - numerical proximity is fine in some cases, e.g. 0129 and 0131 are almost identical as prices or street numbers, but as telephone area codes they could be very far apart.

We extend the type hierarchy by enabling matching methods to be defined for each concept type in the hierarchy. At the top of the hierarchy is a simple equality test; each subtype can have a specialisation or inherit the matching method from its parent. For example, Figure 8 represents a hierarchy of concepts. For the most general, "position", equality (or similarity) is implemented by comparison of the three attributes latitude, longitude and altitude. A ground level location need not compare altitude, since it is guaranteed to be the same for identical locations. Subclasses such as zipcode or location relative to nearby road junctions require completely different methods, and do not inherit from their parent classes. Attaching these procedures to the type hierarchy gives a clean and efficient implementation.

Note that each type has two matching methods - one to be used when the concept is part of a query, the other when it is part of a graph in the knowledge base. This enable us to implement the asymmetric matching required - for example, if event X takes place on Dec 13, 2001 and event Y takes place on Dec13-14, 2001 then clearly they are not the same; however, it would be safe to say that event X could be part of event Y, but not that event Y could be part of event X. The need for asymmetric matching is explored further in the work on semantic unification of fuzzy values [2, 5, 6].

Summary

The vision of a semantic web implicitly includes many aspects which require fuzzy knowledge representation and reasoning, or an equivalent technology. The opportunities are there for the soft computing community - the needs are for an understandable and objective interpretation of membership functions, and efficient, portable, implemented systems which solve real semantic web problems. These include:

☐ the mismatch between the crisp hierarchical structures used to classify objects, and the "fuzzier" real world in which objects may have partial membership in classes
☐ notions of fuzzy equality in data, and semantic equivalence of syntactically different structures
☐ robustness against missing, partial and incorrect data

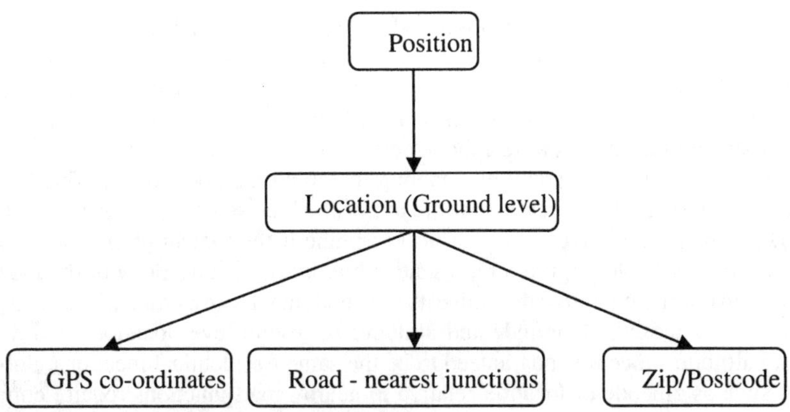

Fig 8   hierarchy of concepts representing geographical location

Knowledge representation by means of RDF maps easily into the conceptual graph framework and Fril++. The use of a concept hierarchy signals a move away from *keyword*-based searching toward *concept*-based searching. The assumption behind current information access techniques is that a user knows terms that appear in documents of interest, and is able to choose query terms that distinguish these documents from others in the collection. Concept-based searching automatically includes some personalisation issues, since users can express the search in their preferred terms which can be converted to the system-based concepts.

The semantic web aims to bring machine and human information processing closer by incorporating knowledge into documents, expressed in a clearly defined representation. We argue that extending this representation to allow fuzzy hierarchical matching will enable better searching and inference, demonstrating the value of soft computing in intelligent information management.

References

1. Baeza-Yates, R. and B. Ribeiro-Neto, *Modern Information Retrieval*. 1999, Harlow, UK: Addison Wesley.
2. Baldwin, J.F., *The Management of Fuzzy and Probabilistic Uncertainties for Knowledge Based Systems*, in *Encyclopedia of AI*, S.A. Shapiro, Editor. 1992, John Wiley. p. 528-537.
3. Baldwin, J.F., T. Cao, T.P. Martin and J.M. Rossiter. *Implementing Fril++ for Uncertain Object-Oriented Logic Programming*. in Proc. *IPMU 2000*. 2000. pp 496 - 503 Madrid.
4. Baldwin, J.F., T.H. Cao, T.P. Martin and J.M. Rossiter. *Towards Soft Computing Object-Oriented Logic Programming*. in Proc. *Ninth IEEE International Conference On Fuzzy Systems (Fuzz-IEEE 2000)*. 2000. pp 768-773 Texas.

5. Baldwin, J.F., J. Lawry and T.P. Martin. *Efficient Algorithms for Semantic Unification.* in Proc. *Information Processing and the Management of Uncertainty.* 1996. pp 527-532 Spain.

6. Baldwin, J.F., J. Lawry and T.P. Martin, *A Mass Assignment Theory of the Probability of Fuzzy Events.* Fuzzy Sets and Systems, 1996. **83**(3): p. 353-367.

7. Baldwin, J.F. and T.P. Martin. *Fuzzy Classes in Object-Oriented Logic Programming.* in Proc. *FUZZ-IEEE-96.* 1996. pp 1358-1364 New Orleans, USA.

8. Baldwin, J.F. and T.P. Martin. *Fuzzy Objects and Multiple Inheritance in Fril++.* in Proc. *EUFIT-96.* 1996. pp 680-684 Aachen, Germany.

9. Baldwin, J.F., T.P. Martin and B.W. Pilsworth, *FRIL Manual (Version 4.0).* 1988, Fril Systems Ltd, Bristol Business Centre, Maggs House, Queens Road, Bristol, BS8 1QX, UK. p. 1-697.

10. Baldwin, J.F., T.P. Martin and B.W. Pilsworth, *FRIL - Fuzzy and Evidential Reasoning in AI.* 1995, U.K.: Research Studies Press (John Wiley). 391.

11. Baldwin, J.F. and S.K. Morton, *Conceptual Graphs and Fuzzy Qualifiers in Natural Language Interfaces.* 1985, University of Bristol.

12. Berners-Lee, T., *Weaving the Web.* 1999, London: Texere. 272.

13. Berners-Lee, T., *Conceptual Graphs and the Semantic Web.* 2001.

14. Berners-Lee, T., *Notation 3.* 2001.

15. Berners-Lee, T., J. Hendler and O. Lassila, *The Semantic Web*, in *Scientific American.* 2001. p. 28-37.

16. Brickley, D., *RDFWeb notebook: aggregation strategies.* 2001.

17. Brin, S. and L. Page. *The anatomy of a large-scale hypertextual Web search engine.* in Proc. *International world wide web conference.* 1998. pp 107-118 Brisbane; Australia: Elsevier Science.

18. Case, S.J., N. Azarmi, M. Thint and T. Ohtani, *Enhancing e-Communities with Agent-Based Systems.* IEEE Computer, 2001. **33**(7): p. 64.

19. Damiani, A.E. and L. Tanca. *Flexible query techniques for well-formed XML documents.* in Proc. *International conference on knowledge-based intelligent engineering systems & allied technologies.* 2000. pp 708-711 Brighton: Institute of Electrical and Electronics Engineers.

20. Damiani, E., L. Tanca and F.A. Fontana, *Fuzzy XML Queries via Context-based Choice of Aggregations.* Kybernetika -Praha-, 2000. **36**(6): p. 635-656.

21. DublinCoreGroup, *Dublin Core Metadata Initiative.* 2001.

22. Erdmann, M., *Formal Concept Analysis to Learn from the Sisyphus-III Material.* 2000.

23. Fensel, D., I. Horrocks, et al., *OIL: An Ontology Infrastructure for the Semantic Web.* Ieee Intelligent Systems and Their Applications, 2001. **16; NUMB 2**: p. 38-45.

24. Goldman, R., J. McHugh and J. Widom, *Lore: A Database Management System for XML.* Doctor Dobbs Journal, 2000. **25**(4): p. 76-80.

25. Gruber, T.R., *A translation approach to portable ontology specifications.* Knowledge Acquisition, 1993. **5**(2): p. 199.

26. Guarino, N., C. Masolo and G. Vetere, *OntoSeek: Content-Based Access to the Web*, in *Creating Robust Software Through Self-Adaptation*, R. Laddaga, Editor. 1999, Ieee Computer Society. p. 70-80.

27. Heflin, J. and J. Hendler. *Semantic Interoperability on the Web.* in Proc. *Extreme Markup Languages.* 2000. pp Monteral: GCA.

28. Ho, K.H.L. *Learning Fuzzy Concepts by Example with Fuzzy Conceptual Graphs.* in Proc. *1st Australian Conceptual Structures Workshop.* 1994. pp Armidale, Australia.

29. Kaski, S., T. Honkela, K. Lagus and T. Kohonen, *WEBSOM - Self-organizing maps of document collections*. Neurocomputing, 1998. **21; ISSUE 1-3**: p. 101-117.

30. Maier, D., *The Theory of Relational Databases*. 1983: Pitman.

31. Martin, T.P. *Searching and Smushing on the Semantic Web - Challenges for Soft Computing*. in Proc. *FLINT 2001 - New Directions in Enhancing the Power of the Internet*. 2001. pp 3-8 Berkeley, CA: University of California, Berkeley.

32. McHugh, J., S. Abiteboul, et al., *Lore: A Database Management System for Semistructured Data*. Sigmod Record, 1997. **26**(3): p. 54-66.

33. Salton, G. and M.J. McGill, *Introduction to Modern Information Retrieval*. 1983, New York: McGraw Hill.

34. Sowa, J.F., *Conceptual Structures*. 1984: Addison Wesley.

35. Sowa, J.F., *Knowledge Representation: Logical, Philosophical, and Computational Foundations*. 1999: Brooks Cole Publishing Co.

36. Staab, S., A. Maedche, C. Nedellec and P. Wiemer-Hastings, *ECAI'2000 Workshop on Ontology Learning*. 2000.

37. Ullman, J.D., *Principles of Database and Knowledge-Base Systems Parts 1 and 2*. 1988: Computer Science Press.

38. van Rijsbergen, C.J., *Information Retrieval*. 2nd ed. 1979, London, UK: Butterworths.

39. Voorhees, E.M. and D.K. Harman, *Overview of the Seventh Text REtrieval Conference (TREC-7)*. Nist Special Publication Sp, 1999: p. 1-24.

40. W3C, *W3C Semantic Web Activity: Resource Description Framework*. 2001.

41. Wuwongse, V. and M. Manzano, *Fuzzy Conceptual Graphs*, in *Conceptual Graphs for Knowledge Representation*, G.W. Mineau, B. Moulin, and J.F. Sowa, Editors. 1993, Springer (LNAI 699). p. 430-449.

Dialogue-based interaction with a web assistant: The ADVICE approach

Ana García-Serrano, David Teruel, Josefa Z. Hernández
Department of Computer Science, Technical University of Madrid (UPM), Spain
{agarcia, dteruel, phernan}@dia.fi.upm.es

Abstract: Searching for and selecting complex products on the web is usually a difficult task for consumers mainly due to the lack of intelligent support or assistance. An advanced solution in an e-commerce setting has to optimize the user search by a customer adapted intelligent assistance that emulates in some way the performance of a human seller. This assistance should comprise customer-adapted suggestion and explanation of product types, features, alternatives and special offers on digital markets.

The *ADVICE approach* allows this enhancement of the interaction with the user by: (a) A fluid communication with the user supported by different media in order to achieve an advanced multimedia user interface. (b) A dialogue-based interaction to help user navigation in the commercial web, supporting a genuine integration of semantics and pragmatics, based on a discourse model with a twofold dimension: informational and intentional. (c) A customized configuration of the offer according to the user needs and the evolution of the dialogue. Moreover, improving system ergonomy can only be place-holder for the system intelligence.

The implemented version to reach a first step on intelligent assistance in e-commerce is based on an agent-based architecture. The prototype includes the Interface Agent to manage the multimedia presentation, the Interaction Agent to support an advanced user-system interaction and the Intelligent Agent incorporating a knowledge-based model of the e-business and domain related knowledge, that supports the reasoning for advise-giving according with the user needs and the dialogue evolution.

This work has been developed in the framework of the European Commission research project ADVICE (IST 1999-11305). The resulting prototype allowed the validation of the global approach for the development of advanced interfaces in web applications.

Introduction

The need of improving the interaction level in the human-system relationship that naturally emerged in an e-commerce scenario, has to be supported within a move from the current catalogue-based customer services to a customer adapted intelligent assistance, emulating in some way the performance of a human seller.

Looking at the state-of-the-art in e-commerce or digital sites endowed with some kind of natural language capabilities or intelligent assistance, four classes of applications can be identified:

- the more simple approach are on-line catalogues explored directly by the users, occasionally with the support of a web map but no particular help from the application.
- on-line catalogues incorporating query forms to be filled by the user according to his requirements that allow a direct access to the products database, improving the navigation with regard to the simple on-line catalogues and supporting a kind of guide for the customer. Examples of this kind of systems are web sites developed for buying second hand cars, houses, etc.
- Meta-search engines that help to find e-commerce sites. For instance, www.askjeeves.com is a search engine that receives a natural language query and provides a set of links containing the related e-sales sites (or web related sites) where the keywords of the user query were found.
- interactive software robots, like ALICE (http://www.alicebot.org/), by Wallace (2000), is a robot that interacts with users in natural language. Includes a set of classified cases (in the same way as the ELIZA work of early natural language developments) and an inference mechanism based in a simple case-based reasoning approach. There are also some templates to produce the answers. The emphasis in the language design is minimalism. Similar chat-bots are designed to talk about specific topics (some of them in e-commerce environments) but without managing a conversation model.

From this short summary about the classes of related applications was outlined an interesting conclusion. It is widely accepted that natural language processing or the mere integration of different communication media is not sufficient for building systems that can engage in a truly co-operative dialogue with humans.

The *ADVICE approach* main goals to succeeded in the enhancement of the interaction with the user are:

- A fluid communication with the user supported by different media (natural language, GUI and 3D character) in order to achieve an advanced multimedia user interface in an e-commerce scenario. The natural language approach is intended to be robust and generic (easily customizable for different e-commerce applications). Thus, available general purpose lexical resources (such as taggers, lexicon, semantic grammars, etc.) are used.
- A dialogue-based interaction to help user during his navigation in the web site. This interaction approach is supported by a discourse model with a twofold

dimension: informational and intentional. The informational approach establishes that the coherence of the discourse follows from semantic relationships between the information conveyed by successive utterances (inference-based approach as major computational tool). The intentional approach claims that the coherence of discourse derives from the communicative intention of both speakers and that mutual understanding depends on the capability to recognize those intentions (following speech act theories). The dialogue-based interaction applied also supports a genuine integration of semantics and pragmatics, since a good analysis of dialogue requires semantic representation (representation of the content of what the participants are saying) and pragmatic information (what kinds of speech acts they are performing - asking/answering a question, making a proposal, ...-, what information is available to each speaker, what is the purpose behind their various utterances).

- A customized configuration of the offer according to the user needs and the evolution of the dialogue.

The global objective of the work presented in the sequel is to describe a knowledge-based approach to the design and deployment of intelligent assistants that complement the current e-shop systems functionality in an e-commerce scenario. This means that the system has to be capable of recognizing who is talking with, which are the particular information needs of the user and then initiate a user-adapted information generation process, i.e. providing the right information to the right user.

The knowledge-based technologies [11,8,13] are being applied in this work to develop an e-commerce prototype using several kinds of domain knowledge identified to (a) search for appropriate products according to the user needs, considering factors such as the client preferences, general characteristics of the products at different levels of abstraction, relations between products and preferences, constraints about certain product configurations, market strategies, etc. and (b) the management of a dialogue-based interaction, considering common dialogues in the e-commerce scenario and previous experiences with "similar" customers regarding to the personal profile, the kind of explanations needed, unexpected utterances etc.

The general architecture [6,12] of the approach defined to support the intelligent assistance on the web, contains three main agents:

- The *Interface Agent* responsible for the multimedia input-output activities of the system. This agent will collect users utterances (English sentences, clicks on the settled items, such as icons, menus, etc.) and transform them into semantic structures (streams of speech acts).
- The *Interaction Agent* is responsible for the adequate management of the interaction between the Interface Agent and the Intelligent Agent, as well as with the user. It means that has to (1) manage the evolution of the conversation in a coherent way, (2) deliver to the Intelligent Agent the query of the user together with relevant information about this user that may influence the

production of the appropriate offer and (3) send to the Interface Agent the question or information to be presented to the user at every moment.
* The *Intelligent Agent* responsible of the generation of the information required by the customer is supported by a knowledge model that contains the reasoning model as well as the domain structure. These agent produces a configurable offer tree that contains the different products that suits the user identified needs. The prune of the tree comes from the new identification of user requirements during the dialogue.

In the following is included, first a description of the dialogue based approach, then is presented the components that allows the intelligent assistance, and a short presentation of the natural language component included in the-interface agent is included. Finally the current working prototype is described and some conclusions are given.

Dialogue based approach

Classical interfaces leave almost all the responsibility of the interaction to the user, instead of sharing the commitment. A dialogue is a full-convene process where both speakers need to be tuned up for the best performance. For a flexible dialogue is required at least one joint commitment (so the speakers can understand one another). These commitments motivate the clarifications and confirmations always present in conversations. Recent IST project Trindi [3,4]. claims for the need of a "common ground" between user and system. IST project Advice joins this line of research prototyping the 'threads model'.

The ADVICE virtual assistant performs a dialogue based interaction supported by the joint intention management and a generic product offer tree provided by the domain model of the product specialist to accomplish the system's participation in the dialogue (pro-active assistance to the user) to improve the success possibilities of the interaction.

The interaction, could be seen as a sequence of alternative interventions of both participants, user and system. A session will be built through several dialogues, eventually, they can be related each other, even nested. That should be seen as a chess game, in which both players make their moves to attain their goals that could imply smaller nested sub goals. In a formal analysis, there could be established some states of the 'dialogue game', and the transition one after the other as steps in that game. Those steps keep a conventional relation, conforming a 'valid dialogue' for both interlocutors through their interaction. Also there is static information all over the dialogue (the context).

From the locutive point of view, both interlocutors play their role by turns, so he dialogue is divided into discourses generated alternatively by user and system. Finally there is a third locutive level the utterance the atomic information piece of the discourse.

A satisfactory management of dialogue requires in general both semantic representation (content of what has been expressed) and pragmatic information. The semantic structures used to link interface and interaction agents in the Advice project are based on Searle's speech acts [17,7]. The set of adapted speech acts are detailed in Table 1.

Table 1. Set of speech acts for the Advice

Courtesy acts	
Salute: < c, a > Farewell: < c, a > Thank: < c, a > Disannoy: < c, a > Empathetic: <c,a> Satisfy: < c, a > Wish: < c, a >	[c]: conventional (formal/informal) [a]: allowable (open/close) <u>Examples</u>: *Nice to see you Ana: salute(i,c)* *Excuse me... :* disannoy(f,c) *Don't worry... :* empathetic(i,c) *Excellent :* satisfy(f,c)

Representative acts	
Inform: <t,m,s,c>	[t]: type (confirmation / data /...) [m]:matter (approve/deny/identity/..) [s]:subject (product/user/system/...) [c]: content (...) <u>Ex</u>: *I'm Ana*: inform(data,identity,user,Ana)

Authoritative acts	
Authorize: <m,a >	[m]:matter (start/ offer / task / ...) [a]: allowable (open / closed) <u>Ex.</u>: *Can I help you?* authorize(task,open)

Directive acts	
Request: <t,m,s,c> Command: <t,s,c>	[t]:type (choice/data/comparison/...) [m]:matter (approve/deny/identity/..) [s]: subject (user/system/...) [c]: content (values...) <u>Ex.</u>: *Who are...?*: request(data,identity,....,) *Show me some saws*: command(search,sys,prod.)

Null Speech	
Null: < >	*Well,... :* null()

The semantic structures (streams of speech acts), used to represent both the content than the intention (pragmatic) of the communication [10]. Therefore, there has to be a component in charge of translating this structures to be understable by the human user and viceversa. This component is the Natural Language processor in the Interface Agent

The Interaction Agent is responsible for the adequate management of the interaction between the Interface Agent and the Intelligent Agent, as well as with the user. It means that it has to (i) manage the evolution of the conversation in a coherent way, (ii) deliver to the Intelligent Agent the query of the user together with relevant information about this user that may influence the selection of the appropriate answer and (iii) send to the Interface Agent the information to be presented to the user at every moment.

The interaction agent manages three different sorts of information from the user participation in the dialogue. First, has to extract the data that shapes the circumstance, the so called static information. Secondly should get the underlying intentions of the dialogue, that is the dynamic information. Finally, it has to attend to the structure of the interaction, for attaining a valid dialogue.

The main components of the interaction agent are lined up with this assortment. Hence, the *Session Model* will deal with the context and the details of the interaction, while the *Dialogue Manager* pay attention to the state of the interaction and checking the coherence preservation with the user.

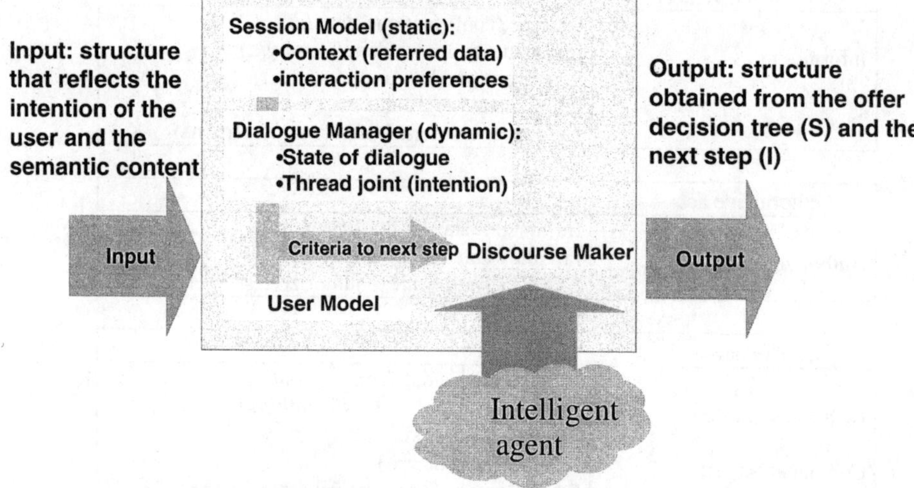

Fig. 1. Main structure of the interaction

On a second division, in order to handle de steps taken by both interlocutors through the interaction, there is a first component which role is to identify a valid dialogue helping to understand upcoming user movements. At the same time provides a set of adequate steps for the system to take. The other constituent of the dialogue manager stands for the intentional processing containing the user thread, the system thread and the thread joint, that supports a common ground for keeping the coherence of the interaction.

Dialogue manager is based in the set of possible states that can take place into a dialogue. The transitions into states are based on the interactions and intentions of the other interlocutor (represented in the thread); and some other events produced by internal actions, like a timer, the incoming of intelligent information, etc. The dialogue manager, depending on the current state and the intentions of the user, changes from a state to another, and each state involve a set of actions, i.e. reply the user in a certain way, ask for external intelligent information, etc. When the dialogue manager decides that a reply to the user must be done, passes this information to the Discourse generator.

Turn	Discourse	Thread	Pattern
System	Welcome to Tooltechnics, I'm the virtual sales assistant.		Greeting
User	Hi there. I'm John Smith.		Greeting
System	Hello Mr. Smith. Nice to meet you. What can I do for you?	Authorize (task)	Greeting Authorize
User	Well, I want a circular saw.	Command (product)	Command
System	What kind of saw do you want to purchase, a pendulum-cover saw or a plunge-cut saw?	Request (product) Request (data)	System requires explanation
User	What's the difference between them?	Request (data)	Question
System	Well, you see...	Solve	System provides explanation

Fig. 2. Dialogue steps: intentional and informational dimensions

The automaton implemented from the analysis of available dialogues that shape the *corpus*. So, as more dialogues the corpus contains as more sophisticated and realistic would be the automaton, so the dialogue capabilities of the system.

The third main component of the Interaction Agent is the *discourse generator*. It has to find a pattern that fits the needed discourse and then to fill up with the context that will provide the session model. When needed some domain dependent information, will construct a request to the Intelligent Agent that will analyze the context and then provide the information requested updating the context. In this process, some events may occur, and could even originate new threads. This would force to restart the discourse generation.

Intelligent assistance on the web

In the current state of knowledge engineering, a knowledge model can be conceived as a hierarchically structured problem-solving model which implies the characterization of several classes of problems to be solved, i.e. tasks to be performed. Our proposal in this direction is to consider the organizational principle that we call the *knowledge-area oriented principle*. This principle establishes that a knowledge model can be organized by a hierarchy of knowledge-areas where each one defines a body of expertise that explains a specific problem solving behavior.

The ADVICE approach has been designed using the KSM (*Knowledge Structure Manager*), a tool and a methodology for developing knowledge based applications [5] to define Knowledge Units as the main components of the application to be implemented.

A *knowledge unit* can be defined as a body of expertise that represents something an expert knows about a particular field. The top-level unit represents the whole model and is decomposed into simpler units that encapsulate the expertise that support the reasoning methods. A KU is the basic building blocks over the knowledge based applications to be defined. These KU does not follow the classical computational division of process and data. On the contrary, internally a KU is composed of two parts: The knowledge of the unit that is structured and can be divided into other KU and a set of tasks that defines what the unit can do, and how to do it.

```
METHOD Get decision tree
ARGUMENTS
     INPUT   p features,a features
     OUTPUT  p tree, a tree,p exp, a exp
DATA FLOW
(Saws specialist) Make decission tree
     INPUT p features, a features
     OUTPUT a tree,p tree,a exp,p exp
CONTROL RULES
START
->   (Saws specialist) Make decision tree,
     END.
```

Fig. 3. Link method for a KU

A particular type of KU is when it can't be decomposed into more KA. Then it is called primary knowledge unit (PKU). These are the lowest level components of the knowledge model. Each PKU is associated with a *primitive of representation.*

These primitives provide, both mechanisms to represent the knowledge of the area into a representation method (rules, patterns, bayesian networks, etc); and the inference methods related with the representation model (backwards chaining, pattern matching, etc).

The inference methods (so-called tasks) of a PKU are defined using these inference methods, that are expressed in a language developed into the KSM tool, the LINK code. For a task is defined the input and output paths (but not the data type of these ones) and the data flow over the different knowledge units.

PKU's tasks are executed over a *knowledge base* (KB) where the knowledge of the area is stored. This knowledge is based on a set of concepts. The definition of these concepts, their structure and relationships with other concepts are described into a *conceptual vocabulary* (CV). Therefore, The vocabulary establishes the basic language, extended and shared by several KU's.

The conceptual Vocabulary is similar to ontology's, but they keep some differences. The main difference between them is that a vocabulary only uses a subset of the elements that defines an ontology. It only uses concepts, attributes, and the ownership and subclass-of relations; whereas an ontology also uses functions, axioms, instances and it can define any kind of relationship.

```
CONCEPT product SUBCLASS OF object.
ATTRIBUTES:
      type {saw, saw blade, rail guide},
      order number (DOMAIN INTEGER).
CONCEPT tool SUBCLASS OF product.
CONCEPT accessory SUBCLASS OF product.
CONCEPT profession SUBCLASS OF object.
ATTRIBUTES:
      tasks (INSTANCE OF task).
CONCEPT task SUBCLASS OF object.
ATTRIBUTES:
      type {cut, sand},
      material {wood, plastic,, aluminium, pvc, paint, chipboard,
laminates, frfplastic},
      object {aperture, door, panel, staircase, floor, joint}
```

Fig. 4. Concel conceptual vocabulary

All these elements are used to design and to implement knowledge based applications using KSM. The methodology has to steps, the *generic model* and the *domain model* generation. The first one models the application in generic terms, not managing elements of a concrete domain, but defining the basic behavior of the system. The domain model inherits the structure and behavior of the generic model, creating a new instance of it for each domain of the application. For example, the Intelligent agent of the Advice system was develop to sell tools. The generic model represents a systems that sells goods, no matter what kind of goods they are. In the domain model, a new instance of the generic one could be create for each tool of the domain.

The generic structure of the knowledge model for the ADVICE Intelligent Agent is subdivided into three Knowledge areas, *Profession Specialist, Tasks Specialist* (specific works with the tool) *and Product Specialist* (Saws). Each area, or specialist, is an expertise of a sub domain of the conceptual model.

Profession Specialist handles high level knowledge about the tasks or works that use to be done in a profession. That means it manage the information of the typical tasks and works that use to develop several professional, as a carpenter, a decorator, etc. The *Task Specialist* contains knowledge relating to these works, and what kinds of tools are needed to accomplish them. In a future evolution of the system, this area could keep information about the way to do this works. So, the system could be able to provide a new type of advice: how to do a particular task, i.e. to paint an old door step by step. Finally, the *Product Specialist* manages knowledge about the features of the tools and accessories, and currently uses that knowledge to match the needs of the user with the products of the catalogue.

Inside of the Product specialist can be found a KU called *Features Abstraction Knowledge*. This unit is the responsible to manage all the abstract or fuzzy values for the attributes for the products. With this KU the system is able to manage values as "big", "cheapest" or "heavy". This abstract values are represented as a set of fuzzy functions. Besides, this functions must take into account the kind of product that this values are refereed to, and the personal aspects of the user, to

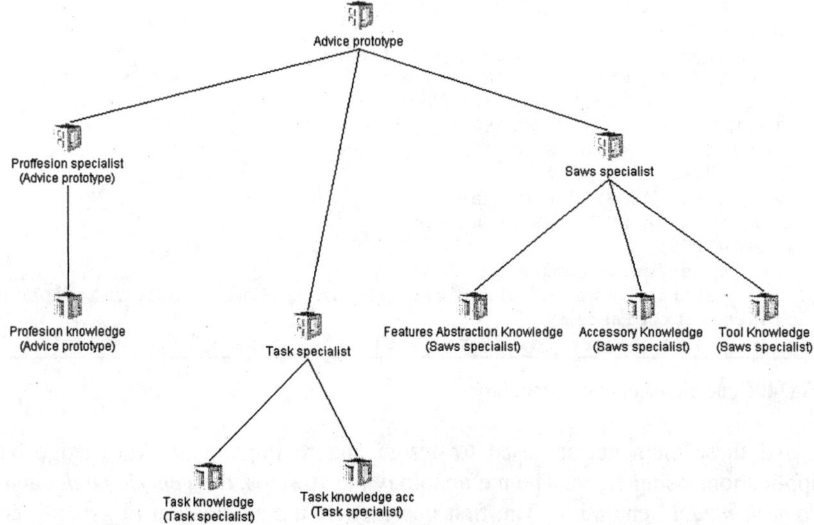

determine the range of numeric values that represent each fuzzy value.

Fig. 5. KSM Model for the Intelligent Agent

The craft domain of the current prototype includes the complete model for one kind of products, the circular saws. The generic knowledge bases identified at the design step, are filled with the specific information of the saws using frames that contains the concepts and attributes specified in the generic vocabulary as well as the specific values to the product. This values and attributes were taken directly from the electronic and paper based product catalogue of a shop. This process could be automated, extracting the needed information and converting it into the correct data structure. In this stage of the development of the system this works has been done manually.

All this information is represented and stored in the KB's as a set of frames. In the previous figure can be seen the description of a product in a frame-like way. Frames are very useful to represent that kind of information because allow the designer to create a single frame to represent each product or each category of product. Moreover, the frame representation helps to make a inference with the KB from the concrete values of some features of the products easily. In the frame definition can be distinguished three parts:

• the name of the frame. This name identifies uniquely the frame in a KB.

- the description of the frame, a list of labeled slots (object – attribute – value) that represents the features of a product and
- the relevance of characteristics. Rules are used to represent which combination of the matched attributes allows the deduction of the whole frame in a given situation.

The most important zones are the description of the frame and the relevance of characteristics. The first one defines the appearance of the frame. It keeps the slots that define what the frame represents. Each slot have an associated label. These labels are used to define the rules of the relevance of characteristics zone. Through a set of rules it's defined the conditions that must carry out the input to make the pattern valid in the pattern matching process. Moreover, with this rules, the process offer a quantitative value of the certain of this result (if the pattern is valid or not). The structure of this rules is as follow.

```
PATTERN atf55EBplus
DESCRIPTION (product)
type = saw [a], name = 'atf55EBplus' [m],
(saw) subtype = plunge [b],
      power = '1200' [c],
      weight = '4800' [d],
      saw blade speed <= 4800 [e],
      saw blade diameter = '160' [f],
      bevel cuts <= 45 [g],
      cutting depth <= 55 [h],
      dust extractor connection = 36 [i],
      systainer = 'yes'            [j],
      rail guide = 'no'            [k],
      electronics = 'yes'   [l],
RELEVANCE OF CHARACTERISTICS
a -> 50%, b -> 50%, c -> 50%, d -> 100%,
f -> 50%, h -> 50%, j -> 50%, k -> 50%,
l -> 50%, m -> 100%,
c,b -> 50%, b,f -> 50%, b,h -> 50%,
c,b,j,k,l -> 100%, b,f,j,k,l -> 100%,
b,h,j,k,l -> 100%, c,d,e,f,g,h,i,j,k,l-> 100%.
```

Fig. 6. Frame based description of a concrete product

- In the IF-zone of the rule represent those conditions. Is a list of the slot's labels with two operators: AND (,) and OR (;).
- The THEN-zone represents the certainty value with which the pattern will be valid if the condition is true.

So, the rule a,b,c → 90% means that if the slots a, b and c are true (they fit with the user requirements) the pattern is valid with a 90 percent of certainty. Given that several rules can be used in a current situation, a simple uncertainty model computes the matching degree of the frame. The matched frames are organized into a tree structure according with the domain knowledge. The nodes with alternatives have attached the attribute-value pairs which is expected will allow the discrimination of the alternatives according with next steps of the dialogue.

This rules, which solve if a frame is valid or not, could also be useful to reach a more complex mechanism for the deduction than with traditional frame-based one. Changing the slots' weights, this feature could get more importance into the decision when the product fits the user's requirements.

The frame based representation and inference method that has been used is different from the traditional ones, with pre and post conditions. However, this type of representation could also be obtained with the frame-based KSM primitive.

The output of the intelligent agent is basically a tree structure that represents the products and product categories of the catalogue, obtained with the frame-based inference mechanism explained, so also it's needed to simulate the pruning process. That way, if a node is cut off, its children must be also cut off. This simulation is obtained thanks to the rules of the relevance of characteristics zone. A frame-based node contains the rules of its ascendants, so if any of the predecessor is invalid, the children should be invalid too. Moreover, it's needed the opposite way, if all the children of an internal node are invalid, this node should be cut off. So, the slots of an internal node contains all the possible values of a feature for all the children.

The Interaction Agent send a request message to the Intelligent Agent. In this message, the Interaction Agent places all the information it has obtained from the interaction with the user. This information is relative to the characteristics that must have the tool the user wants. The Intelligent Agent then start a reasoning process, and it builds an offer tree with the products that fulfil the identified user requirements. This tree is configured according to the frames from the KB that has been validated in the pattern matching process explained before. Then the Interaction Agent explores the tree and can find a solution node or selects from the alternative nodes the pairs to generate the next step in the dialogue. This information is sent to the Interface Agent that generates a new question to the user (pro-activity of the intelligent assistant) or produce an explanation showing some information.

If more information is needed from the user, that its to say, the tree still have alternative nodes, the system ask it to the user. This information should give values to unknown features of the products. When the required information from the user is received and returned to the Interaction agent, two options could be taken. The first is that the Interaction Agents has enough information to prune the tree by itself, and take the opportune decisions with this new tree. The second is to compile again all the information about the user's requirements and pass to the Intelligent Agent to create a new tree. The first option depends on the amount of extra information that the Intelligent Agent include into the decision tree. It can include information about what features are more appropriate to ask the user for.

Natural Language Processing

One of the main aspects of the buying-selling interaction in the web is the capability of the web site of generating some kind of trust feeling in the buyer, just like a human shop assistant would do in a person-to-person interaction. By understanding the buyer needs, is being able to give him technical advice, assisting him in the final decision are not easy things to achieve in a web selling site. Natural language (NL) techniques can play a crucial role in providing this kind of enhancements [16].

Another good motivation to integrate natural language technology in this kind of sites is to make the interaction easy to those people less confident with the Internet or even computer technologies. Inexperienced users feel much more comfortable expressing themselves and receiving information in natural language rather than through the human standard ways.

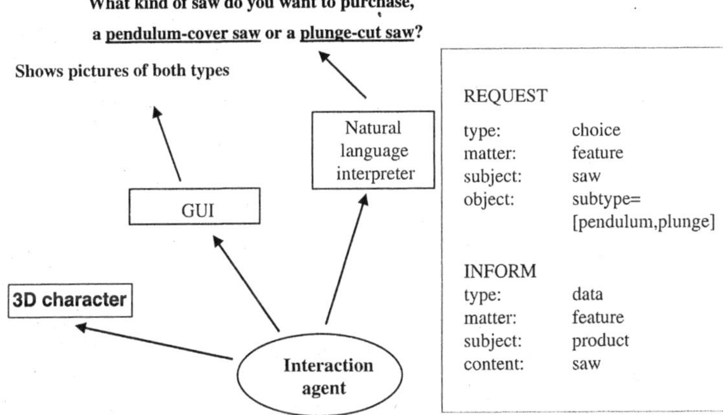

Fig. 7. Understanding Process

The Interface Agent in ADVICE project includes the NL Interpreter and Generator components. The input user utterances are interpreted in order to obtain a feature-typed semantic structure (one or more due to ambiguity) that contains speech act information and some features about the relevant items of the user utterance [9, 10].

At this moment, two interpretation strategies are defined. Firstly, message extraction techniques useful in specific domains are used in ADVICE, implemented by means of semantic grammars reflecting e-commerce generic sentences and idioms, sublanguage specific patterns and keywords. Secondly, if the pattern matching analysis does not work successfully, a robust processor that makes use of several linguistic resources (Brill tagger [1], WordNet [15], EuroWordNet [18] and a Phrase Segmenter[13]) integrates syntactic and semantic analysis in different ways.

The complexity of the Natural Language applications makes almost compulsory to get profit of the existing resources that are available even though these resources were not full compatible with our requirements. The complexity of the Natural Language applications makes almost compulsory to get profit of the existing resources so are using a knowledge-based methodology in order to adapt and reuse existing English resources [14, 9].

The lexicon is structured in three classes of words: general vocabulary, e-commerce vocabulary and domain specific vocabulary. The grammar rules contain no-terminal symbols that represent the domain concepts (tool, model, task, accessory and so on) and terminal symbols that represent the lexicon entries (vocabulary) in our application (saw, sander, to buy, to need and so on).

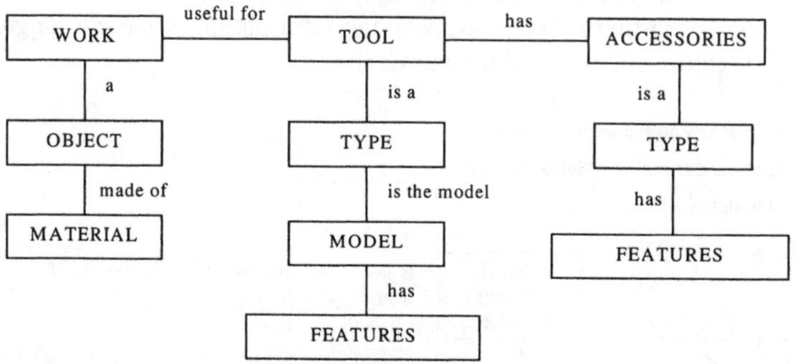

Fig. 8. Ontology to fill the product-related lexicon

Three complementary techniques are used for the NL analysis and interpretation [9]. Each one presents different degree on precision and coverage in their results. The first technique uses a syntactic grammar. In this grammar are represented the prototypical and other sentences that are obtained from the analysis of the corpus. If the user utterance fits perfectly with one of the rules of the grammar, the interpretation is correct. So, the precision of this technique is very high. However this technique could not interpret some utterances correctly if they differ slightly from the templates.

In order to improve the coverage of the analyzer, two techniques are added to the NL agent. The first one is the usage of relaxed grammars. That means that the phrase could be interpreted correctly if it keeps the structure of the template although it'd contain additional non-relevant information. This technique improves the coverage, although some relevant information from the user's utterance could be omitted. The last technique consists on identifying basic concepts related on the domain into the phrase. This technique identifies the keywords on a phrase and offers an interpretation based only on this concepts. This technique is less accurate than the others, but almost every time obtains a result. Finally, one of the three interpretation is chosen depending on the correctness of the interpretation.

Concerning the generation of answers, a domain-specific template-based approach is currently used. The templates used to generate natural language answers to the user can be two typed: with arguments (if they require arguments to fill the slots in) or without them (for instance, agreements, rejections and topic movements).

In the current working prototype, templates do not contain issues concerning the User Model although they are ready to cover them in next step of the project. The user features that are considered in this first version are: Some templates include a special argument (Expertise Level) that causes different levels of explanation in the system answers displayed to the user (pop-up links). These

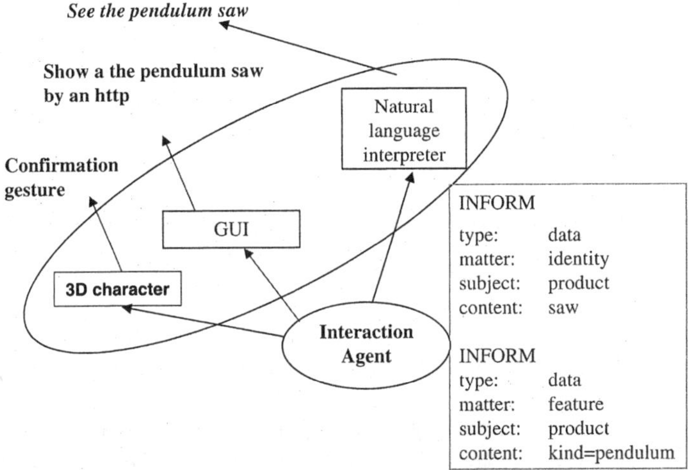

explanations are in a glossary containing each domain term (tool class, accessory and so on) together with its different explanations; Moreover, each template includes several possibilities of answer. Thus, the system does not generate always the same answer under the same conditions in order to achieve natural dialogues.

Fig. 9. Generation Process

Working Prototype

This section is devoted to present some details of the performance of current working prototype for the dialogue example that first interaction steps are shown in figure 9. The domain of the application for this prototype has been reduced to a interaction where a customer's aim is to try to buy a product, more concretely a circular saw. The user starts the interaction introducing himself. Afterwards, the user tells the system what he wants from it. In this case he wants to purchase a circular saw.

In the previous interactions no intelligent information is needed. The Interaction Agent manages the user's intentions directly. But when it receives from the Natural Language Agent the speech acts related to "I want a circular saw", it start a new intelligent task, which implies an interaction with the Intelligent Agent.

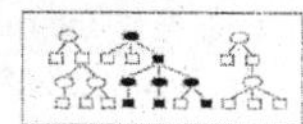

Fig. 10. Some advice-customer interaction steps

The Interaction Agent send to the Intelligent Agent all the information it has obtained form the user till this moment. In this case, it sends the information about the product the user wants to purchase, that it's to say, the circular saw. With this information, the Intelligent Agent configures an offer tree where the nodes represents the products that fulfil the user requirements. These valid products are the result of the pattern matching process with the content of the Knowledge Bases.

From the utterance (or user cliks on the saw alternative):
   "I think I need a pendulum-cut saw"

• Identification of the comunicative acts:
      null, inform(data, identity, product, saw)
      inform(data,feature,product,type=pendulum-cut)

• Identification of a new state into the (current) question-answer
dialogue pattern:
      "solve" (the previous state was "request(product)")

• Identification of the topic:
      product type pendulum-cut (add to the session model).

Fig. 11. Informational contents of an utterance

Moreover, when the offer tree is built, the Intelligent Agent attaches to the internal nodes with alternatives (with two or more paths below them) the best discriminating feature to solve the fork. The Interaction Agent will use this feature to ask the user for more information, in order to make more concrete its advice to the user. This features allows the Interaction Agent to make intelligent decision about the next step on the dialogue without having to ask the Intelligent Agent for it. This reduces substantially the response time of the system. In this prototype the complexity of this task is very little. The Intelligent Agent manages just one feature for each fork. So, the dialogue is not very flexible. In the future versions, this aspect will be improved. The idea is to create a new KB to manage this information. Moreover, the Intelligent Agent should have an additional functionality, or task, for the dynamical selection of the best discriminate feature in each situation.

In next step in the dialogue the user answers the system question : *"What kind of saw do you want: a pendulum or a plunge cut saw?"*. The figure 11 contains all the information extracted from the answer: *"I think I need a pendulum cut saw"*.

The processing steps after user answer are presented in the sequel. The user answer contains an 'inform act' that is stored in the session model. The dialogue manager changes the dialogue state from 'require clarification' to 'solve task'. The thread will close system 'asking about type' element, and is again the user's 'request product' element next to be reach. The discourse maker fails in constructing a response with the final solution, because the tree of solutions has yet several of them. Hence, adds a new element in the system thread: 'request more data'. The discourse maker now is able to act sending to the Interface Agent the question referring next bifurcation in the offer tree. Finally, the action taken by the system will be used for updating the dialogue state (again to 'require data' to further step processing).

In the current prototype, the inter-agent message communication is supported by sockets. There are two kinds of messages (speech-acts for the communication between the NL Agent and the Interaction one; and queries-trees for the communication with the Intelligent agent), but both are encoded using XML. The Intelligent Agent was developed in C++ and Java and the Interaction Agent as well as the NLP components in Ciao Prolog [2]. Figure 12 shows the conceptual architecture of the ADVICE prototype. As can be seen, the NL compponents are embedded into the Interface Agent. This component is responsible of managing the input and output devices: GUI, NL and the 3D Avatar. Both the input devices than the output ones must be coordinated to offer a coherent interaction with the user.

Conclusions

The work presented in this paper corresponds to a real-world experience with a complex problem, the development of an intelligent virtual assistant. Main goals in this open and active research area can be summarized as follows:

- integration of advanced understanding and expression capabilities with the intelligent configuration of answers and the control performed by the dialogue

modules leading to the generation of the right information in the right way
and moment.

- use of a knowledge-based methodology to obtain a robust NL-based iteration
  and scalability issues by the reuse of existing resources (English) and the
  performance of keyword-based analysis and shallow parsing techniques.
- integration of cognitive modules developed with the technologies mentioned
  above with avatar and shop-system technologies.

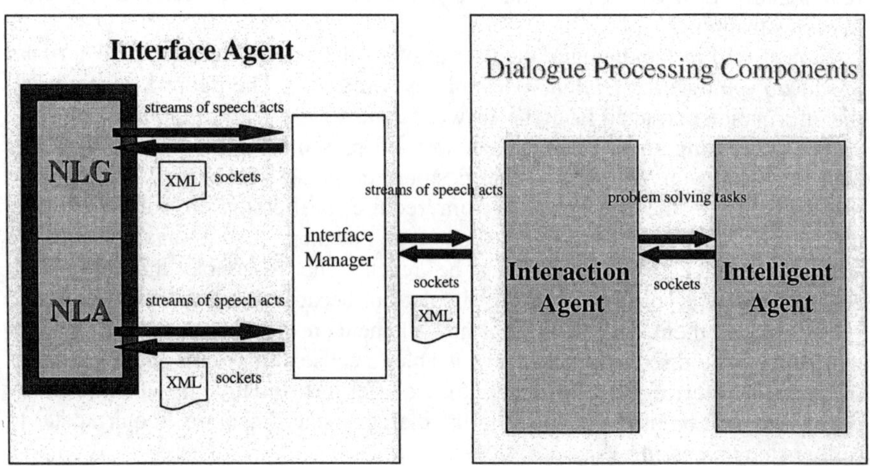

Fig. 12. Architecture of the current prototype

The work presented in this paper was focused on the application of knowledge
engineering technologies to face problems still not solved in the existing
applications, i.e. support of coherent natural language dialogues held in a
particular conversational framework and dynamic configuration of user-adapted
answers.

The designed solution, based on the use of structures of problem-solving
methods, has shown to be successful mainly because of the capability of offering
an understandable view of the reasoning and the incorporated knowledge. Second
main advantage comes from the integration of an advanced human-computer
understanding and expression capabilities with the intelligent configuration of
answers that leads to the generation of the right information in the right way and
moment in this kind of scenarios.

The first round of this work finished with a first prototype using new
technologies and standards:

- The agent-based architecture of the complete system is based on specialised
  agents to achieve the needed scalability and modularity of the components.

- The agents intra-communication is a new version of the speech acts approach used in conversation models to deal with intentional and semantic representation of the message contents. Has been implemented in XML and RMI interfaces (partially following the EU net AgentLink standards)
- The model of the problem solving steps applied by a person used to provide customer service is an hybrid solution using a frame-based representation of the products and a set of rules to generate generic offer trees to be one of the needed elements of the lay-out of the dialogue.
- The concrete domain for the prototype is BME Cat-compliant (EU standard for e-commerce applications). Two products were completely described for twenty different models and accessories available.
- The intelligent dialogue is supported by a technique capable to model joint commitments during the dialogue, to perform pro-active system participation as well as to manage the whole process to a high performance of the system.
- The NL interpreter and generator were developed following a corpus-based methodology for a domain-dependent language (to enhance the human-computer dialogue). Several resources developed in previous EU-projects were re-used (more details can be found in [14])

This work was developed in the framework of the VFP of the European Commission project ADVICE (IST-1999-11305). The ADVICE prototype allowed the validation of the global approach and the identification of the news improvements needed in this kind of advanced interfaces in web applications. The approach is currently being used and extended to deal with a new domain and operational framework in the on-going EC project VIP-ADVISOR (IST-2001-32440).

Acknowledgements

We thank the Intelligent Systems Research Group (ISYS at the UPM) and to the ADVICE consortium for their support during the two years of the ADVICE European Project (IST 1999-11305) (http://www.isys.dia.fi.upm.es/advice).

References

[1] Brill E (1994) Some advances in rule-based part of speech tagging. Proceedings of the Twelfth National Conference on Artificial Intelligence (AAAI-94), Seattle, Wa. , pp. 722-727, 1994
[2] Bueno F, Cabeza D, Carro M, Hermenegildo M, López P, Puebla G (1999). *The Ciao Prolog System: Reference Manual. The Ciao System Documentation Series* Technical Report CLIP 3/97.1, The CLIP Group, School of Computer Science, Technical University of Madrid.

[3]   Cohen PR, Levesque HJ (1991) *Confirmation and Joint Action*, Proceedings of International Joint Conf. on Artificial Intelligence, 1991

[4]   Traum, DR, Bos, J., Cooper, R., Larsson, S., Lewin, I., Matheson, C., Poesio, M., (1999). *A Model of Dialogue Moves an Information* Trindi Project (LE4-8314)

[5]   Cuena, J., and Molina, M. (1997). KSM: An Environment for Design of Structured Models. In S.G. Tzafestas (ed.), Knowledge Based Systems: Advanced Concepts, Techniques & Applications. World Scientific Publishing Company.

[6]   Silva J and Demazeau Y (2002) Vowels Co-ordination model, AAMAS'02, ACM pp:15-19

[7]   García-Serrano A and Peñas A (1999) *Interpretación de mensajes en un entorno de comunicación libre: Una aplicación a las conversaciones de correo electrónico.* Technical Report FIM/110.1/IA/99. Technical University of Madrid,

[8]   García-Serrano A, Martínez P and Teruel D (2001) *Knowledge-modelling techniques in the e-commerce scenario,* IJCAI Workshop on E-Business and the Intelligent Web, Seattle (USA)

[9]   García-Serrano A, Martínez P and Rodrigo L (2001) *Adapting and extending lexical resources in a dialogue system.* ACL 01 Workshop on Human Language Technology and Knowledge Management, Toulouse (France), July 2001

[10] García-Serrano A, Rodrigo L and Calle J (2002) *Natural Language Dialogue in a Virtual Assistant Interface.* Proceedings of the 3rd International Conference on Language Resources and Evaluation (LREC 2002). (Spain)

[11] Hernández J Z, and García-Serrano A (2000) *On the use of knowledge modelling techniques for knowledge management: A perspective*, EKAW Workshop on Common approaches in knowledge management.

[12] Jennings N, (1999) *Agent-oriented software engineering,*LNAI XXXX

[13] Martínez P and García-Serrano A (2000) *The role of knowledge-based technology in language applications development.* Expert Systems with Applications 19 ,pp 31-44

[14] Martínez P, García-Serrano A, Calle J and Rodrigo L *(2001) An agent-based design for a NL-interaction for intelligent assistance in the e-commerce scenario,* IEEE International Workshop on Natural Language Procc and Knowledge Eng. (Arizona)

[15] Miller GA, Beckwith R, Fellbaum, C, Gross D, Miller K (1993) *Introduction to WordNet: An On-line Lexical Database.* Princeton University, New Jersey

[16] Oviatt S and Cohen P, *What comes naturally ,(2000)* Comm. of ACM, Vol 43, N 3

[17] Searle JR, (1969). Speech Acts: an essay in the philosophy of language. Cambridge Univ. Press.

[18] Vossen P, Bloksma L, Rodríguez H, Climent H, Calzolari N, Roventini A, Bertagna F, Alonge A, Peters W (1998) The EuroWordNet Base Concepts and Top Ontology. Version2. EuroWordNet (LE 4003) Deliverable.

Personalized Search Agents Using Data Mining and Granular Fuzzy Techniques

Yu Tang[1], Yan-Qing Zhang[1], Abraham Kandel[2], T.Y. Lin[3], and Y.Y. Yao[4]

1 Georgia State University, Atlanta, GA 30303 USA
2 University of South Florida, Tampa, FL 33620 USA
3 San Jose State University, San Jose, CA 95192 USA
4 University of Regina, Regina, Saskatchewan, Canada S4S 0A2

Abstract. In a traditional library system, a search result will be exactly the same if different users with different preferences use the same search criteria. It is obvious that the traditional library system cannot provide high QoS (Quality of Service) for different users. To solve this problem, a personalized library search agent technique is proposed based on data mining technology. By mining the training data sets, the attributes that are related to borrowing tendency of users are analyzed, and then users are divided into different groups. The SLIQ (Supervised Learning In Quest) algorithm is used to mine data in this Web-based personalized library search agent system successfully. Simulations have shown the personalized library search agent can generate personalized search results based on users' preferences and usage. Therefore, a user can use the personalized library search agent to get quick result. The fuzzy Web search agent and the granular Web search agent are proposed to deal with uncertainty and complexity of huge amounts of Web data. In general, Computational Web Intelligence (CWI) can be used in the personalized search agent to improve QoS for Web users.

1 INTRODUCTION

In our time, with the Internet boom many libraries provide on-line service. People can search what they want through the Internet. Using the traditional search agent, every user is treated as the same way even if they have the different backgrounds – they have the different job, different favorite, different age and so on. But for the system, they have no difference, the search result are exactly same. Sometimes, such system can't provide high quality of service. Suppose a user is interested in graphics, he wants to find some book about java used in graphics. If he only types 'java' as keyword to search for books, the results displayed on the screen may begin with the books about java used in other fields such as java servlet, java server page and java language and so on, which he doesn't like. In this case, he has to take more time to find the books that he really needs. In the worst case – the books he needs appear on the end of the list – he maybe looses patience before he finds the books. It is obvious that the traditional searching system is time-consuming. To solve this problem, in the traditional library system, user has to give more detail information to narrow the search result. Our goal is to develop a personalized search agent to enable users get the different sequence of search result according to their personality. Users can get the different sequence of search results even if the same search criteria are used. For example, a user belongs to group A. In this group, users tend to borrow books about network. Suppose he types 'java' as keyword to search for books, the books listed in the firstly page will be the books about network with 'java' in their titles and then the other books will display. By this way, users can find what they want quickly than the traditional way. Our approach is to employ data mining techniques.

There is another problem during the searching. That is the searching based on exact quantities is a not suitable for all situations. Sometimes, the users want to seek a hotel at a proper price. But proper price may not be an exactly number (say $49.99). Under this situation, we should use fuzzy terms like around $50 and about 120 miles for fuzzy search. In general, a linguistic search agent is an ideal system that can use flexible linguistic terms and different languages. In addition, granular computing can be used to design pure granular search agents. For example, rough sets can be used to design a rough search agent. The rough search agent may provide more relevant results by using rough sets and data mining techniques. The interval search agent can select possible search results based on interval computing [9]. Clearly, fuzzy computing and granular computing can enhance the QoI (Quality of Intelligence) of the smart search agent.

Computational Web Intelligence (CWI) is a hybrid technology of Computational Intelligence (CI) and Web Technology (WT) dedicating to increasing QoI of e-Business applications on the Internet and wireless networks [8]. Fuzzy computing, neural computing, evolutionary computing, probabilistic computing, granular computing, rough computing, data mining, personalization and intelligent agent technology are major techniques of CWI [7][8][9]. Currently, seven major research areas of CWI are (1) Fuzzy WI (FWI), (2) Neural WI (NWI), (3) Evolutionary WI (EWI), (4) Probabilistic WI (PWI), (5) Granular WI (GWI), (6) Rough WI (RWI), and (7) Hybrid WI (HWI). Here, relevant FWI and GWI are described. FWI has two major techniques that are (1) fuzzy logic and (2) Web Technology. The main goal of FWI is to design intelligent fuzzy e-agents that can deal with fuzziness of data, information and knowledge, and also make satisfactory decisions like the human brain for e-applications effectively. GWI has two major techniques that are (1) granular computing and (2) Web Technology. The main goal of GWI is to design intelligent granular e-agents that can deal with granulation of data, information and knowledge for e-applications effectively.

In this chapter, a library search agent using data mining techniques is proposed. Since every user has his/her own personality and borrowing history, querying the same database by the same way will get the different results that may be meaningful and useful for the given user. Employing the data mining techniques, users can be divided into different groups, by this way, the system not only can provide the default list of books for every user according to the group that the user assorted, buts also it can provide different sequence of search result for different kind of users. The advantage of using such system is time saving and providing more satisfying service. In addition, Techniques of FWI and GWI can be used to design fuzzy granular Web search agents to deal with uncertainty and complexity of huge amounts of Web data.

2 DATA MINING OVERVIEW

In the past three decades, we depend on the statistics to analysis the collect data and we get success in some fields. But there is a drawback: the statistics used to analysis data starts with a hypothesis about the relationship among the data attributes, and then prove or disprove that hypothesis. If the data with a lot of attributes, this hypothesis-and –test methodology is time spending.

Another element makes the thing worse: with the development of techniques, the ability of the computer to store data is increasing. We can now store and query terabytes and even petabytes of data in one management system. The explosion of stored data requires an effective way to analysis data and to get the useful and meaningful information. It is clearly that using the statistics to analyze such massive amount of data is impossible.

For these reasons, we need to develop a new means to analysis data. Fortunately, power of computation gets great improvements while the increasing of the power of store. Meanwhile, artificial intelligence (AI) also develops. The algorithm of AI is opposed to statistical techniques; it can automatically analyze data and build data models that make us to understand the relationships among the attributes and class of data. As described in [2], [4], [5], [6]. This algorithm employs "test-and-hypothesize" paradigm instead of "hypothesize-and-test" that used in statistics.

2.1 Basic Concept of Data Mining

Data mining has been defined as fellows:
   "the process of exploration and analysis, by automatic or semi-automatic
   means, of large quantities of data in order to discover meaningful patterns
   and rules." [2].

Generally, data mining is the process of analysis data from different perspectives and summarizing it into useful information. It starts with raw data and gets the results that may be insights, rules, or predictive models. Usually, such process is work on larger relational database.

Because data mining can automate the process of finding predictive trends and behaviors in large database and automate the discovery of previously unknown patterns that may be missed by expert for they are hidden in data, using this technique, we can make proactive, knowledge-driven decisions.

Data mining is the result of a long time process of research on statistics, artificial intelligence, data visualization, machine learning and so on, and it gets its techniques and algorithms from these fields, so it has much in common and has some difference with them:

1. Data mining has common with machine learning in the study of theories and algorithm for system that extract patterns and models form data. But data mining focuses on the extension of these theories and algorithm to the problem of finding special patterns that may be interpreted as useful or interesting knowledge in large sets of data while machine learning focuses on how to reproduce the previous situation and make generalizations about new cases by analyzing the previous examples and their results.

2. Data mining also has in common with statistics in the modeling data and handling noise. But it differs from traditional statistics. Data mining is data driven, it uses test-and-hypothesis paradigm while statistics is human driven, it uses hypothesis-and-test paradigm. Moreover, sometimes they have the different goal: data mining is human centered and focuses on the human-computer interface research but statistics is more interested in getting logical rules or visual representations.

2.2 Process of Data Mining

Data mining process is interactive and iterative, it often starts with a large, arbitrary data set and with as few assumptions as possible. The initial data are treated as if there is no information available, the system must extract potential rules or patterns from that data, and then use algorithm to choose among them. This technique that used to get the important information is data modeling.

Modeling is simply the act of building a model in one situation where you know the answer and then applying it to another situation that you don't know [6]. The way that computer built the model is like the way that people build the model. First, computer is loaded a lot of data include variety of situations and their results, and then the system runs through all of data and extracts the characteristics of the data; finally, it builds the model by using this information. Once the model is built, it can be used to give the answer for the similar situations.

Data mining analyzes the relationships and patterns; it implements the any of the following types of function:

- Classification: Stored data items are mapped or classified into several predetermined exclusive groups by a function. The members in the same group are as "close" as possible to each other and the members in the different group are as "far" as possible from one another where distance is measured with respect to specific variable(s) that the system try to predict [6].
- *Regression*: Stored data items are mapped into a real-valued prediction variable by a particular function.
- *Clustering*: Stored data items are divided into different groups according to the logical relationship or consumer preferences. The members in the same group are as "close" as possible each other and the members in the different group are as "far" as possible from one another.
- *Summarization*: A report/documentation or a compact description for a subset of data is consolidated.
- *Dependency modeling*: A model that describes significant dependencies between variables is found by the particular methods. It exists at two levels: the structure level and the quantitative level.
- *Change and deviation detection*: The significant changes in the data from the historic pattern or normative values are discovered.

Data mining process includes many steps. Brachman and Anand (1996) gave a practical view of such process. The main steps are following:

1. *Analyzing problem*: This step involves analyzing the business problem, understanding the application domain, the relevant prior knowledge, and what the result of data mining the end-user wants to get.
2. *Preparing data*: This step involves creating a target data set, data cleaning and preprocessing, data reduction and projecting. In this step, system will select a data set or a subset of variables or data samples, then it will have some basic operations on the selected data such as noise removing, necessary information collecting, and will transform the selected data to the format required by the data mining algorithms.
3. *Choosing the data mining task*: In this step, the aim of data mining process will be decided, the possible aims of the process could be classification, regression, clustering, or others.
4. *Choosing the data mining algorithms*: In this step, the appropriate algorithm for searching for the pattern will be selected. It includes selecting the appropriate model and appropriate parameters that may match the particular data mining method.
5. *Generating pattern*: In this step, system will generate the pattern by using rule induction (automatic or interactive) and the selected algorithm. The pattern could be in a particular representational form or a set of such representations: classification rules or trees, regression, clustering or others.
6. *Interpreting pattern*: In this step, pattern will be validated and interpreted, it is possible to return to the previous step for further iteration.
7. *Consolidating knowledge*: In this step, pattern will be deployed, and the guideline or reports will be produced. The related knowledge will be incorporated into the real-world performance system, or simply reported to interested parties.

8. *Monitoring pattern*: This step assures that data mining strategy is correct. The historic patterns are regularly monitored against new data to detect the change in this pattern as early as possible.

2.3 Methods of Data Mining

There are a variety of data mining methods, many of them have been used for more than a decade in specialized analysis tools, but their capabilities are evolving now and more powerful than before. The followings are some popularly used techniques:

- *Artificial neural networks*: It is a method that inspired by human brain. It builds a non-linear predictive model that learns through training and resemble biological neural network in structure. This method was developed by several groups of researchers.
- *Decision trees and rules*: It is a method that uses shaped-tree to represent decisions. Decision trees grouped data into set of rules that are likely to have a different effect on a target variable. The trees and rules have a simple representational form, so it easy to be understood by users. Specific decision tree methods include Classification and Regression Trees (CART) and Chi Square Automatic Interaction Detection (CHAID). They provide a set of rules for the unclassified new dataset, and get the predictive answer for it.
- *Genetic algorithms*: It is a method that used in a design that based on the concepts natural evolution. It uses processes such as genetic combination, mutation and natural selection. It is an optimal technique.
- *Nonlinear regression and classification*: It is a method that includes a set of techniques for prediction, it fits linear and non-linear combinations of basic functions to combination of the input variables. This method is powerful in representation, but it is very difficult to interpret.
- *Example-based method*: It is a method to represent the model by the approximate example from the database. In this method, the representation is very simple, but it requires a well-defined distance metric for evaluation the distance between the data points.
- *Nearest neighbor method*: It is a method that used to classify each record in database into different sets. For every record, it will be put in the dataset that combination of the class of the k records most similar to it in a historical dataset. This method is also known as k-nearest neighbor technique.
- *Data visualization*: It is a method to represent the data visually. It illustrates the visual interpretation of complex relationships in the multidimensional data. In order to make the representation understand easily by users, it often uses graphics tools to represent the data relationship.
- *Relational Learning Models*: It is a technique that uses the more flexible pattern language of first-order logic. It is less restricted compare to decision-trees and rules methods. The relational learner can easily find formulas such as X = Y. This method is also known as inductive logic programming.

2.4 Trends of Data Mining

Today, data mining applications are used successfully on all size system for mainframe, PC platform and client/server. There are some external trends that will drive it to get more progress:

- *Size of database*: It is the most fundamental external trend that pushes the data mining technique going ahead. There is explosion of digital data in our days, and most of data are accessed via network. The more data needs to be processed and maintained, the more powerful techniques and system are needed.
- *Computation power of computer*: As the size of database becomes larger and larger, queries become more complex and the number of queries becomes greater, the power of computation is important.
- *Development of hardware*: Data mining techniques require numerically and statistically intensive computation, the increasing memory and processing speed enable the techniques solve the problems that were too large to be solved before. The development of hardware can drive this technique become more powerful and useful.
- *Speed of network*: As mentioned above, much of data are accessed via network, so the speed is very important for data mining techniques. The next generation Internet (NGI) will connect sites at OC-3 (155 Mbit/sec) speeds or higher that 100 times faster than today's speed. With this speed, correlating distributed data sets using current algorithms and techniques becomes possible.

Since the data mining is affected by the above elements, perhaps the followings are the three most fundamental tendencies:

- *Scaling algorithm to larger database*: Most data mining algorithms today are memory bound, so the technique can't get success on large data set until the amount of data fits into main memory. For the increasing of amount of data, the complexity of data, and complexity of queries, the algorithms must be developed.
- *Extending algorithm to new data type*: Most data mining algorithms work on flat data today, but data type is more complex than before, such as there are collection-valued and object-valued attributes. The data mining technique needs to develop to satisfy require of these new data type. In addition, the algorithm for semi-structured data and unstructured data also need to develop.
- *Development distributed algorithm*: Most of data mining algorithms of today require the data to be mined on the same location and in the same structure. But usually most data is distributed. The important trend of data mining is the algorithm can work with distributed data.

3 CLASSFICATION ALGORITHMS

3.1 Algorithms

- *ID3 algorithm*: It is a decision tree building algorithm. It determines the classification of data by testing the value of the properties and builds the tree in a top down fashion. It is a recursively process.
- *C4.5 algorithm*: It is an algorithm that recursively partitions the given data set to generate a classification decision tree. It considers the entire possible test that can split the data set and then select the best test. The decision tree uses Depth-first strategy. This algorithm was proposed first by Quinlan in 1993.
- *SLIQ(Supervise Learning In Quest) algorithm*: It is a decision tree classifier designed to classify large training data [1]. It uses a pre-sorting technique in the tree-growth phase. The decision tree uses Breadth-first strategy. This algorithm was proposed and developed by IBM's Quest project team. The details of this algorithm will be introduced in the next chapter.
- *Nave-Bayes algorithm*: It is a simple induction algorithm. It assumes a conditional independence model of attributes given the label. It was firstly proposed by Good in 1965 and developed by Domingos and Pazzani. *Nearest-neighbor algorithm*: It is a classical algorithm. It has options for setting, normalizations and editing. It was firstly proposed by Dasarathy in 1990 and developed by Aha in 1992 and Wettschereck in 1994.
- *Lazy decision tree algorithm*: It is a tree building algorithm. It builds the "best" decision tree for every test instance. This algorithm was proposed by Friedman, Kohavi and Yun in 1996.
- *Decision table algorithm*: It is a simple but useful algorithm. It uses a simple lookup table to select the feature subset.

The classification algorithms have much in common with traditional work in statistics and machine learning. It describes a model that based on the features present in a set of training data for each class in the database. The advantage for the algorithms is clearly: it is easy to understand for users and easy to implement on all kind of system. But the drawbacks are also obviously: if there are millions of data in the database or each data has a large number of attributes, the time for implementing will be huge, and the algorithm is not realistic.

3.2 SLIQ Algorithm

The SLIQ (Supervised Learning In Quest) algorithm is used to classify the training dataset. It is introduced in [3].
- Basic principle of SLIQ

As many other classic classification algorithms, the SLIQ also can be implemented in two phases: tree building phase and tree pruning phase. Because it is .t for both numerical and categorical attributes, there are a few differences in handling the two kinds of attributes. In the tree building phase, it uses a pre-sorting technique in the tree-growth phase for numerical attributes for evaluating splits while it uses a fast subsetting algorithm for categorical attributes for determining splits. This sorting procedure is integrated with a breadth-first tree growing strategy to enable classification of disk-resident datasets. In the pruning phase, it uses a new algorithm that based on the MDL (Minimum Description Length) principle and gets the results in compact and accurate trees.

- Details of the algorithm

SLIQ algorithm is .t for both numerical and categorical attributes. In this system, the attribute history we will consider are numerical and the others are categorical.

Phase of building tree: In this phase, there are two operations happen. First operation is to evaluate of splits for each attribute and to select the best split; Second operation is partition the training dataset using the best split. The algorithm is described as following:

```
MakeTree(Training Data T)
      Partition (T);
      Partition (Data S)
      if(all records S are in the same class) then return;
      Evaluate splits for each attribute A
      Use the best split to partition S into S1 and S2;
      Partition (S1);
      Partition (S2);
```

Before we analysis the numerical attributes, we partition the dataset by attributes -favor and job. Both of them are categorical. Let S(A) is the set of possible values of the attribute A, the split for A is of the form A S', where S' is subset of S. The number of possible subset for an attribute with n possible value is $2 ** n$. If the cardinality of S is large, the evaluation will be expensive. Usually, if the cardinality of the S is less than a threshold, MAXSETSIZE (the default value is 10), all of the subsets of S are evaluated. Otherwise, we use the greedy algorithm to get the subset. The algorithm starts with an empty S' and adds one element of S to S' that be the best split, these process will be repeated until there is no improvement in the splits.

For the attributes history, we pre-sorted first. Because in this system, we suppose there are four kinds of books in library, we divided the history into 4 parts, each part is for the number of one kinds of book that the users had borrowed. That means there are 4 numerical attributes need to be considered. To achieve this pre-sorting, we used the following data structure, we created a separate list (historyList[][]) for each attribute of the training dataset. history[][0] store the attribute value, history[][1] store the according index in the dataset. Then we sorted these attributes list. After the attributes list is sorted, we processed the splitting. The algorithm is given below,

EvaluateSplits()

    Step 1: for each attribute A do
                traverse attribute list of A;
    Step 2: for each value v in the attribute list do
                find the corresponding entry in the class list,
                and hence the corresponding entry class and the
                leaf node (say l) updates the class histogram in the leaf l;
    Step 3: if A is numeric attribute then
                compute splitting index for test(A $_i$= v);
    Step 4: if A is a categorical attribute then
                for each leaf of the tree do
                        find subset of A with best split.

For the numerical attribute, we need to compute the splitting index. In this algorithm, we use gini index (L.Breiman et.al.), which proposed by Wadsworth and Belmont. gini(T) defined as

$$gini(T) = 1 - \Sigma(pj * pj) \quad (1)$$

In this formula, T is a dataset that contains set of examples from n classes; pj is the relative frequency of class j in dataset T.

To calculate all of the ginin indexes for each attribute value, we compute the frequency for each class in the group first, and then found the best one for split the group. Because the value between vi and vi+1 will divide the list into two same parts, we choose the midpoint as the split point. For each group, one part is the examples that the value of attribute less than or equal to the split point, the other part is the examples that the value of attribute larger than the split point. We did the split as the same way one by one attribute until the node is the pure node (that is to say, all the examples in the node are the same class).

Phase of pruning tree: In this phase, the initial tree that built by using the training data will be examined and the sub-tree with the least estimated error rate will be chosen. The strategy is based on the principle of Minimum Description Length (MDL). It includes two parts: Data encoding and model encoding, comparison of the various sub-tree of T.

- Advantages of SLIQ

SLIQ is an attractive algorithm for data mining for its advantages:

1. The pre-sorting technique used in tree building phase and the MDL principles used in tree pruning phase make the result exhibits the same accuracy characteristics while the executes time is much shorter and the tree is smaller.
2. It can get the higher accuracies by classifying larger (disk-resident) datasets that can't be handled by other classifiers.
3. 3. It can scale for large data sets and classify datasets irrespective the number of records, attributes and classes.

4 SYSTEM DESIGN

Although data mining techniques have been used in scientific and business field successfully for tracking behavior of individuals and groups, processing medical information, selecting market, forecasting financial trends and many other applications for several years, their uses in library system are limited. Many people argue that the current data mining technique is not appropriate for library system because of its lack of standards, its unproven in library and the big technical hurdles remain. With the development of the size of database, we have to admit that the traditional catalogs can't satisfy the user's need, but the efficient new way is not discovered now. So, in this system, we try to find an alternative way to access it: to save time and make user more satisfied.

Our system is an attempt to propose a new way for searching in library. During designing this system, we broke the project into 5 different phases. In the traditional library's searching system, all the users will get the same sequence of the search result if they use the same query to the same database, but they have own favorite field and their own need, so they are may be not interested in the books that will listed on the firstly pages.

In this system, data mining technique is used to analysis the information about the users. There are many attributes in the user's information. Here, our major interesting is the user's personality (includes his/her favorite field and profession) and the user's borrowing history. Users are classified into different classes by this information. In the same class, all the users mostly tend to borrow the same kind of books.

For example, a user's favorite fields are graphics and network, his profession is programmer, and in his borrowing history, the number of books about graphics he borrowed is 40% of the total number of books he borrowed, the number of the books about network he borrowed is 25% of the total number of books, and the number of books about program language is 30% of the total number. For the information about this specified person, using the mining result, we may classify him into the class GPGN (the priority of the kind of books for the member of this class is: graphics, program language, network and the others). So, when he uses the keyword search and types the keyword "java", all the books displayed on the screen will begin with the books about graphics with "java" appearing in the title, then all the books about program language with "java" appearing in the title, and then all the books about network with "java" appearing in the title and the other books. By this way, the user can find his/her wanted books faster than the traditional way.

In the data preparing phase, data are prepared for mining. There are a lot of records in the database and many attributes for each record, not only we need to select a relative small data set, but also we need to select the attributes that have effect on borrowing tends about the record. For the data about the user, we needn't consider all of the attributes because not all of attributes have impacts on the tendency of borrowing such as address, email, social security number and password and so on. After analyzing a lot of data, we found the impactions come from these features: the favorite fields, the borrowing history, the profession and the age of the person. To make the problem simple, we just consider the three of them: the favorite fields, the borrowing history and the profession of the user in the dataset.

SLIQ algorithm is used to generate pattern in this system, it is described in details in the last chapter. We get the decision tree by SLIQ algorithm, and we need to incorporate this knowledge. We will obtain the features for each pure node from the decision tree and put the results in the database, a table contains class type and the class features. When we need to decide the type of one member, we can get this information from database and make a decision.

As we analysis before, the goal of using data mining techniques is providing the faster and more satisfactory service for users. In this system, we can approach this goal by classifying the members into different class type and giving the different search result sequence for the different class members. The mining result is used in the two phases:

For the member whose class type is unknown, we can identify him by his/her information and the result of data mining. For the member whose class type has been decided, because his borrowing history is changing, or his profile may be changed, we can update his class type according to these changes. In this system, when the member logout, we will classify the member's class type, so the logout process is a time consuming, but in the searching process, it saves time for member to find the book he most need. The personalize search agent will help user find what he wants quickly in two ways: it provides the default list for every user. According to the type of the user assorted to, system will automatically give the top 10 popular books of that kind that the user may most interested in. The other way is it will provide different search result for the different type of users.

In the application, with the increasing of the number of members and the changes of the member's class type, the initial decision tree may be lost its accuracy. It is necessary to mining data again. We will run the mining process after a period time to try to keep its accuracy; it is time consuming but can saving time in searching process.

It is obvious that in this system user can get the more useful search result quickly. Compare to traditional library system, it is time saving and provides more satisfying service.

Tables 1 and 2 are used to show the difference of search results between traditional search agent and personalized search agent:

Table 1. Search Result By the Personalized Agent

	Job	Favor	Search result
User A	Student	Programming language, e-business	e-business, programming language, graphics, system management
User B	Engineer	graphics	Graphics, system management, programming language, e-business

	Job	Favor	Search result
User A	Student	Programming language, e-business	Programming language, graphics, system management, e-business
User B	Engineer	graphics	Programming language, graphics, system management, e-business

5 FUZZY WEB SEARCH AGENTS

For clarity, here we use fuzzy logic as a basic tool to design a fuzzy-logic-based Web search agent (a fuzzy Web search agent in short) for better QoS of Web search. The novel technique proposed here can use not only traditional fuzzy-keyword-based search method but also fuzzy-user-preference-based search algorithm so as to get more satisfactory personalized search results for a particular user. In this sense, if user A and user B type in the same search key words with fuzzy operators such as fuzzy AND or fuzzy OR, user A and user B will get two different search results because user A has a different profile from user B. Clearly, personalized fuzzy Web search agent is more powerful than traditional fuzzy Web search engine because a user's profile is taken into account. Therefore, the traditional fuzzy Web search engine is a bottom building block of the personalized fuzzy Web search agent. In general, the personalized fuzzy Web search agent consists of the basic Fuzzy Web Search Engine (FWSE), the Personalized Fuzzy DataBase (PFDB), and the final Fuzzy Fusion System (FFS).

A fuzzy relevancy matrix is used to show similarity between two fuzzy search terms. A problem is the fuzzy relevancy matrix could be very large. To reduce complexity of the fuzzy relevancy matrix, a personalized fuzzy relevancy matrix is used to actually design a personalized fuzzy Web search algorithm. Since a user has a small number of frequently used search words, the personalized fuzzy relevancy matrix will be small. The personalized fuzzy relevancy matrix is updated dynamically based on the PFDB that is also updated periodically by mining the user's Web usage and preferences. The personalized fuzzy Web search algorithm is described logically as below:

Begin

Step 1: Use input key words and operators to match the personalized fuzzy relevancy matrix, and find out relevant key words ranked by similarity;

Step 2: Use these key words to do regular Web search to find out candidate results;

Step 3: Use personal profile in the PFDB to select a small number of personalized results from the candidate results based on ranked personal preferences.

Step 4: Display the final results in the ranked order.

End

6 GRANULAR WEB SEARCH AGENTS

Granular computing (GrC) is a label of theories, methodologies, techniques, and tools that make use of granules in applications [15][16][17][18]. Basic ingredients of granular computing are subsets, classes, and clusters of a universe [15][16]. There are many fundamental issues in granular computing, such as granulation of the universe, description of granules, relationships between granules, and computing with granules.

In designing a GrC based Web search agent, we will focus on the three important components of a search system, a set of Web documents, a set of users (or a set of queries, as user information needs are typically represented by queries), and a set of retrieval algorithms. For text based Web documents, they are normally represented through a set of index terms or keywords. A GrC based agent will explore potential structures on these four sets of entities, in order to improve efficiency and effectiveness of Web search. Granulation of the set of documents has been considered extensively in the design of cluster-based retrieval systems, in order to reducing computational costs [11][13]. In this approach, a collection of documents is divided into clusters such that each cluster consists of similar documents. A center is constructed for each cluster to represent all the documents in that cluster. A hierarchical clustering of documents is produced decomposing large clusters into smaller ones. The large clusters offer a rough representation of the document. The representation becomes more precise as one moves towards the smaller clusters. A document is then described by different representations at various levels. Hence, a cluster-based retrieval system implicitly employs multi-representation of documents.

Retrieval in the system is carried out by comparing a query with the centers of the larger clusters. If the center of the current cluster is sufficiently close to the query, then the query will be compared against the centroids of the smaller clusters at a lower level. In other words, if it is concluded that a document is not likely to be useful using a rough description, then the document will not be further examined using more precise descriptions. Different retrieval methods strategies may also be employed at different levels. It is important to realize, however, that the use of document clustering only reduces the dimensionality of the document collection while the dimensionality of index terms remains the same. That is, the same number of terms is used for the representation of cluster centers regardless of the level in the document hierarchy.

The notion of constructing a term hierarchy to reduce the dimensionality of terms has been studied [10][12]. A main consideration is the existing trade-off relationship between the high dimensionality of index terms and the accuracy of document representation. One may expect a more accurate document representation by using more index terms. However, the increase of the dimensionality of index terms also leads to a higher computational cost. It may also be argued the addition of index terms may not necessarily increase the accuracy of document representation as additional noise may be added. Recently, Wong et al. [14] suggested granular information retrieval. It is explicitly demonstrated that document clustering is an intrinsic component of term clustering. In other words, term clustering implies document clustering. In a term hierarchy, a term at a higher level is more general than a term at a lower level. A document is then described by fewer more general terms at a higher level, while is described by many specific terms at a lower level. Retrieval in a term hierarchy can be done in a manner similar to retrieval in a document hierarchy. There are many advantages to our proposed approach of granular information retrieval. As already mentioned, the proposed method reduces the dimensionality of both the document and term spaces. This provides the opportunity to focus on a proper level of granulation of the term space. In general, the method provides a model for developing knowledge based intelligent retrieval systems.

In a similar way, we can granulate the set of users. If queries or user profiles are represented in a similar form as documents, the process is much simpler. Granulation of users can be done either by pure hierarchical clustering or through concept hierarchy. A hierarchical structure may also be imposed on the document retrieval functions. Many retrieval functions have been developed for information retrieval, including exact Boolean matching, co-ordination level matching, fuzzy logic matching, inner product, and cosine similarity measure. Obviously, these functions do not share the same computational complexity and accuracy characteristics. For example, the coordination level matching is less expensive to compute than the cosine similarity measure, while at the same time being less accurate. At the higher levels of the term hierarchy involving more general descriptions, a simpler less expensive retrieval function may be used. On the contrary, a more expensive retrieval function can be used at the lower levels of the term hierarchy.

In summary, GrC based retrieval agents will explore the structures of several of entities in a retrieval system through granulation. In particular, different granulated views can be developed, and an agent chooses suitable views to achieve the best results. The framework of granular Web search agents allows multi-representation of Web documents and users, as well as multi-strategy retrieval. The challenging issues will be the granulation of documents, terms, users and retrieval algorithms, the representation of various objects under different granulated views, and the selection of suitable granulated views. It is expected that granular Web search agents will be a potential solution to many difficulties involved in Web search.

7 CONCLUSIONS

In this system, we have successfully used data mining techniques to save search time for users. This system can provide more satisfied service for users. It is an improvement based on the existed library system and is a successful example that data mining used in practice. More important, it proves the possibility of using the data mining technique in library system. We are sure data mining technique will be used in more fields and become more popular in business.

There are more works need to do to improve the library system. We can consider all the attributes that will impact on the borrowing trend of the users. If we can do that, the accuracy of decision tree will be increased. The other possible work is that we can use data mining technique in catalog. This will save more time for user in searching. With the development of the data mining techniques, the system will become smarter and smarter; it can make users more satisfied.

In the future, advanced intelligent techniques such as soft computing based data mining [18], CWI, NWI, EWI, and RWI will be used in the personalized library search agent system to continue to improve QoS of a library system and other information systems.

References

1. Agrawal, A.Arning, T.Bollinger, M.Mehta, J.S hafer, R.Srikant(1996) The quest data mining system, Proc. of the 2nd Int'l Conference on Knowledge Discovery in Databases and Data Mining, Portland, Oregon.
2. Berry, Michael J.A , Lino., Gordon (1997) Data mining techniques for marketing, sales and customer support.
3. Manish Mehta, Rakesh Agrawal and Jorma Rissanen: SLIQ: A Fast Scalable Classifier for Data Mining.
4. http://sac.uky.edu/ stang0/DataMining/WhatIs.html What is Data Mining.
5. http://www.anderson.ucla.edu/faculty/jason.frand/teacher/technology/datamining.html Data Mining: What is Data Mining?
6. http://www3.shore.net/ kht/text/dmwhite.html An introduction to data mining.
7. Tang Y. and Zhang Y.-Q. (2001) Personalized Library Search Agents Using data Mining Techniques, Proc. of FLINT2001, 119–124.
8. Zhang Y.-Q. and Lin T.Y. (2002) Computational Web Intelligence (CWI): Synergy of Computational Intelligence and Web Technology, Proc. of FUZZIEEE2002 of World Congress on Computational Intelligence 2002: Special Session on Computational Web Intelligence, 1104–1107.
9. Zhang Y.-Q., Hang S., Lin T.Y., and Yao Y.Y. (2001) Granular Fuzzy Web Search Agents, Proc. of FLINT2001, 95–100.
10. Chen H., Ng T., Martinez J., Schatz B. (1997) A Concept Space Approach to Addressing the Vocabulary Problem in Scientific Information Retrieval: An Experiment on the Worm Community System, Journal of the American Society for Information Science, 48, 17–31.
11. Rasmussen E. (1992) Clustering Algorithms, In: Frakes, W., Baeza-Yates, R. (Eds.): Information Retrieval: Data Structures and Algorithms, Prentice Hall, Englewood Cliffs, USA, 419–442.
12. Jones K.S. (1971) Automatic Keyword Classification for Information Retrieval, Butterworths, London, UK.
13. Willett (1988) Recent Trends in Hierarchic Document Clustering: A Critical Review, Information Processing and Management, 24, 577–597.
14. Wong S.K.M., Yao Y.Y., and Butz C.J. (2000) Granular information retrieval, Soft Computing in Information Retrieval: Techniques and Applications, Crestani, F. and Pasi, G. (Eds.), Physica-Verlag, Heidelberg, 317–331.

15. Yao Y.Y. (2000) Granular computing: basic issues and possible solutions, Proceedings of the 5th Joint Conference on Information Sciences, 186–189.

16. Yao Y.Y. and Zhong N. (1999) Potential applications of granular computing in knowledge discovery and data mining, Proceedings of World Multiconference on Systemics, Cybernetics and Informatics, 573–580.

17. Lin T.Y. (1999) Granular Computing: Fuzzy Logic and Rough Sets, Computing with words in information/intelligent systems, L.A. Zadeh and J. Kacprzyk (eds), Springer-Verlag.

18. Y.-Q. Zhang, M. D. Fraser, R. A. Gagliano and A. Kandel (2000) Granular Neural Networks for Numerical-Linguistic Data Fusion and Knowledge Discovery, Special Issue on Neural Networks for Data Mining and Knowledge Discovery, IEEE Transactions on Neural Networks, **11**, 658–667.

Business Operation Intelligence

Eric Shan[1] Fabio Casati[2] Umesh Dayal[2] Ming-Chien Shan[2]
[1]Department of EECS, University of California, Berkeley, California
[2]Hewlett-Packard Laboratories, Palo Alto, California

1 Introduction

Is my business performing? This is the fundamental question facing every business manager, and enterprises have advanced a growing dependence on IT technology to devise a solution to that very question.

In the 80's, companies started to leverage IT technology to improve the efficiency of their business system execution. During this early IT adoption stage, companies enhanced IT technology mainly to increase human productivity and, for that reason, concerned themselves with ensuring that their underlying network and computers were under normal execution. The issue of whether a business was performing was equivalent to asking the question: *Is my IT infrastructure performing?*

In the 90's, companies came to realize that functionally excellent application packages were crucial to optimizing each facet of business operation. To support this need, vendors developed various application modules to take care of each key operation segment of the whole value chain. For example, CRM packages supported front-end, customer, and market-facing operations; SCM packages managed planning and forecasting operations; ERP packages executed accounting, billing, and inventory operations; and LTM packages coordinated transportation arrangements, warehousing, and distribution center management operations. Consequently, the question became: *Is every factor of my enterprise being well-executed by apposite software applications?*

In the late 90's, enterprises started to recognize the necessity for a holistic integration and coordination of their operations and made a significant move from a data-centric approach to a process-centric approach. To automate the end-to-end business operation processes, vendors developed technology, such as message broker and workflow, to facilitate companies in linking and directing the execution of all relevant applications. Vendors also provided performance monitoring tools to oversee the execution of these processes. The fundamental question had grown into: *Does my underlying technology provide effective linkage and efficient execution of the complete end-to-end business operation cycle?*

Today, with the emergence of the global economy and the rapid growth of various new business operation models, companies around the world are dramatically changing the way they conduct business. Many companies have embarked on aggressive initiatives to provide the needed flexibility and dynamics for their business operations, only to find that they are unable to achieve their objectives due to a lack of certain core capabilities. In order to truly facilitate prompt decision-making by line-of-business managers or corporate executives, a new demand has emerged for a level of abstraction that takes IT data and places it in the proper business context suitable for business managers.

Hence, IT-oriented measurements, such as the number of processes executed per hour, the average duration of each individual activity, and the details of application failures and exception rates, should be intelligently transformed into practical and valuable information to a business manager, i.e., the number of accepted purchase orders grouped by week, customers, or suppliers, the status of deliveries, and the details on which customer or supplier interactions are profitable. Ideally, we would like to provide:

sales managers with information regarding whether an order can be fulfilled based on suppliers' credibility and CM partners' capacity, whereas, so far, we can inform them only on the IT-based execution status of SCM and MRP applications, and

finance managers with information regarding the stability of account receivables being undertaken to facilitate their cash flow management, whereas, currently, we can merely provide the maximum number of concurrent users allowed for the ERP applications under the current workload.

In other words, having a network or machine up 99.999% of the time is an IT goal, but it does not necessarily mean that a mission critical process, such as order fulfillment management or customer service delivery, is functioning. With companies striving to attain zero latency across entire value chain executions, it is critical for business managers to know when a business operation, not a particular IT element, is in danger.

In a recent article, the NY Times reported that the inability of companies "to focus effectively on the core things that drive their business" has spurred increasing interest in "a digital dashboard or cockpit that presents not just the key business data, but also the reasons behind the data," where "these things are all about making sure we react before things get out of hand." Gartner Group also recognized this need and coined the term BAM (Business Activity Monitoring) for it. Gartner Group now anticipates that by 2004, BAM will be one of the top four initiatives in enterprises where faster response is the key to performance and expects the BAM composite market to grow from about $115 million in 2001 to a whopping $2607 million in 2005. Consequently, vendors are starting to build digital dashboards to provide business managers with business statistics, trends, and figures that allow them to effortlessly gauge the health of a company's operations. This "feeling the pulse of your business" requirement, coupled with the complexity of today's business operation, translates our original question into another new form: *Do I have total visibility over my business operation processes so that all relevant information is presented in the proper business context, helping me to make real-time decisions?*

While total visibility provides many benefits, it is not an end in itself. To achieve an even higher level of value in modern business operation systems, any product solution is expected to be able to guide companies and act upon the operation information to effect action across the enterprise and beyond. For example, besides identifying low-quality executions and explaining why they happened, the system should be able to predict when they may happen again and automatically take preventative action. It should also be able to dig up hidden knowledge regarding the execution details of business operations and perform analysis to suggest or automate changes regarding business operation process optimization, partner/supplier relationship management, and resource utilization. Now, more than ever, the economic pressure and the fast pace of today's businesses are forcing companies to look for tools that will be able to constantly align their operation executions with their changing operation goals and environments. These demands translate our original question into its final form: *How does my business operation system behave in coping with the fast-changing world in terms of dynamic operation correction, disadvantaged situation prediction and prevention, and operation self-optimization?*

Having realized that the availability of these functions can mean the difference between success and failure for an enterprise, it clearly lends to the need for a new layer in the business operation software stack. We will more aptly refer to this layer as the Intelligent Business Operation Management (IBOM) layer as depicted in Fig. 1.

2 The overall IBOM operation scenario

The operation scenario of the IBOM layer involves numerous interactions throughout various components in this layer. The major steps for IBOM deployment and execution are depicted in Fig. 2.

First, based on corporate business balanced scorecards, business managers and analysts must identify a set of Key Business Performance Indicators (KBPIs) that will facilitate their monitoring of business operations, decision-making, and responses. This step, which may take several iterations of refinement, is crucial to

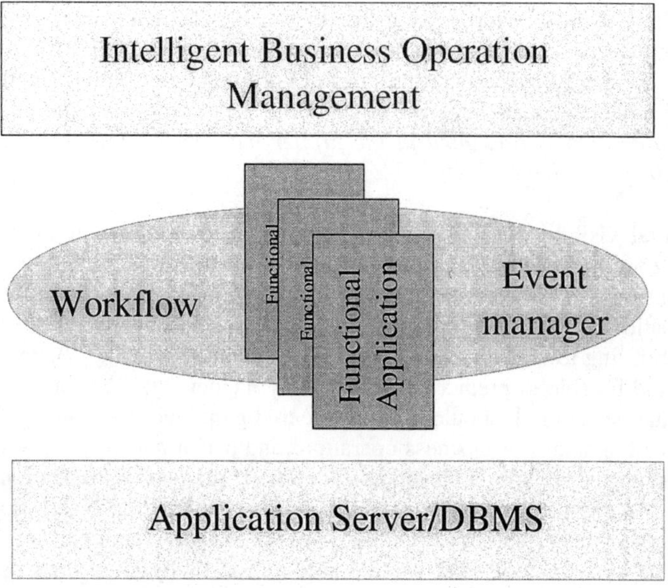

Fig. 1. Business Operation Software Stack

the deployment of an IBOM system because it properly specifies the requirements of the manager and helps to ensure they match with the IT realization of those requirements. Each identified KBPI will then be described by metrics that define the business context and data format in which the business performance measurements will be presented. As stated earlier, in most cases, the business-oriented information delivered by the defined metrics will be very different from the IT-oriented data collected by the underlying operation system in both semantics and syntax. Therefore, a mapping mechanism that transforms the IT-oriented data into the proper business performance measurement defined by a metric is required. These mappings are usually developed jointly by the business managers and corporate IT-staff. This is indicated by step 1a in Fig. 2.

It is also very useful to capture certain types of business operation knowledge in a repository for automation. Provided with the best business practice experience, business managers will then be able to identify the set of data usually needed for collection to help immediately resolve an issue when a specific event happens or describe the actions to be taken as the first step when a particular business operation condition occurs. This kind of knowledge can be collected and stored as a form of Business Operation Context Templates (BOCTs) to facilitate the re-use of their valuable experience and to possibly execute automatic responses, as described later. This is indicated by step 1b in Fig. 2.

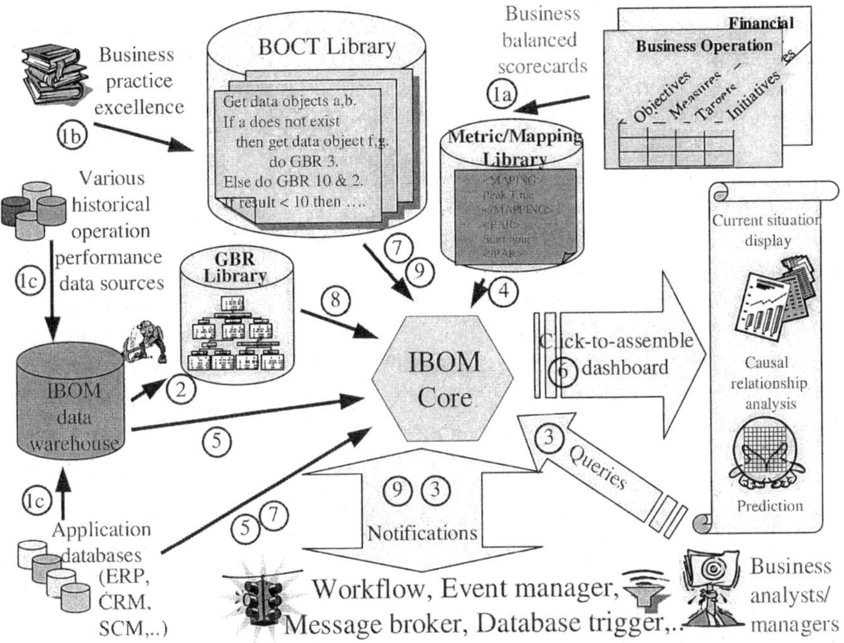

Fig. 2. Overall IBOM operation scenario

To compute the metric-based information supporting KBPI, the system needs inputs from various operation and performance data sources. In addition to its current data kept on-line, each business operation system will also archive its past operation and performance data in a repository. Usually, this historical data lies in logs, such as the workflow process instance execution log, event history log, message exchange log, database trigger activity log, and other legacy operation system logs. This will allow us to take advantage of past experience to predict trends or to proactively respond to a current operation's condition so as to prevent that operation from worsening.

However, this raw data may be in different formats and/or with different semantics, which limits the direct use of the data. Due to its original use, it is very likely that this raw data is not kept in a form facilitating the computation of the metrics in which business managers are interested. It is desirable to apply the standard ETL (extract, transform, and load) process to convert the data into a unified data format supporting those computations. This constitutes the IBOM data warehouse as indicated by step 1c in Fig. 2.

Furthermore, data analysis and mining can be conducted on this data to derive additional operation knowledge such as operation behavior patterns for the prediction of future trends, formulas for execution time estimation of certain activities, or rules for decision-and-action suggestions. A Generic Business Rule (GBR) library will be needed for the storage of this useful knowledge and for assisting in the retrieval of it as well. This is indicated by step 2 in Fig. 2.

When a business manager or analyst wants to query the execution status of certain operations or when a notification has been delivered, IBOM will compute the requested metrics and display the results in the corresponding KBPIs. The notification reflects an operation condition that needs immediate attention and is usually raised by a workflow exception, an occurrence of an event, a newly delivered message, or a database trigger execution. In addition, other relevant data may be computed or certain corrective actions may be invoked based on the operation knowledge kept in the GBR library and the BOCT library. This is indicated by step 3 in Fig. 2.

To compute the metrics, we will first retrieve the relevant mappings from the metrics library and the needed operation and performance data from the IBOM data warehouse and/or the databases of the operation systems. Then, the appropriate metric mappings will be applied on the raw data to generate the required KBPIs. Depending upon their job functions, different business managers may want to have the KBPIs presented in different formats and layouts, i.e., they would probably like to assemble their own personal dashboards with different configurations of the various metrics indicators, such as in pie chart, bar chart, or other preferred forms and displays. These are indicated by steps 4, 5, and 6 in Fig. 2.

To facilitate a business manager's decision-making process, the manager must have relevant data available. One function of a BOCT is to capture this needed information for the system. Therefore, the appropriate BOCTs will be identified and retrieved from the library and used as the basis to compute the needed information from various business operation data sources. This is indicated by step 7 in Fig. 2.

In the case that a business manager would like to view some trends or that the IBOM itself needs to conduct a prediction on certain mission-critical operations, the knowledge kept in the GBR library will be used to derive the information. This is indicated by step 8 in Fig. 2.

Lastly, the business operation needs to be optimized. This is usually done separately as a "re-engineering" task at a different time. However, certain types of corrective and preventative actions can be done automatically without human intervention. Again, the business operation knowledge kept in BOCTs will be referenced to ensure the quality of the business operations. For instance, a BOCT may specify that the appropriate operation procedure for a company to handle customer orders is as follows: When a specific inventory drops to a certain level, it will trigger the check of the contracted suppliers' status. If none of them shows an immediate availability of a reasonable quantity of the needed material, a purchase order process will be initiated to buy the material from a marketplace. If this order cannot be fulfilled within a certain period, it is best to stop the current order and disable the process from accepting any new orders. This is indicated by step 9 in Fig. 2.

To summarize the IBOM lifecycle, we outlined the major phases of operation and their transitions in Fig. 3. It is worthwhile to note that the IBOM process is a continuous effort. For example, new metric mappings are constantly under construction to support new KBPIs, the GBRs need to be validated and revised periodically to ensure their quality, and the BOCTs need to be developed and enhanced over time to reflect changes in business environments. On the other hand, it is expected that a pilot project with the objectives of developing a small set of GRBs and BOCTs and of beginning to demonstrate the value of IBOM could be completed within three to six months.

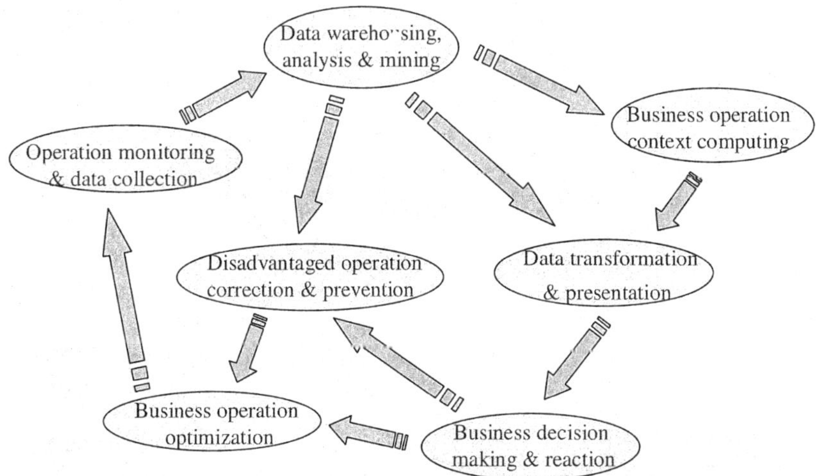

Fig. 3. Lifecycle of the IBOM operation

3 IBOM's implementation and challenges

In this section, we will highlight the design and implementation of an IBOM system and identify the key research challenges in achieving this level of operation excellence. The overall architecture of an IBOM system is depicted in Fig. 4. We will take a closer look at possible implementations for each IBOM component in the following subsections.

3.1 Business metrics and mappings

Each set of metrics may represent the performance of a business entity (e.g., a supplier or inventory) or operation (e.g., a process or service) or even an objective (e.g., cash flow level or profit level). Metrics may be defined in a hierarchy to facilitate the roll-up and drill-down of information browsing, allowing business managers to view different business perspectives at different abstraction levels. Once defined, we need to develop the mappings that transform the collectable IT-oriented

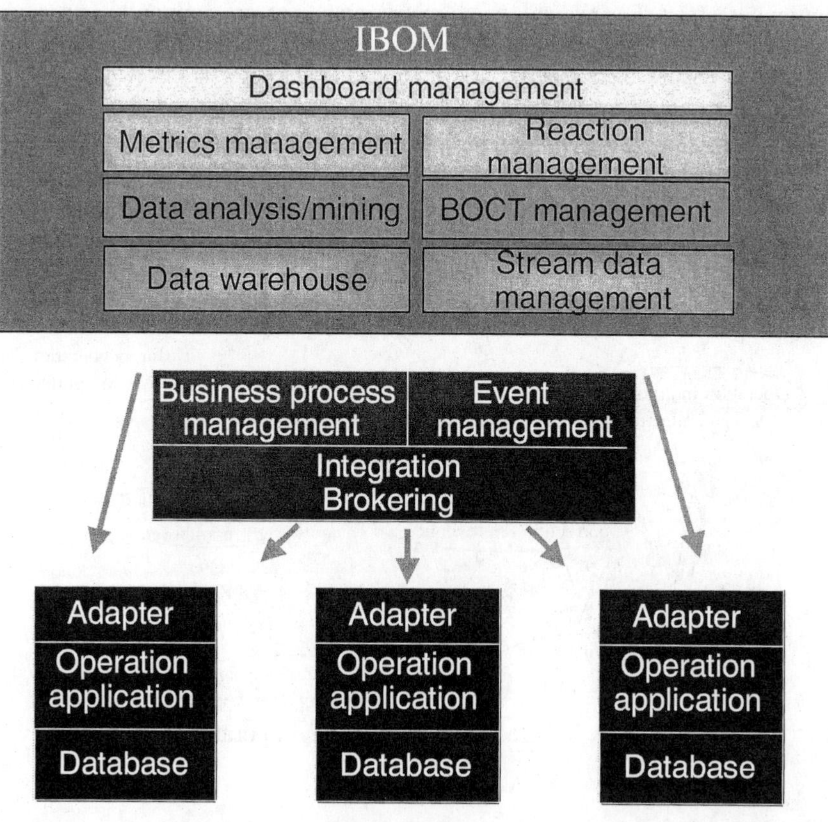

Fig. 4. Overall IBOM system architecture

measurements (i.e., the raw data) into the target business-oriented performance indicators (i.e., the data metrics).

A mapping can be defined as simply as an SQL view definition to filter or aggregate the raw data, or as a Java program that can perform any comprehensive information transformations. Let's illustrate this through a simple example. Suppose business processes of 3 different types (A, B, C) are usually carried out in company XYZ. Due to the unique nature of each of these 3 processes, their peak times may be different. For example, most of the type A (travel expense report) processes are initiated on Monday (when people come back from business trips) during regular working hours (9am–5pm), the processes of type B (purchase order) are usually initiated evenly across the whole working week (i.e., Monday through Friday, 10am–3pm), and the processes of type C (customer service request) are usually triggered during weekends (Saturday through Sunday, 9am–9pm).

Suppose that the business managers want to focus their monitoring of these processes on their average response times during only the peak time intervals. Thus, they would require performance data charts that solely target the data corresponding to specific time intervals since the definitions of peak time are different for these 3 types of processes. In this case, business managers would prefer to see performance metrics as shown in Fig. 5.

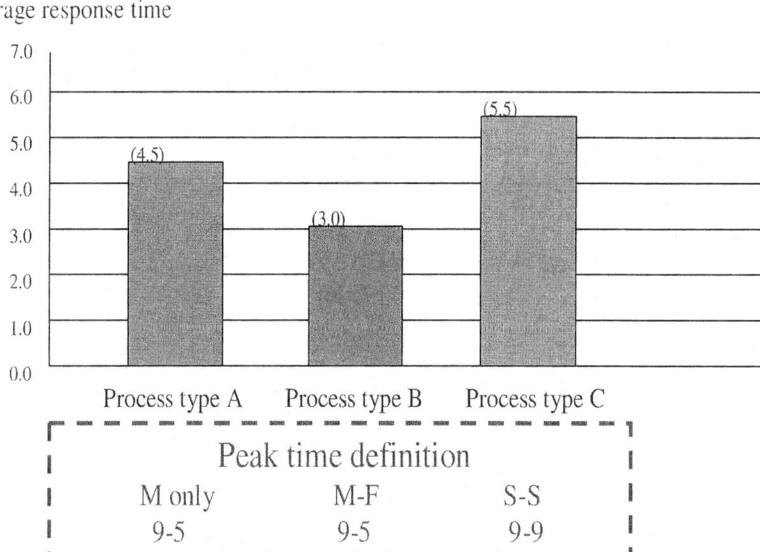

Fig. 5. Metrics for peak time performance

In this presentation of the desired business metrics, the complexity of the different peak time definitions is hidden. Instead, a unified view of peak time performance is presented to give them a direct access to the information they need. The mapping itself can be defined as an XML file:

```
<METRIC>
<NAME>Peak time execution performance</NAME>
<DESCRIPTION>Process performance during peak time</DESCRIPTION>
<TYPE>NUMERIC</TYPE>

<MAPPING>
<MAPPING_BODY>
     SELECT AVG(execution -time) FROM $t WHERE &a <= starting-time <= &b
</MAPPING_BODY>

<APPLYING_CONTEXT>
          <ENTITY-TYPE>Process type A </ENTITY_TYPE>
<MAPPING_PARAMETERS>
<PAR>
        <PAR_NAME>Performance-data-table</PAR_NAME>
        <PAR_VALUE>Process-A-data</PAR_VALUE>
<\PAR>
<PAR>
        <PAR_NAME>Start time</PAR_NAME>
        <PAR_VALUE>(Monday, 9)</PAR_VALUE>
</PAR>
<PAR>
        <PAR_NAME>End time</PAR_NAME>
     <PAR_VALUE>(Monday, 17)</PAR_VALUE>
</PAR>
</APPLYING_CONTEXT>

<APPLYING_CONTEXT>
        <ENTITY-TYPE>Process type B </ENTITY_TYPE>
<MAPPING_PARAMETERS>
<PAR>
        <PAR_NAME>Performance-data-table</PAR_NAME>
        <PAR_VALUE>Process-B-data</PAR_VALUE>
<\PAR>

<PAR>
        <PAR_NAME>Start time</PAR_NAME>
        <PAR_VALUE>(Monday, 9)</PAR_VALUE>
</PAR>
<PAR>
        <PAR_NAME>End time</PAR_NAME>
        <PAR_VALUE>(Friday, 17)</PAR_VALUE>
</PAR>
</MAPPING_PARAMETERS>
</APPLYING_CONTEXT>
. . . . . . . . . .
</METRIC>
```

As another example, we can define Java functions that map the local purchases of certain goods into one of the three cost categories (low, normal, or high) as shown in Fig. 6. These mappings may involve a computation based on the local price, shipping cost, custom duty, or other relevant factors. Such an indicator provides business managers with a quick idea of the cost when they evaluate a purchase order. More complex mappings can be defined in a similar manner.

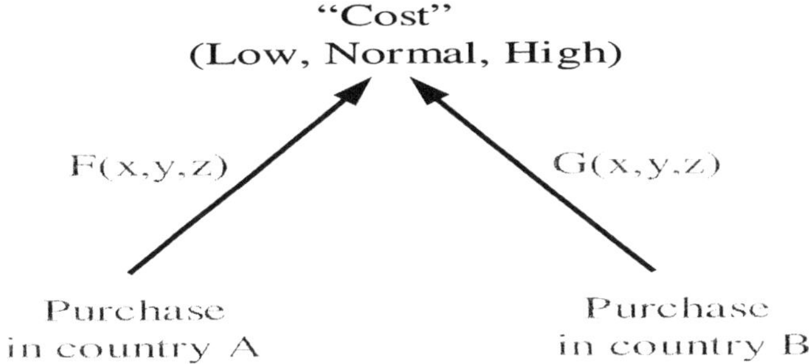

Fig. 6. Metrics for cost

Furthermore, the metric calculation can be defined on top of other metric values, such as computing the correlation or aggregation among these metric values. This will also support the metrics' hierarchy management, helping to synchronize the data across multiple metrics. The remaining challenge is to develop a tool that facilitates the capturing of mapping knowledge and automates the generation of the mappings between the raw data and the data metrics.

3.2 Operation data warehouse and steam data management

The raw data is usually kept by the operation systems in their own databases to serve their own applications. The IBOM system needs to collect them from various sources into a single repository. Since most of the data is kept in a form serving some other functions, it is very likely not to be in a form suitable for IBOM's business analysis. It will need to conduct the usual ETL (extract, transform, and load) process to clean and re-organize the data. The challenge is on how to define a proper schema (the fact tables and dimension tables) for this data warehouse to facilitate the metric value computation and data mining process. This heavily depends on the type of business analysis to be conducted and the kind of raw data available. An example of a business process-centric schema is shown below, which aims at the analysis of business process execution performance based on workflow log data.

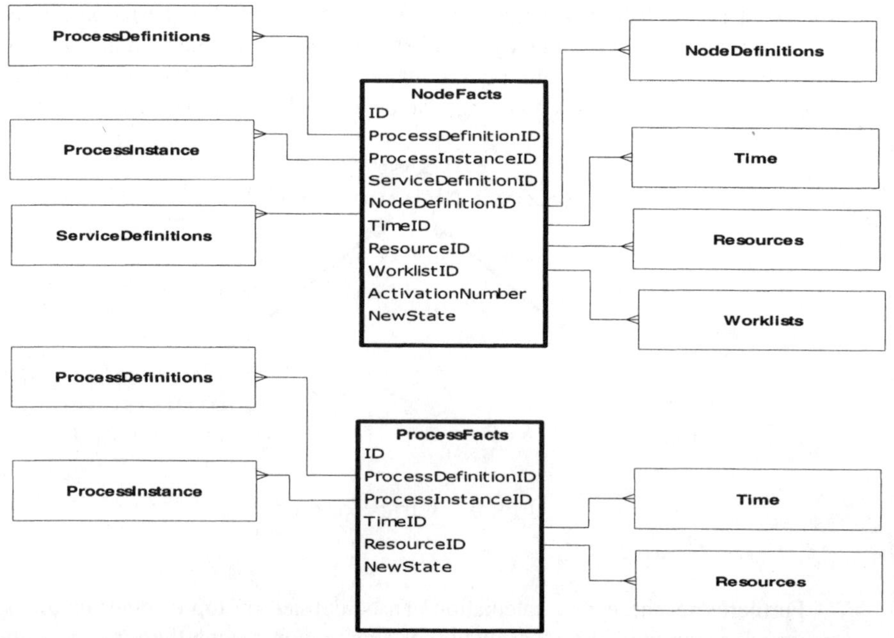

Bigger challenges remain on how to retrieve the data from multiple heterogeneous data sources in an efficient way and on how to manage real-time stream data input to update the IBOM warehouse and any derived information. The former requires heterogeneous DBMS technologies like the one developed in the Pegasus project, while the latter may be able to leverage technology developed by stream data management projects like the Telegraph project.

3.3 GBR and BOCT libraries

Now, let's pursue mechanisms to facilitate the leveraging of past experience of business operations. As usual, data mining technologies will be applied to discover valuable hidden knowledge from the operation data and derive various advisory and predictive aids, such as decision trees for event prediction, associated rules for suppliers' behavior, and statistics-model-based formulas for total cost estimation. These various advisory aids will be stored in a generic rule library and invoked to predict operation trends and/or occurrences of certain events or exceptions, to calculate the estimation of certain operation costs or shipping delays, or to provide suggestions regarding certain operation conditions or alternatives. The following shows an example of a decision tree generated during this step to be used to predict the completion time of an expense approval process at different steps.

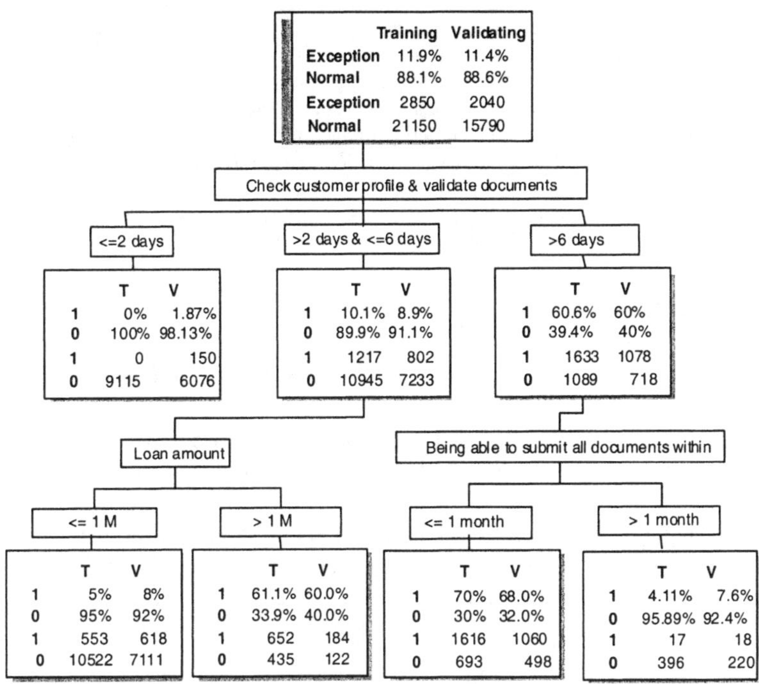

There are many ways to mine the operation data. The challenge is how to identify the most effective tool for the job. Many algorithms, such as decision trees, support vector machines, and boosting, have been used to select the business attributes controlling the prediction. Choosing which algorithm should be applied to support each specific business analysis, however, is still an open research issue.

Note that the operation knowledge derived by data mining technology represents only a small portion of IBOM's stored "experience." There are many other situations where accumulated business practice wisdom will help business managers to accomplish their job better and faster. As mentioned before, certain types of the business operation knowledge can be captured in a form like the BOCT. A sample BOCT is depicted in the diagram below.

We introduce the BOCT because we consider it the most important new mechanism to capture business operation knowledge and enable the automatic execution of that knowledge. As a matter of fact, the BOCT can serve multiple other purposes: (i) describe what relevant data may be needed to provide a better context of information for a situation, (ii) suggest what GBRs should be applied to derive additional information, (iii) specify the proper sequence to follow to resolve

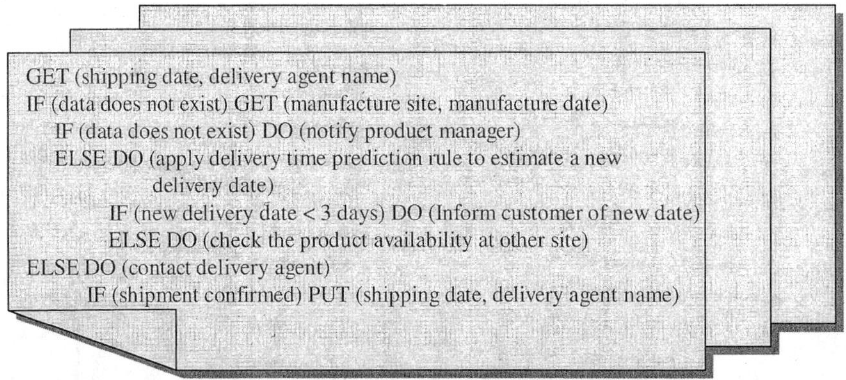

an issue, and (iv) indicate what reaction to cause to compensate for unexpected events or which correction to apply to prevent disadvantageous outcomes.

A sample display by a BOCT is shown below. It provides information in an effective context for managers to manage their businesses.

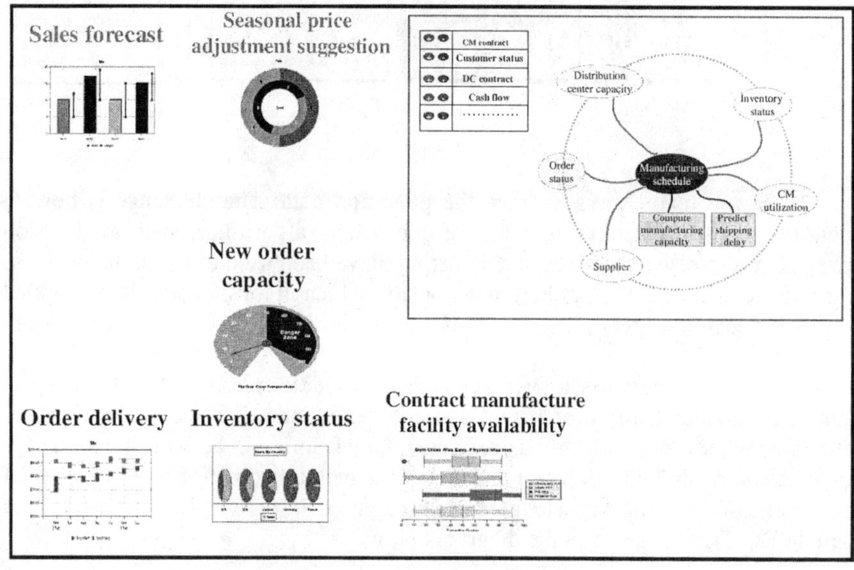

As shown in this example, on one display, business managers can view the base line information (e.g., inventory level and CM available capacity) as well as aggregated information (e.g., the new order taking capacity that is computed based on the base line data). It will also be able to show the auxiliary information calculated by our GBRs (e.g., the sales forecast and suggested seasonal product price). The most interesting part is the business context information display that, based on the appropriate BOCT, will automatically collect all information related to the current situation and present it in the proper means shown below.

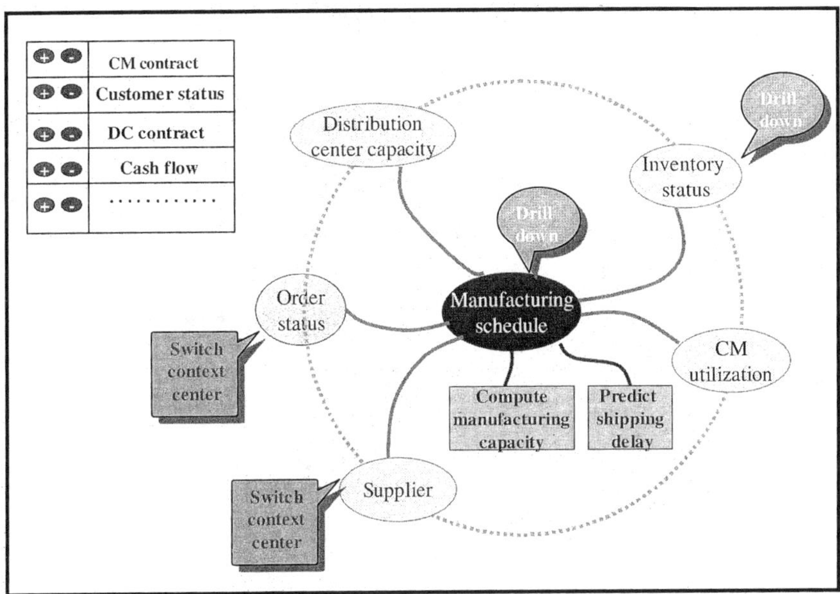

Actions are also indicated so that they can serve as reminders to business managers or be invoked automatically. Information items can be added to or deleted from the display to support individual needs. During different stages of a task, a business manager can switch between different business information contexts supported by different BOCTs to provide the right information and right advice at the right time.

Since the BOCT is a new concept, many associated issues need to be addressed. The first major challenge lies on how to design an appropriate BOCT language. The BOCT language will be quite different from the workflow definition language and the rule engine language. Based upon its execution nature and application, it needs to be very simple but still be able to support all of the basic business operation scenarios. Our first implementation supports a BOCT language consisting of only five key operators (GET, PUT, IF, ELSE, DO). Therefore, it will have tremendous execution speed (since the language is simple) yet still offers the flexibility supporting most BOCT level operation scenarios (since each operator is very generic).

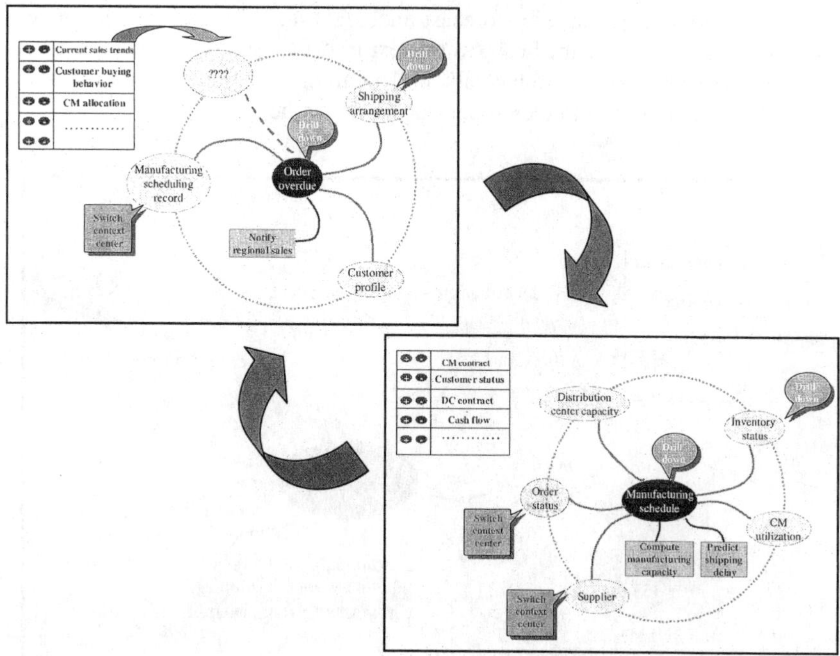

The second major challenge lies in how to design a tool to generate the appropriate data marts supporting the BOCT information context in a real-time fashion. Remember, the size of these data marts will be much smaller than usual and may consist of a few data items in certain cases. However, the requested data may be stored among multiple systems and clearly requires the combination of heterogeneous database query processing technology and data semantics conciliation technology to produce the required mini data marts. In addition, the new data stream management will add another dimension of complexity to the BOCT information context management.

3.4 Business process management

Even though the business process management is not considered as part of an IBOM system, it is worthy to understand its value in an IBOM-based solution.

Without a business process management system to capture the end-to-end operation flow, it will be quite difficult to develop a complete picture regarding the overall operation flow and will increase the difficulty of analyzing the relationships, dependencies, interaction sequences, and consequences among different business activities.

Let's review the case of a RosettaNet PIP-based operation. Usually, multiple RosettaNet PIP transactions need to be carried out in a certain sequence to complete one entire business operation as shown below.

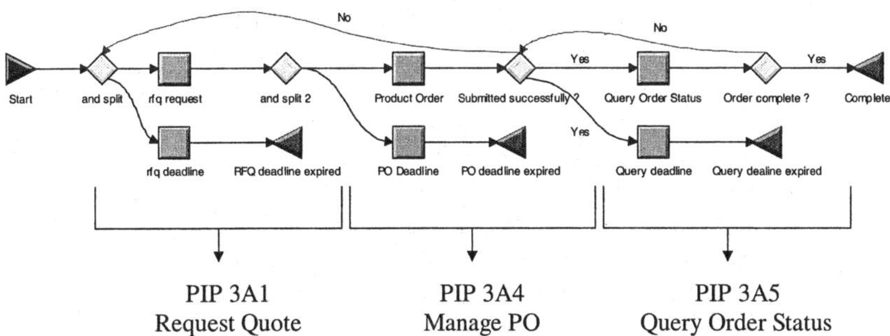

Without a business process system in the background to thread these PIP transactions, the data collected will only be able to present a set of fragmented operation scenarios. This leaves the IBOM system in a very difficult situation to develop an overall understanding of the underlying operation flow and will not be able to accomplish certain analyses. On the other hand, an IBOM solution with a business process management component will provide an overall execution framework of the business operation, thereby facilitating the analysis. Both scenarios are shown below for comparison.

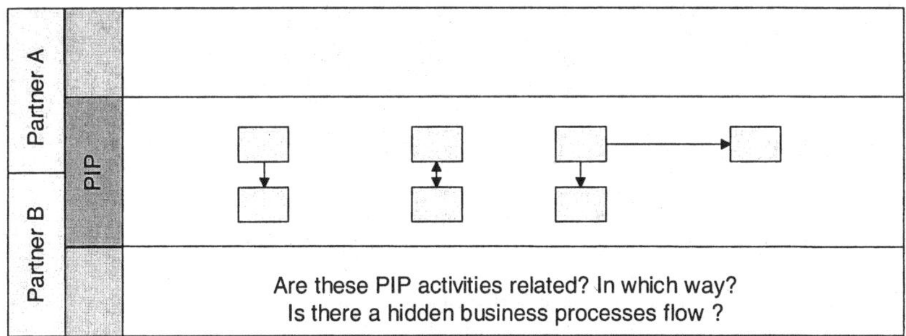

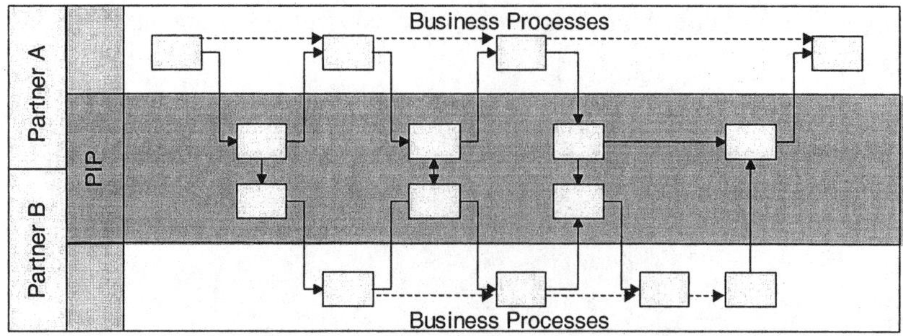

In addition, most of the performance data collected by various operation systems may be focused on their own interests in the operation. It is likely that business managers will need data outside the individual system's interest to perform their tasks. The business process management system provides a facility for IBOM to collect this additional data. For example, at the beginning or completion of each activity within a business process, the invoked application's parameters, return data, and other operation-related data can be logged. This provides an easy way to collect the right set of data in the right context at the right place and time. This will dramatically simplify the context computations for BOCT support.

4 Conclusion

To survive in today's competitive business environment, companies need an IBOM-based operation system to cope with the changing world. The line of response does not stop at the information display. It needs to take action to respond to certain critical situations in a real-time manner. The successful deployment of an IBOM system requires the combination of multiple disciplines, including IT management and business operation management. A long-term goal is to feed back the results to achieve the self-optimization objective. The mechanism we introduced, i.e., the BOCT, is very likely to play a key role as we move along this direction.

References

1. "How do I link IT metrics to business performance?", Technical report, META Group, Inc. Stamford, CT, April 2002.
2. Angela Bonifati, Fabio Casati, Umesh Dayal, and Ming-Chien Shan, "Warehousing Workflow Data: Challenges and Opportunities", Proc. of 27th Int. Conf. on VLDB, Rome, Italy, September, 2001.
3. Erika Brown, "Analyze This", Forbes magazine, August, 2002.

4. Fabio Casati, Umesh Dayal, and Ming-Chien Shan, "Business Process Cockpit", Proc. of 28th Int. Conf. on VLDB, Hong Kong, China, August, 2002.

5. Fabio Casati and Ming-Chien Shan, "Semantic monitoring and analysis of business processes", Proc. of Int. Conf. On EDBT, Prague, Czech Republic, March, 2002. (Lecture Notes in Computer Science, Springer-Verlag.)

6. Fabio Casati, Umesh Dayal, and Ming-Chien shan, "E-Business Applications for Supply Chain Automation: Challenges and Solutions", Proc. of ICDE Conference 2001, Heidelberg, Germany, April, 2001.

7. Fabio Casati and Ming-Chien Shan, "Process Automation as the Foundation for E-Business", Proc. Of the 26th Int. Conf. on Very Large Databases, Cairo, Egypt, September, 2000.

8. Sirish Chandrasekaran, Amol Deshpande, Mickael Franklin, Joseph Hellerstein, Wei Hong, "TelegraphCQ: Continueous Dataflow Processing for an Uncertain World", UCB Technical Memo, September 2002.

9. Weimin Du and Ming-Chien Shan, "Query Processing in Pegasus", Object Oriented Multidatabase Systems, Information Management, edited by Ahmed Elmargarmid, Prentice Hall, 1995.

10. Jane Griffin, "Information strategy: The XIE and zero latency: Partners in the new era of BI", DM Review magazine, January 2002.

11. Daniela Grigori, Fabio Casati, Umesh Dayal, and Ming-Chien shan, "Improving Business Process Quality through Exception Understanding, Prediction, and Prevention", Proc. of 27th Int. Conf. on VLDB, Rome, Italy, September, 2001.

12. Mark Hellinger and Scott Fingerhut, "Business activity monitoring: EAI meets data warehousing", eAI Journal, July 2002.

13. David McCoy, "Business activity monitoring scenario: The new competitive advantage", Gartner report, June 2002.

14. Dennis Powell and Tori Boutte, "Agile e-business", The Business Integrator Journal, Second Quarter, 2002.

15. Philip Russom, "Analytic apps meet BPM", Intelligent Enterprise magazine, September, 2002.

16. Mehmet Sayal, Fabio Casati, Umesh Dayal, and Ming-Chien Shan,"Extending Workflow Technology to support B2B (RosettaNet) interactions", Proc. of ICDE Conference 2002, San Jose, California, February, 2002.

17. Ming-Chien Shan, 'HP Workflow Research: Past, Present, and Future, Proc. of NATO ASI on Workflow systems and Interoperability, Springer-Verlag, Turkey, August, 1997.

18. Ming-Chien Shan, Weimin Du, Jim Davis and William Kent, "The Pegasus project - A Heterogeneous Information Management System", Modern Database Systems, edited by W. Kim, Addison-Wesley Publishing Comp., September 1994.

19. Bob Tedeschi, "Digital Cockpits track a corporation's performance", The New York Times, July, 2002.

Evaluating E-Commerce Projects
Using an Information Fusion Methodology

Kurt J. Engemann

Department of Information and Decision Technology Management, Iona College
New Rochelle, NY 10801, USA
kengemann@iona.edu

Holmes E. Miller

Department of Business Administration, Muhlenberg College
Allentown, PA 18104, USA
homiller@muhlenberg.edu

Ronald R. Yager

Machine Intelligence Institute, Iona College
New Rochelle, NY 10801, USA
yager@panix.com

Abstract. E-commerce is changing the business landscape, but applications carry investment risks. E-commerce applications redefine business models ranging from retailing to trading stocks to originating mortgages. Organizations will be making significant investments in e-commerce applications that may or may not realize their potential payoff. In this paper we propose a methodology to evaluate investments in e-commerce applications. We frame investment issues in the e-commerce environment and present qualitative models to augment quantitative investment analysis. We introduce the power average as an aggregation operator for information fusion in group decision-making in e-commerce project selection. We outline a collaborative methodology to arrive at a group consensus involving individual qualitative assessments. We present a scoring model for individual applications, and a portfolio model for groups of applications.

1. Introduction

Although executives have been making information technology (IT) investment decisions for decades, evaluating IT investments continues to pose problems that concern both the difficulty in evaluating the investments relative to their payoff stream [11] and the effectiveness of the investments themselves relative to business objectives and productivity impact [16]. Evaluating IT investments presents several problems that are not as problematic as when assessing the payoff of more traditional assets. Classical financial investment evaluation methods, such as net present value, require hard numbers, while much of the payoff in IT is indirect and qualitative. Evaluating portfolios of e-commerce projects expands the complexities to additional dimensions because of technological and business uncertainties underlying e-commerce itself [17]. As organizations invest in IT to position themselves in the world of electronic commerce, there is greater desire to measure the payoff of IT investment.

This paper focuses on methods for assessing portfolios of e-commerce projects. The idea of portfolios of IT projects has gained wide acceptance [13], albeit without the quantitative elements associated with financial instruments whose returns are easier to measure [12].

In many instances committees provide input and decide on the development and selection of new information technology. There is a need to combine or fuse the opinions of individuals in a group to reach a consensus regarding the evaluation of a potential application. We describe a new tool, the power average, to aid and provide more versatility in the information fusion process. As we shall describe, the power average is a new measure in which the values being aggregated support or provide power to each other. We use the power average throughout our methodology for evaluating e-commerce project portfolios. We outline a collaborative decision making methodology to aggregate individual qualitative assessments in order to establish group consensus.

2. Dimensions for Assessing E-Commerce Projects

Information technology investment decisions must balance the investment's initial and ongoing costs with present and future payoffs. IT costs are more dynamic than costs of other technologies, and benefits often are difficult to estimate at the time the decision is being made [9]. IT exists in an environment of rapid technological change, and applying traditional investment methodologies may ignore key variables and lead to poor decisions [4, 14]. Despite IT's potential benefits, many IT investments have had mixed reviews regarding their impact on the bottom-line [3] and measuring their effectiveness by relating IT to organizational performance variables is challenging [10, 15].

Specific methodologies for selecting an IT investment often involve a detailed analysis, and differ among organizations. Generally IT payoffs and costs are analyzed along the dimensions of functionality, competitive advantage, and interaction with existing information technologies. Functionality is the most amenable to quantification, and often the only one used when analyzing the IT investment. New capabilities from a competitive perspective are not as evident, and may be more difficult to quantify because of their strategic nature. IT risks often are addressed only as an afterthought to the traditional analysis of IT payoffs and costs. Doing so results in an incomplete perspective. Methodologies for evaluating IT investment should account for IT risk, including risks concerned with maintaining business continuity, and risks associated with information security [5].

At its broadest level, e-commerce can mean any use of electronic technology in any aspect of commercial activity [2,7]. Many technologies can be used in support of e-commerce, highlighting the fact that it is more than just an Internet-based phenomenon. Electronic commerce includes key technologies and procedures including: streamlining processes, interconnectivity, Internet, electronic data interchange, electronic funds transfer, e-mail, security, electronic document management, workflow processing, middleware, bar coding, imaging, smart cards, voice response, and networking [8,19,20].

Four attribute dimensions to consider when evaluating e-commerce investments are customization effects; technology effects; security and legal issues; and operational issues.

Customization Effects - The Internet facilitates customization both in marketing and pricing. One-on-one marketing will become even more prevalent in the future. Examples of one-on-one marketing include web pages personalized to the specific customer visiting the site and advertising tailored to specific individuals. One-on-one marketing will require e-commerce applications to contain increasingly sophisticated software, database, and telecommunications capabilities. In addition, the need to keep current will increase both direct costs and indirect costs including risk while the rewards will be the additional revenues flowing from the one-on-one marketing efforts. One-on-one pricing, such as Internet bidding and on-line auctions, challenges the concept of fixed prices for goods and services. One-on-one pricing may change both the per unit price used to calculate benefits as well as the pattern of demand for those products.

Technology Effects - Technology effects in e-commerce are similar to technology's effects on other IT investments. Direct costs result from hardware, software and telecommunications and direct benefits flow from customer usage. Indirect costs and benefits stem from an uncertain future and are even more immeasurable due the greater uncertainties of e-commerce. The main differences occur in the speed with which the technologies are developed and implemented, and the uncertainties created by the interplay of the technologies and e-commerce customers. The speed of development and speed of service condenses the maintenance and enhancement cycle to make existing applications better.

Security and Legal Issues - E-commerce raises new security concerns, both because of the ease with which potential perpetrators may access e-commerce systems and do harm, and relative lack of security sophistication of many users.

This affects both the likelihood of security incidents and also the perceived impact of those incidents when they do occur. E-commerce applications must be accessible, and that accessibility and ease-of-use creates opportunities for hackers or a cyber-thief. Moreover, the network itself creates the opportunity for widespread damage. In addition to information security, e-commerce applications need to ensure continuity of service. E-commerce is new and the legal environment governing e-commerce transactions still is evolving. Many issues -- for example, intellectual property rights, rules governing commercial transactions, enforcement standards, and privacy and free-speech issues -- are still being formalized.

Operational Issues - E-commerce, while in many respects "electronic," still has "manual" operational elements that must be integrated into a seamless system. In many cases that integration involves a physical distribution of goods. In other cases, the link is electronic with non e-commerce systems, for example, booking airline or hotel reservations, or answering customer e-mail inquiries. The success of the e-commerce business model may hinge more on operational linkages than on the customer-Internet connection. For business-to-business applications this is particularly important. For example, in supply-chain applications requiring just-in-time delivery of parts, manufacturing and distribution must work seamlessly with electronic linkages. Moreover, the increasingly tight expectations for timely service and high quality product, places even more requirements on vendors and blurs supply-chain roles.

E-commerce investments can be evaluated along three dimensions: criticality, quantitative, and qualitative. An application's criticality is determined first, and refers to the degree to which the type of application must be implemented. Classical quantitative ROI methods can be used to evaluate business-to-business and business to consumer e-commerce applications. Quantifiable costs and benefits exist for all four categories discussed above -- for example, software costs for one-on-one marketing and pricing, telecommunications costs for rapid response, reduced item costs from detailed product searches, firewall costs for system security, and costs integrating physical distribution with e-commerce front ends. Each specific e-commerce application would have a roster of costs and benefits that could be quantified.

A characteristic of e-commerce, however, is the wide scope of qualitative elements. Beyond explicit productivity measures, many IT investments create customer value along dimensions such as timeliness, customer service, and product quality [1]. Business performance is separate from productivity and customer value when assessing IT's impact. Some dimensions of business performance involve how e-commerce applications extend a company's market reach, provide a platform for new business opportunities, create higher visibility for the firm and products, and are related to many security and legal issues. Analyzing the qualitative aspects of an e-commerce investment can be linked with more traditional quantitative analysis and an assessment of the investment criticality to form an overall approach. The qualitative assessments can be converted into quantitative ratings that can then be used in a scoring model.

Often groups do such evaluations, where individual evaluations need to be combined. In the next section we introduce the power average as a tool useful in fusing such information.

3. Information Fusion using the Power Average

In most organizations, people collaborate in making major decisions. This is particularly true with information systems steering committees evaluating potential e-commerce applications. Collaborative computing technologies support group decision making in different ways, enabling teams to share opinions, data, information, knowledge, and other resources. In many instances groups need to combine or fuse the opinions of individuals in the group to reach a consensus. The method of information fusion, which is used, plays a key role in the final deliberation of the group. For example, in combining individual ratings that represent evaluations on the worthiness of a potential project, a group may use any of several measures of central tendency (e.g. mean, median, mode). The method of information fusion chosen clearly directly influences the result, and each method possesses characteristics that may make its application appealing as well as disadvantageous.

We describe a tool, the power average, to aid and provide more versatility in the information fusion process. As we shall describe, the power average is a new measure in which the values being aggregated support or provide power to each other. We propose the power average as a useful tool in aggregating individual assessments of e-commerce applications both in an evaluation model for a single project and in a portfolio model for a group of projects.

The Power Average (P-A) is an aggregation operator that takes a collection of n values and provides a single value. The P-A operator is defined [18] as follows:

$$P\text{-}A(a_1, ..., a_n) = \frac{\sum_{i=1}^{n} (1 + T(a_i))a_i}{\sum_{i=1}^{n} (1 + T(a_i))}$$

where $T(a_i) = \sum_{j \neq i} \mathrm{Sup}(a_i, a_j)$.

Sup(a,b) is defined as the support for element a from element b. Typically we shall assume that Sup(a, b) satisfies the following three properties:

1) $\mathrm{Sup}(a, b) \in [0, 1]$
2) $\mathrm{Sup}(a, b) = \mathrm{Sup}(b, a)$
3) $\mathrm{Sup}(a, b) \geq \mathrm{Sup}(x, y)$ if $|a - b| \leq |x - y|$.

Thus closer values provide more support for each other. It is convenient to denote

$$V_i = 1 + T(a_i) \quad \text{and} \quad w_i = \frac{V_i}{\sum_{j=1}^{n} V_j}.$$

The w_i are a proper set of weights, $w_i \geq 0$ and $\Sigma_i \, w_i = 1$. Using this notation we have:

$$P\text{-}A(a_1, ..., a_n) = \Sigma_i \, w_i \, a_i,$$

Therefore the power average is a weighted average of the a_i. Note, however, that it is a non-linear weighted average because the w_i depends upon the arguments.

The power average is a generalization of the mean. If $Sup(a, b) = 0$ for all a and b, then

$$P\text{-}A(a_1, ..., a_n) = \Sigma_i \, a_i/n.$$

More generally when all the supports are the same, the power average is equivalent to the mean. If $Sup(a, b) = k$ for all a and b, then

$$T(a_i) = k \, (n - 1) \text{ for all i and } P\text{-}A(a_1, ..., a_n) = \Sigma_i a_i/n.$$

The power average is commutative; it doesn't depend on the indexing of the arguments. The power average remains the same for any permutation of the arguments. The power average is bounded since $w_i \geq 0$ and $\Sigma_i \, w_i = 1$. Thus:

$$Min_i \, [a_i] \leq P\text{-}A(a_1, ..., a_n) \leq Max_i[a_i]$$

Therefore the power average is idempotent, because if $a_i = a$ for all i, then $P\text{-}A(a_1,..., a_n) = a$. A property typically associated with an averaging operator that is not generally satisfied by the power average is monotonicity because the w_i depend upon the arguments. Monotonicity requires that if $a_i \geq b_i$ for all i then $P\text{-}A(a_1, ..., a_n) \geq P\text{-}A(b_1, ..., b_n)$. The fact that monotonicity is not generally satisfied by the power average makes its use appealing in certain situations. For example, if an argument became too large its effective contribution to the power average could diminish (such as an outlier) and the power average could decrease.

The power average has some of the characteristics of the mode operator. We recall that the mode of a collection of arguments is equal to the value that appears most frequently. We note that the mode is bounded by the arguments and commutative, and is not generally monotonic. For example: $Mode(1, 1, 3, 3, 3) = 3$ and $Mode(1, 1, 4, 5, 6) = 1$, here we increased three arguments and the mode decreased. In the case of the mode we are not aggregating values, rather the mode must be one of the arguments. In the case of power average we are allowing blending of values.

We shall consider some useful parameterized formulations for expressing the Sup function. The determination of the values of the parameters may require the use of some learning techniques.

A binary Sup function can be expressed as follows: if $|a-b| \leq d$ then $Sup(a, b) = K$ and if $|a-b| > d$ then $Sup(a, b) = 0$. A natural extension of this is a partitioned type support function. Let K_i for $i=1$ to p be a collection of values such that $K_i \in [0, 1]$ and where $K_i > K_j$ if $i < j$. Let d_i be a collection of values such that $d_i \geq 0$ and where $d_i < d_j$ if $i < j$. We now can define a support function as:

$$\text{if } |a - b| \leq d_1 \text{ then } Sup(a, b) = K_1$$
$$\text{if } d_{j-1} < |a - b| \leq d_j \text{ then } Sup(a,b) = K_j \qquad \text{for } j = 2 \text{ to } p-1$$
$$\text{if } d_p - 1 < |a - b| \text{ then } Sup(a, b) = K_p$$

Inherent in the above type of support function is a discontinuity as we move between the different ranges.

One useful form for the Sup function that provides a continuous transition, is:

$$Sup(a, b) = K \, e^{-\alpha(a-b)^2}$$

where $K \in [0,1]$ and $\alpha \geq 0$. We easily see that this function is symmetric and lies in the unit interval. K is the maximal allowable support and α is acting as an attenuator of the distance. The larger the α the more meaningful differences in distance. We note here that $a = b$ gives us $Sup(a,b) = K$, and as the distance between a and b gets large then $Sup(a, b) \to 0$.

Using this form for the support function we have

$$P\text{-}A(a_1, ..., a_n) = \frac{\sum_{i=1}^{n} (1 + T(a_i))a_i}{\sum_{i=1}^{n} (1 + T(a_i))}$$

where $T(a_i) = \Sigma_{j \neq i} K \, e^{-\alpha(a_i - a_j)^2}$. Denoting $V_i = 1 + T(a_i)$ we express:

$$P\text{-}A(a_1, ..., a_n) = \Sigma_i w_i \, a_i$$

where $w_i = = \dfrac{V_i}{\sum_{j=1}^{n} V_j}$.

We will now apply the power average in an evaluation model of e-commerce applications.

4. Evaluation of Qualitative Dimension

We suggest a model for the qualitative dimension of an e-commerce project that relates the *attributes* of customization effects, technology effects, security and legal issues, and operational issues with the *customer and business value categories* of customer service, product/service quality, timeliness, market reach, new business opportunities, visibility, and business risk [6]. A score representing the qualitative dimension for a potential application would be determined by having a group of knowledgeable people first discuss the application, then having the individuals provide scores, and then combining the scores using the power average. Potential applications would be evaluated for each *customer and business value category* considering relevant *attributes* of the application. For example, for a retail customer interface application, some discussion points related to *customer service* include:

Customization effects
- menus tailored to customer purchases
- screen appearance user dependent
- customized pricing

Technology effects
- customer technology not constrain use
- available to broad range of user technology
- advanced technology utilized

Security and legal issues
- chance of customer fraud
- credit card information confidential
- personal purchase history confidential

Operational issues
- link to inventory system
- track order status
- on-line queries about product quality

Each member of the evaluating committee would provide a rating between 0 and 20 (best), and the individual committee member's scores would be combined using the power average. This would be done for all customer *and business value categories*. Table 1 illustrates a committee's resulting scores for three potential applications.

Table 1: Scoring of Qualitative Factors

Customer & Business Value Categories	Weight	Application 1	Application 2	Application 3
Customer service	.25	7	15	12
Product/service quality	.20	10	13	14
Timelines	.15	4	17	8
Market reach	.10	8	15	4
New business opportunities	.10	7	17	14
Visibility	.05	9	11	8
Business risk	.15	8	11	10

We propose a collaborative methodology to arrive at a group consensus involving individual qualitative assessments. The power average provides a useful method for determining an MIS steering committee's consensus value for each element of Table 1. To arrive at the ratings we propose a methodology in which each committee member individually provides scores and a group support system automatically calculates the committee's consensus, using the power average as the aggregation operator. The normalized weights are also determined using the power average.

The power average's non-monotonic behavior makes the operator useful when the values being aggregated provide a degree of support to each other. This is ordinarily the way committees operate, with individuals receiving various level of support. Consider the power average of ten elements, five of which are ten's and five of which are five's. In this case the simple average is 7.5 and for any choice of K and α the power average also is 7.5.

Figure 1 displays the non-monotonic behavior of the power average as we change one of the values originally equal to 10. For illustrative purposes we used K = 1 and α = 0.3. We see that as we decrease the value, it joins the cluster of fives, decreasing the P-A faster than the simple average. In the case of increasing the value, initially the power average begins to decrease, exhibiting non-monotonicity, instead of increasing as does the simple average. This decrease is a reflection of the fragmentation of the cluster at 10, it is losing its power because it lost a member and the cluster at five has gained in power more than compensating for the increase in value. This decreasing in the P-A continues as we increase the element until it reaches fourteen at which time we see a reversal and now the P-A starts increasing   At this point the increase in value begins overcoming the loss of power. This non-monotonic behavior makes the power average an appealing aggregation operator for combining individual committee member's scores.

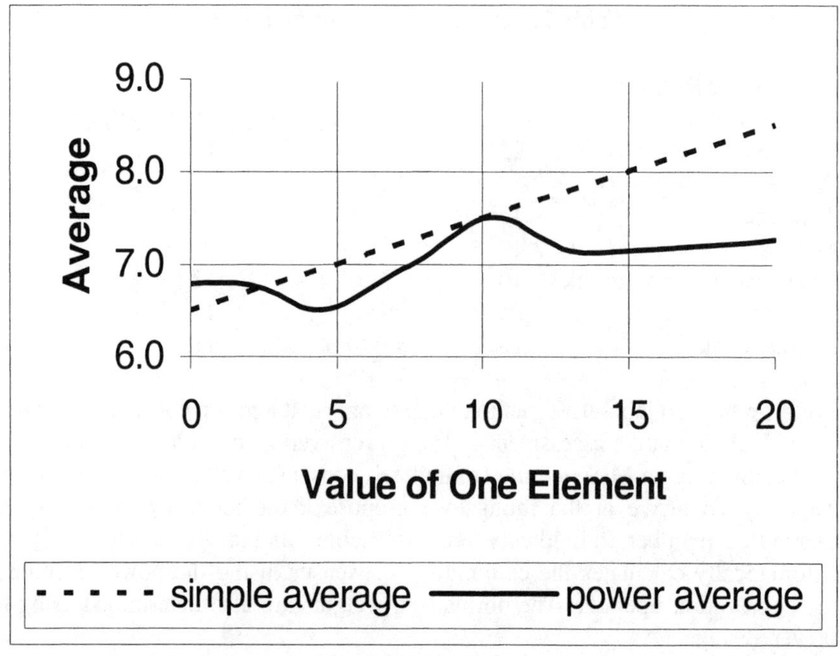

Fig. 1. Non-monotonicity of the Power Average

To obtain a score for project i representing its qualitative factors, let:

P_{ij} = score for project i for *customer and business value category* j,

W_i = weight for *customer and business value category* i, where $0 \leq W_i \leq 1$ and $\Sigma_i\ W_i = 1$,

$S_i = \Sigma_j W_i P_i$ = score for project i.

In this illustration $S_1 = 7.\,5$, $S_2 = 14.3$, $S_3 = 10.7$.

For each project being evaluated, the value S_i would be calculated and used in conjunction with results from the criticality and quantitative financial analyses to arrive at a decision regarding the e-commerce investment.

5. Optimization Model for a Project Portfolio

Managers are interested in the portfolio of projects of all e-commerce applications comprising the business in addition to evaluating individual applications. Mathematical programming can be used to develop a model to optimally select a portfolio of projects. We propose a model that seeks to maximize the aggregate sum of the qualitative dimension scores for the projects, subject to constraints that specify:

- One project of each critical "type" must be selected, where for example "type" may be order tracking, which may be accomplished by means of any one of several applications
- Budget constraints are satisfied
- The aggregate net present value of the returns from the project portfolio must exceed some specified floor limit N.

We use the following notation for the model:

S^k_i = the qualitative dimension score associated with implementing project i of type k. This is obtained via weighted ratings based on applying the power average procedures discussed above,

x^k_i =1 if project i of type k is implemented, otherwise 0,

c^k_i = the cost of project i of type k,

B = budget constraint for total of all projects

N^k_i = the net present value of implementing project i of type k

N = minimum aggregate net present value

The mathematical program follows:

$$\text{Maximize } \Sigma_k \Sigma_i S^k_i x^k_i$$

Subject to:

$$\Sigma_i x^k_i = 1, \quad \text{for all } k$$
$$\Sigma_k \Sigma_i c^k_i x^k_i \leq B$$
$$\Sigma_k \Sigma_i N^k_i x^k_i \geq N$$
$$x^k_i = 0,1.$$

Table 2 gives the data and optimal results for a sample problem containing seven types of projects and three alternatives for each type. For example, alternative 1 of project type 1 has a qualitative dimension score of 7.5, and project alternatives 2 and 3 have scores of 14.3 and 10.7, respectively.

		Project Alternative		
		1	2	3
	1	7.5	14.3	10.7
Project	2	15.5	11.8	9.6
Type	3	5.2	7.3	6.4
	4	4.1	8.8	5.9
	5	10.9	16	16.8
	6	8.8	5.6	9.4
	7	10.5	9.7	8.6

QUALITATIVE PROJECT DIMENSION SCORE

		Project Alternative		
		1	2	3
	1	0	1	0
Project	2	1	0	0
Type	3	0	1	0
	4	0	1	0
	5	0	1	0
	6	0	0	1
	7	1	0	0

1 IF PRJT i OF TYPE k SELECTED
0 OTHERWISE

		Project Alternative		
		1	2	3
	1	2.7	4.9	6.3
Project	2	8.4	8.7	9.9
Type	3	5	6.5	9.1
	4	2	6	4
	5	2.5	7.7	8.2
	6	3	3.6	4.2
	7	7	9	9

PROJECT COSTS

		Project Alternative		
		1	2	3
	1	7.5	8.3	11
Project	2	7.9	12.5	11.4
Type	3	9.5	10.8	10.9
	4	8.1	11.4	12
	5	13.6	16.9	15
	6	9	12	15
	7	6.5	18.3	13

PROJECT NPV

Used Budget	44.7
Allowed Budget	45

Solution NPV:	76.8
Minimum NPV	75

The optimal solution obtained via Excel Solver:

Do project 1 of types 2 and 7

Do project 2 of types 1, 3, 4 and 5

Do project 3 of type 6

Optimal objective function value: 81.8

TABLE 2: DATA AND RESULTS:
SAMPLE PORTFOLIO OPTIMIZATION

6. Conclusion

In evaluating e-commerce investments, the application's criticality, quantitative performance and qualitative performance should be considered. In this paper we presented a modeling approach to support this analysis. Qualitative aspects are particularly important in e-commerce environments. We presented a scoring model to help determine the qualitative value of individual projects. We introduced the power average as an aggregation operator, which allows argument values to support each other. We applied the power average to a collaborative group decision-making situation in which individual evaluations of e-commerce projects were aggregated. We presented an optimization model in e-commerce environments to select applications for project portfolios. Future steps include refining the methodologies and testing them on applications occurring in practice.

7. References

1. Brynjolfsson E. (1994). Technology's true payoff. Information Week, October 10, 34-36
2. Carter J (2002), Developing E-Commerce Systems, Prentice-Hall, Upper Saddle River. NJ
3. Chircu A, Kauffman R. (2000). Limits to value in electronic commerce-related IT investments. Journal of Management Information Systems, vol 17, no 2, pp 59-80, Fall-2000
4. Dos Santos B; Sussman L (2000). Improving the return on IT investment: the productivity paradox. International Journal of Information Management, vol 20, no 6, pp 429-440, Dec. 2000
5. Engemann K, Miller H (1999). Evaluating information technology investment: a methodology for managing risk. Measuring Information Technology Investment Payoff: Contemporary Approaches, (ed. M. Mahmood and E. Szewczak) Idea Group Publishing, pp. 321-342
6. Engemann K, Miller H, Yager R (2001). Information fusion in group support systems using the power average. Advances in Decision Technology and Intelligent Information Systems, Volume II (ed. K. Engemann and G. Lasker), pp. 7-11, The International Institute for Advanced Studies in Systems Research and Cybernetics, Windsor, Canada
7. Fingar P, Kumar H, Sharma T (2000). Enterprise E-Commerce. Meghan-Kiffer Press: Tampa, Fl
8. Finin T, Grosof B (1999). Artificial Intelligence for Electronic Commerce: Papers from the AAAI Workshop. AAAI Press, Menlo Park, CA.
9. Irani Z, Love P (2000/2001). The propagation of technology management taxonomies for evaluating investments in information systems. Journal of Management Information Systems, vol 17, no 3, pp 161-176, Winter 2000/2001
10. Kayworth T, Chatterjee D, Sambamurthy V (2001). Theoretical justification for IT infrastructure investments. Information Resources Management Journal, vol 14, no 3, pp 5-14, Jul/Sep-2001

11. Mahmood M, Mann G (1993). Measuring the organizational impact of information technology investment: an exploratory study. Journal of Management Information Systems, Summer, vol 10, no 1, pp 97-122
12. Markowitz H (1952). Portfolio selection. Journal of Finance, 7, pp 77-91, March 1952
13. McFarlan F (1981). Portfolio approach to information systems. Harvard Business Review, September-October 1981
14. Riggins F (1999). A framework for identifying web-based electronic commerce opportunities. Journal of Organizational Computing and Electronic Commerce, vol 9, no 4, pp 297 310
15. Shafer S, Byrd T (2000). A framework for measuring the efficiency of organizational investments in information technology using data envelopment analysis. Omega (Oxford), vol 28, no 2, pp 125-141, Apr. 2000
16. Tzu-Chuan C; Dyson R (2000). Managing strategic IT investment decisions: from IT investment intensity to effectiveness. Information Resources Management Journal, vol 13, no 4, pp 34-44, Oct./Dec.-2000
17. Turban E, Lee J, King D, Chung H (2000). Electronic Commerce: A Managerial Perspective. Prentice Hall: Upper Saddle River, NJ
18. Yager R (2001). The power average operator. IEEE Transactions on Systems, Man and Cybernetics Part A, 31, pp 724-730
19. Yager R (2000). Targeted e-commerce marketing using fuzzy intelligent agents. IEEE Intelligent Systems, Nov/Dec, pp 42-45
20. Yager R, Pasi G (2001). Product category description for web-shopping in e-commerce. International Journal of Intelligent Systems vol 16, pp 1009-1021

Dynamic Knowledge Representation for *e*-Learning Applications

M. E. S. Mendes and L. Sacks

Department of Electronic and Electrical Engineering, University College London, Torrington Place, London WC1E 7JE, UK

Abstract. This chapter describes the use of fuzzy clustering for knowledge representation in the context of *e*-Learning applications. A modified version of the *Fuzzy c-Means* algorithm that replaces the Euclidean norm by a dissimilarity function is proposed for document clustering. Experimental results are presented, which show substantial improvement in the modified algorithm, both in terms of computational efficiency and of quality of the clusters. The robustness of this fuzzy clustering algorithm for document collections is demonstrated and its use for on-line teaching and learning in a complete end-to-end system is explored.

1. Introduction

In recent years, network-based teaching and learning (*e*-Learning) has become widespread, with bespoke solutions by individual institutions and standardizing initiatives for learning technologies [14]. The work presented in this chapter is in relation with the CANDLE[1] project, a European Commission funded project that is developing an *e*-Learning service mainly focused on providing a flexible learning environment and on delivering sharable and reusable course material in the area of Telematics, within a standards based framework.

---

[1] <u>C</u>ollaborative <u>A</u>nd <u>N</u>etwork <u>D</u>istributed <u>L</u>earning <u>E</u>nvironment: http://www.candle.eu.org/

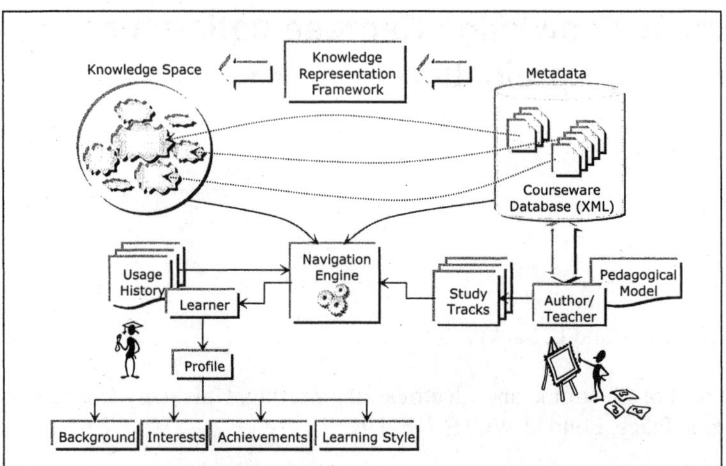

Fig. 1. Adaptive knowledge-based content navigation for flexible *e*-Learning

Although *e*-Learning systems lack face-to-face interaction between teachers and students, they have the major advantage of enabling people to access learning facilities regardless of their location and whenever it is most convenient to them. Large networked repositories of learning material may be accessed by students, but it is necessary to narrow down the available resources to a particular individual based on the learning context, *i.e.*, to take into account learning objectives, pedagogical approaches and profile of the individual learner. This may range from relatively rigid training objectives through to exploratory or research oriented interactions [25] - the system should be flexible enough to accommodate these different learning scenarios. Towards the more exploratory end of the learning spectrum, tools such as an adaptive navigation engine are required to determine which documents are the most relevant for a given student, who wants to learn a particular subject.

In Fig.1, a framework for adaptive knowledge-based content navigation is depicted. The student's exploration of the content database is constrained by the pedagogical approach defined by the teacher. In a more instructional approach there would be a low degree of freedom as the student would have to follow a specific sequence of links. More flexible approaches should also be supported. However, giving the freedom to explore the whole database would not be a good option because of the ease with which learners can stray away from the original learning goals. A solution is to have a group of related material organized in an abstract knowledge space so that when the student wants to learn more about a given topic, the navigation engine can follow the knowledge space relationships between documents to derive the links relevant to the students' interests and objectives.

To achieve this, there needs to be a way to classify and organize learning materials in terms of knowledge domains. The proposed approach to this problem is to use fuzzy clustering to dynamically discover document relationships. Such approach is explored in the section 1.2. The remaining sections of this chapter describe the background and experiments with fuzzy clustering for knowledge representation. In section 1.3, details about document representation schemes, distance functions, clustering algorithm and performance evaluation measures are presented. A modified version of the *Fuzzy c-Means* [4] clustering algorithm is also introduced. In section 1.4, the experiments with fuzzy clustering are presented and analyzed. The integration of the dynamic knowledge representation with adaptive *e*-Learning systems is described in section 1.5, and the conclusions are presented in section 1.6.

2. Fuzzy Clustering for Knowledge Representation

An abstract knowledge space can be built in several ways. A popular emerging approach is to develop an ontology of the domain, defining a set of concepts and relations between those concepts. In CANDLE, each piece of course material is tagged with metadata according to an extended version of IEEE LOM (Learning Objects Metadata) model [13]. Metadata is descriptive information about resources such as: title, authors, keywords, technical requirements, conditions of use, and such like. CANDLE's metadata model includes a category to classify courseware according to a pre-defined taxonomy (concepts) to form an ontology of the Telematics domain. By tagging each piece of content with specific concepts, the relationships defined in the ontology can be used to locate related course material. This approach is a major feature of the more general enterprise of the Semantic Web [2,11].

The rich semantic information captured by the ontology facilitates the search of material both for authors and learners. Authors may start from one point in the ontology and follow the appropriate links to locate material with the relationships required for their new courses. The learner may use the ontology for navigation and possibly automated location of content – by following a set of relationships relevant content can be located. The key question with this approach is which ontology to use. Two problems can be foreseen. On the one hand, different experts in a given field are likely to disagree on the correct ontology. On the other hand, in fields like engineering or telematics the true ontology quickly changes through time as the fields develop. Hence, the deployment and maintenance efforts are costly. Similar limitations have been pointed out in the context of the Semantic Web [8]. This provides the motivation to develop a process for dynamic ontology discovery.

Our proposal is to employ fuzzy clustering techniques for discovering the underlying knowledge structure and identifying knowledge-based relationships between learning materials. Fuzzy clustering methods fit in the broad area of pattern recognition. The goal of every clustering algorithm is to group data elements according to some similarity measure so that unobvious relations and structures in the data can be discovered. Information retrieval is one of the fields where clustering techniques have been applied, with the objective of improving search and retrieval efficiency. The use of clustering in this area is supported by the so-called *cluster hypothesis* [24], which assumes that documents relevant to a given query tend to be more similar to each other than to irrelevant documents and hence are likely to be clustered together. Clustering has also been proposed as a tool for browsing large document collections [9] and as a post-retrieval tool for organizing Web search results into meaningful groups [34]. The organization of online learning material into knowledge domains can be seen as a document clustering problem, as every piece of content is represented as an XML document [5], containing the descriptive metadata. Clustering algorithms can thus be used to automatically discover, sometimes unobvious, knowledge-based content relations from the metadata descriptions.

Agglomerative hierarchical clustering algorithms are perhaps the most popular for document clustering [31]. Such methods have the advantage of providing a hierarchical organization of the document collection but their time complexity is problematic in view of partitional methods such as the *k-Means* algorithm [19] (also quite used for document clustering). These types of algorithms generate hard clusters in the sense that each document is assigned to a single cluster. With fuzzy clustering multiple memberships in different clusters are allowed. Such algorithms have not been widely explored for document clustering although a study in [16] indicates that the *Fuzzy c-Means* algorithm can perform at least as well as the traditional agglomerative hierarchical clustering methods.

Given that the goal is to discover the best representation for the actual ontology of the domain, fuzzy clustering techniques offer substantial advantages over their hard clustering counterparts. And the reasons are the following:

- One of the limitations with pre-defined ontologies is the need for consensual definition of the right ontology. For most systems, knowledge can be ambiguous, can vary between experts and evolves through time. Therefore, a fuzzy knowledge space is more likely to represent the "true" underlying knowledge structures.
- The process of tagging learning content with taxonomic terms is the author's attempt to define the subjects associated with each material. This classification process is intrinsically imprecise, and may differ between authors in their understanding of the keywords and relationships between parts of the system.

The concept behind fuzzy clustering is that of fuzzy logic and the theory of fuzzy sets [33], which provides the mathematical means to deal with uncertainty. Fuzzy clustering brings together the ability to find relationships and structure in data sets with the ability to handle uncertainty and imprecision.

3. Background for the Clustering Experiments

The relevant background for the fuzzy clustering experiments is given in this section. Each step of the knowledge discovery process is depicted in Fig. 2.

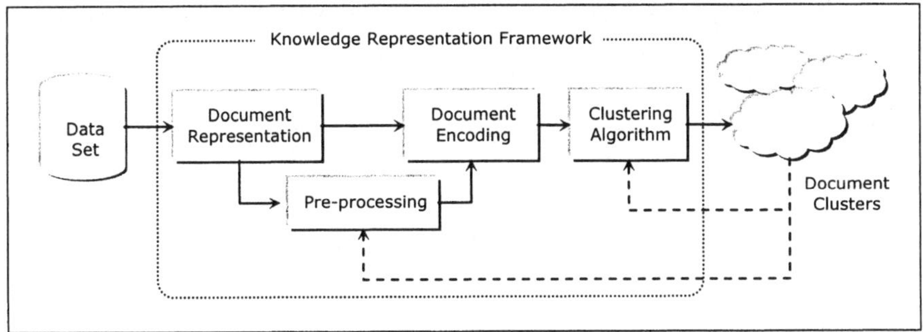

Fig. 2. Knowledge representation framework

Given a test document collection each document has to be numerically encoded for handling by the clustering algorithm. A suitable document representation has to be generated and some pre-processing may be applied for feature selection. The discovered document relationships are then analyzed to assess the performance of the clustering algorithm. Each of these steps is described in the following sub-sections.

3.1 Document Representation

In order to apply a clustering algorithm to a document collection, there needs to be an appropriate document representation scheme. In the classical models of information retrieval (IR) documents are represented by a set of indexing terms in the form of k-dimensional vectors [1]:

$$x_i = [x_{i1}\ x_{i2}\ ... x_{ik}]\qquad(1.1)$$

where k is the total number of terms and x_{ij} represents the weight of term j in document x_i. The indexing terms are usually obtained through automatic indexing and they represent the words contained in the document collection after stop words removal (*i.e.* discarding insignificant words like 'a', 'and', 'where', 'or') and stemming (*i.e.* the removal of word affixes such as 'ing', 'ion', 's') [23,26].

3.2 Pre-processing

In general the total number of indexing terms for a given document collection will be very high. Some of those terms are usually insignificant for document retrieval and also for clustering purposes and they can be excluded by pre-processing, thus reducing the dimensionality of the problem space.

The significance of a term is usually assessed based on its inverse document frequency (IDF), *i.e.* the number of documents that contain the term. *Entropy* and *specificity* are two examples of significance measures [26]. The *entropy* measure is derived from Claude Shannon's Information Theory [28], with the premise that the more informative a term is the lower its *entropy*. *Entropy* is defined as:

$$H_j = \sum_{i=1}^{N} p_{ij} \cdot \log_2 \frac{1}{p_{ij}}, \quad \text{with } p_{ij} = \frac{f_{ij}}{F_j} \text{ and } F_j = \sum_{i=1}^{N} f_{ij} \qquad (1.2)$$

where p_{ij} is the probability of term j appearing in document i and F_j is the sum of the frequencies of term j (f_{ij}) in all documents (N). Terms that appear in most documents exhibit high *entropy* and may be regarded as noise in the clustering process and thus, they may be discarded.

The *specificity* measure identifies terms that only appear in a small percentage of documents (very specific terms) or terms that appear in almost every document (too general). Such terms are considered irrelevant and can be discarded. The *specificity* of a term is defined as follows:

$$sp = \log(N / n_j) \qquad (1.3)$$

where N is the total number of documents in the collection and n_j the number of documents containing term j. These measures provide metric scales for keeping or discarding indexing terms. The sensitivity of the pre-processing will be analyzed in depth in section 1.4.2.

3.3 Document Encoding

The term weights x_{ij} that compose the document vectors as in Eq. (1.1) can be obtained in various ways. In the *Boolean model* of IR binary term weights are used, *i.e.* $x_{ij} \in \{0,1\}$, that just acknowledge the presence or absence of terms in the document. Alternatively, in the *Vector model* of IR the weights are a function of the terms frequency [1]. This model has been selected for document encoding discussed here.

In the *Vector model*, the simplest term weighting scheme, know as *tf*, uses the term frequencies as weights – $x_{ij} = f_{ij}$. Another scheme that has proved to perform well, attributes high weights to terms that occur frequently in a small number of documents (*i.e.* terms that have high specificity) [27]. Such weighting scheme is known as *tf×idf*. The weights are usually given by the formula in Eq. (1.4) or variations of this formula.

$$x_{ij} = f_{ij} \cdot \log(N / n_j) \qquad (1.4)$$

Longer documents contain more words and consequently higher term weights. But in IR systems, documents with different lengths should have equal chances of being retrieved. For this reason, vector length normalization is usually applied (*i.e.* $\|x_{ij}\|=1$). For clustering applications, such normalization should also be made.

3.4 Clustering Algorithm

Considering the requirements of the knowledge representation framework, the *Fuzzy c-Means* algorithm (FCM) [4] is used. It is perhaps the most popular fuzzy clustering technique and it generalizes a popular document clustering algorithm - the hard *k-Means* algorithm [19]. The later produces a crisp partition of the data set whereas the former generates a fuzzy partition: documents may be assigned to several clusters simultaneously. A description of the algorithm follows.

Fuzzy c-Means Clustering Algorithm

Given a data set with N elements each represented by k-dimensional feature vector, the *Fuzzy c-Means* algorithm takes as input a ($N{\times}k$) matrix X=$[x_i]$. It requires the prior definition of the final number of clusters c ($1<c<N$), the choice of the fuzzification parameter m ($m>1$) and the selection of a distance function $\|\cdot\|$, the most common being the Euclidean norm.

The algorithm runs iteratively to obtain the cluster centers (or prototypes) – V=$[v_\alpha]$: ($c{\times}k$) – and a partition matrix – U=$[u_{\alpha i}]$: ($c{\times}N$) – which contains the membership of each data element in each of the c clusters. Both the cluster centers and the partition matrix are computed to minimize the following objective function:

$$J_m(U,V) = \sum_{i=1}^{N}\sum_{\alpha=1}^{c} u_{\alpha i}{}^m \left\| x_i - v_\alpha \right\|^2 = \sum_{i=1}^{N}\sum_{\alpha=1}^{c} u_{\alpha i}{}^m d_{i\alpha}{}^2 \qquad (1.5)$$

The FCM algorithm starts with a random initialization of the partition matrix subject to the following constraints:

$$1.\ u_{\alpha i} \in [0,1],\ \forall_{\alpha \in \{1,\dots,c\}}\ \forall_{i \in \{1,\dots,N\}} \qquad (1.6)$$

$$2.\ \sum_{\alpha=1}^{c} u_{\alpha i} = 1,\ \forall_{i \in \{1,\dots,N\}} \qquad (1.7)$$

$$3.\ 0 < \sum_{i=1}^{N} u_{\alpha i} < N,\ \forall_{\alpha \in \{1,\dots,c\}} \qquad (1.8)$$

At each iteration t the grades of membership and the cluster centers are updated according to Eqs. (1.9) and (1.10) respectively:

$$u_{\alpha i} = \sum_{\beta=1}^{c} \left(\frac{\left\| x_i - v_\alpha \right\|^2}{\left\| x_i - v_\beta \right\|^2} \right)^{-\frac{1}{(m-1)}} = \sum_{\beta=1}^{c} \left(\frac{d_{i\alpha}{}^2}{d_{i\beta}{}^2} \right)^{-\frac{1}{(m-1)}} \qquad (1.9)$$

$$v_\alpha = \sum_{i=1}^{N} u_{\alpha i}{}^m \cdot x_i \Big/ \sum_{i=1}^{N} u_{\alpha i}{}^m \qquad (1.10)$$

The algorithm ends when the termination criterion is met ($\| U^{(t+1)} - U^{(t)} \| < \varepsilon$) or the maximum number of iterations is achieved ($t+1 > t_{max}$).

Selection of the Distance Function

Most clustering algorithms group related data elements based on some notion of distance/proximity between elements. Hence, the choice of a particular distance function should reflect the nature of the data set.

Although Euclidean distance is usually applied in the FCM algorithm, it is not the most suitable for measuring the proximity of documents vectors, which tend to be very sparse and high-dimensional. To justify this statement we give the following example. Let us consider two documents x_A and x_B that are indexed with a set of k terms T. Let us also assume that most terms in T, say k', appear neither in x_A nor in x_B. Also that x_A and x_B have no terms in common. Since the two document vectors agree in k' dimensions in which they both have zero term frequencies, their Euclidean distance may be relatively small when in fact x_A and x_B are totally dissimilar. Hence, the problem with the Euclidean norm is that the non-occurrence of the same terms in both documents is handled in similar way as the co-occurrence of terms.

In the *Vector model* of information retrieval the degree of similarity between query vectors and document vectors is usually evaluated as the inner product of the two vectors [1]:

$$S(x_\alpha, x_\beta) = \langle x_\alpha, x_\beta \rangle = \sum_{j=1}^{k} x_{\alpha j} \cdot x_{\beta j} \qquad (1.11)$$

Considering the case when both x_α and x_β are normalized to unit length (*i.e.* $\|x_\alpha\| = \|x_\beta\| = 1$) this similarity function S represents the cosine of the angle between the document vector x_α and the query vector x_β - reason why it is also referred to as the *cosine measure*. This measure exhibits the following properties:

$$0 \leq S(x_\alpha, x_\beta) \leq 1, \forall_{\alpha,\beta} \qquad (1.12)$$

$$S(x_\alpha, x_\alpha) = 1, \forall_\alpha \qquad (1.13)$$

A simple transformation in Eq. (1.11) can be made to obtain a dissimilarity function D:

$$D(x_\alpha, x_\beta) = 1 - S(x_\alpha, x_\beta) = 1 - \sum_{j=1}^{k} x_{\alpha j} \cdot x_{\beta j} \qquad (1.14)$$

In this case,

$$0 \leq D(x_\alpha, x_\beta) \leq 1, \forall_{\alpha,\beta} \qquad (1.15)$$

$$D(x_\alpha, x_\alpha) = 0, \forall_\alpha \qquad (1.16)$$

Referring back to the example of the two documents, x_A and x_B, this function would result in the maximum value, *i.e.* $D(x_A, x_B) = 1$, indicating total dissimilarity.

Relation between Euclidean Distance and Dissimilarity

The Euclidean distance is an inner product induced norm defined by the following equation:

$$d_{AB} = \|x_A - x_B\| = \langle x_A - x_B, x_A - x_B \rangle^{1/2} = \left[\sum_{j=1}^{k} (x_{Aj} - x_{Bj})^2 \right]^{1/2} \qquad (1.17)$$

When x_A and x_B are normalized vectors it follows that

$$\langle x_A - x_B, x_A - x_B \rangle = \langle x_A, x_A \rangle - 2\langle x_A, x_B \rangle + \langle x_B, x_B \rangle = 2 - 2\langle x_A, x_B \rangle \qquad (1.18)$$

which reveals that the squared Euclidean distance between two unit length vectors is directly proportional to the dissimilarity function defined in Eq. (1.14):

$$d_{AB}^2 = 2 \cdot D(x_A, x_B) \qquad (1.19)$$

The previous remark might suggest that applying the *Fuzzy c-Means* algorithm to a document collection, with documents represented as normalized term vectors, will produce equivalent results either using the Euclidean norm or the dissimilarity function. But this is not the case because the cluster prototype vectors are not normalized in the original algorithm (see section 1.4.2 for a comparative analysis). Therefore, the squared Euclidean distance between document vectors and cluster prototypes will not follow Eq. (1.19).

3.5 Hyperspherical Fuzzy c-Means Algorithm

We have applied the dissimilarity function introduced in the previous section for clustering normalized document vectors using the *Fuzzy c-Means* approach. The objective function in Eq. (1.5) has been modified by replacing the squared norm with the function defined in Eq. (1.14):

$$J_m(U,V) = \sum_{i=1}^{N} \sum_{\alpha=1}^{c} u_{\alpha i}{}^m D_{i\alpha} = \sum_{i=1}^{N} \sum_{\alpha=1}^{c} u_{\alpha i}{}^m \left(1 - \sum_{j=1}^{k} x_{ij} \cdot v_{\alpha j}\right) \qquad (1.20)$$

A new update expression for the clusters prototypes has to be defined in order to use the dissimilarity function such that properties (1.15) and (1.16) hold. This implies that the cluster prototype vectors need to be normalized[2]. Consequently, a constraint for the optimization of J_m is introduced:

$$S(v_\alpha, v_\alpha) = \sum_{j=1}^{k} v_{\alpha j} \cdot v_{\alpha j} = \sum_{j=1}^{k} v_{\alpha j}{}^2 = 1, \forall_\alpha \qquad (1.21)$$

Using the method of the Lagrange multipliers [3] it is possible to minimize Eq. (1.20) subject to constraints, as this method converts constrained optimization problems into unconstrained ones. Besides constraint from Eq. (1.21), constraints

---

[2] Since the presentation of their paper [20] at the FLINT 2001 Workshop, the authors came across a paper by Klawonn and Keller [15] which contains a similar modification of the FCM algorithm. A later publication by Miyamoto [21] also explores the normalization of the cluster prototypes.

(1.6), (1.7) and (1.8) still need to be regarded. Since the optimization is made taking U and V separately, minimizing the function in Eq. (1.20) with respect to $u_{\alpha i}$ (v_α fixed) leads to a similar result of that in Eq. (1.9), the only difference being the replacement of $d_{i\alpha}^2$ and $d_{i\beta}^2$ by $D_{i\alpha}$ and $D_{i\beta}$,. The expression for $u_{\alpha i}$ is now:

$$u_{\alpha i} = \sum_{\beta=1}^{c} \left(\frac{D_{i\alpha}}{D_{i\beta}} \right)^{-\frac{1}{(m-1)}} = \sum_{\beta=1}^{c} \left(\frac{1 - \sum_{j=1}^{k} x_{ij} \cdot v_{\alpha j}}{1 - \sum_{j=1}^{k} x_{ij} \cdot v_{\beta j}} \right)^{-\frac{1}{(m-1)}} \tag{1.22}$$

To minimize Eq. (1.20) with respect to v_α ($u_{\alpha i}$ fixed) with constraint (1.21), the Lagrangian function is defined as:

$$L(v_\alpha, \lambda_\alpha) = J_m(U, v_\alpha) + \lambda_\alpha \cdot [S(v_\alpha, v_\alpha) - 1]$$

$$= \sum_{i=1}^{N} u_{\alpha i}{}^m (1 - \sum_{j=1}^{k} x_{ij} \cdot v_{\alpha j}) + \lambda_\alpha (\sum_{j=1}^{k} v_{\alpha j}{}^2 - 1) \tag{1.23}$$

where λ_α is the Lagrange multiplier. To convert the optimization problem into an unconstrained problem the derivative of the Lagrangian function is taken,

$$\frac{\partial L(v_\alpha, \lambda_\alpha)}{\partial v_\alpha} = \frac{\partial J_m(U, v_\alpha)}{\partial v_\alpha} + \lambda_\alpha \cdot \frac{\partial [S(v_\alpha, v_\alpha) - 1]}{\partial v_\alpha} = 0 \tag{1.24}$$

that is equivalent to,

$$- \sum_{i=1}^{N} u_{\alpha i}{}^m x_i + 2\lambda_\alpha v_\alpha = 0 \Leftrightarrow v_\alpha = \frac{1}{2\lambda_\alpha} \cdot \sum_{i=1}^{N} u_{\alpha i}{}^m x_i \tag{1.25}$$

and by applying constraint from Eq. (1.21) follows,

$$\sum_{j=1}^{k} v_{\alpha j}{}^2 = \left(\frac{1}{2\lambda_\alpha} \right)^2 \cdot \sum_{j=1}^{k} \left(\sum_{i=1}^{N} u_{\alpha i}{}^m x_{ij} \right)^2 = 1 \Leftrightarrow \frac{1}{2\lambda_\alpha} = \left[\sum_{j=1}^{k} \left(\sum_{i=1}^{N} u_{\alpha i}{}^m x_{ij} \right)^2 \right]^{-1/2} \tag{1.26}$$

Finally, replacing $\dfrac{1}{2\lambda_\alpha}$ in Eq. (1.25) leads to:

$$v_\alpha = \sum_{i=1}^{N} u_{\alpha i}{}^m x_i \cdot \left[\sum_{j=1}^{k} \left(\sum_{i=1}^{N} u_{\alpha i}{}^m x_{ij} \right)^2 \right]^{-1/2} \tag{1.27}$$

The modified algorithm runs similarly to the original FCM differing only on the expressions to update v_α. It can be shown that the new expression for v_α represents a normalization of the original v_α as defined in Eq. (1.10).

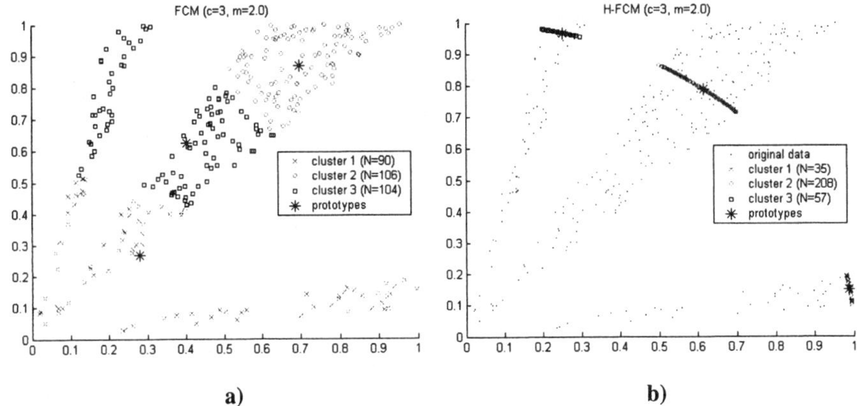

Fig. 3. Clustering of points in a bi-dimensional space using: **a)** FCM and **b)** H-FCM with normalized data vectors

For future reference, the modified algorithm will be labeled as *Hyperspherical Fuzzy c-Means* (H-FCM), since both data vectors and cluster centers lie in a *k*-dimensional hypersphere of unit radius. To illustrate this, consider a bi-dimensional data set that can be partitioned into three clusters of highly similar data elements (*i.e.* any pair of elements in the cluster has a high *cosine measure*). In Fig. 3, the spatial distribution of the data elements is shown and the clusters discovered with a) FCM and b) H-FCM can be discriminated by the different plot markers.

The H-FCM plot shows both cluster prototypes and normalized data elements located in a circle of unit radius. H-FCM distributes the data elements among the clusters following a criterion of minimum dissimilarity to the cluster prototypes, whereas with FCM this does not happen. The FCM graph and the data in Table 1 show that cluster 1 includes elements which are quite dissimilar ($D = 0.7016$).

As a final note: applying FCM for document clustering might lead to high pairwise dissimilarities within clusters, which means that documents sharing very few terms might be grouped together. Such problem makes the case for using the *Hyperspherical Fuzzy c-Means* algorithm in preference. Results of a performance comparison between FCM and H-FCM are presented in section 1.4.2.

Table 1. Maximum (*max*), average (*avg*) and standard deviation (*stdev*) of the pairwise dissimilarity between elements within each cluster

Dissimilarity	FCM			H-FCM		
	max (D_{ij})	avg (D_{ij})	stdev(D_{ij})	max(D_{ij})	avg(D_{ij})	stdev(D_{ij})
Cluster 1	0.7016	0.2099	0.0629	0.0047	9.97×10^{-4}	4.63×10^{-4}
Cluster 2	0.0273	0.0046	0.0022	0.0307	5.20×10^{-3}	2.20×10^{-3}
Cluster 3	0.1586	0.0382	0.0138	0.0052	6.40×10^{-4}	3.72×10^{-4}

3.6 Performance Evaluation

The performance of clustering algorithms is generally evaluated using internal performance measures, *i.e.* measures that are algorithm dependent and do not contain any external or objective knowledge about the actual structure of the data set. This is the case of the many validity indexes for the *Fuzzy c-Means* algorithm. When there is prior knowledge on how clusters should be formed, external performance measures (algorithm independent) can be used to compare the clustering results with the benchmark. The next sub-sections cover these two types of evaluation measures.

Internal Performance Measures: Validity Indexes for the FCM

As discussed in section 2, the document clusters should be fuzzy so that uncertainty and imprecision in the knowledge space can be handled. However, there needs to be a compromise between the amount of fuzziness and the capability to obtain good clusters and meaningful document relationships. It is known that increasing values of the fuzzification parameter - m - lead to a fuzzier partition matrix. Thus, this parameter can be adjusted to manage this compromise. Establishing the appropriate values for m requires the use of a validity index.

There are several validity indexes for the *fuzzy c-means* algorithm that are used to analyze the intrinsic quality of the clusters. A simple cluster validity measure that indicates the closeness of a fuzzy partition to a hard one is the *Partition Entropy* (PE) [4], defined as:

$$PE = -\frac{1}{N} \sum_{i=1}^{N} \sum_{\alpha=1}^{c} u_{\alpha i} \log_a (u_{\alpha i}) \qquad (1.28)$$

The possible values of *PE* range from 0 – when U is *hard* – to $\log_a(c)$ – when every data element has equal membership in every cluster ($u_{\alpha i} = 1/c$). Dividing *PE* by $\log_a(c)$ normalizes the value of *PE* to range in the [0,1] interval.

The *Xie-Beni* index [32] evaluates the quality of the fuzzy partition based on the compactness of each cluster and separation between cluster centers: the assumption is that the more compact and separated the clusters are the better. This index is defined as follows:

$$S_{XB} = \frac{\sum_{i=1}^{N} \sum_{\alpha=1}^{c} u_{\alpha i}{}^{m} \left\| x_i - v_\alpha \right\|^2}{N \cdot \min_{\substack{\varphi \neq \gamma \\ \varphi, \gamma \in [1,c]}} \left\| v_\varphi - v_\gamma \right\|^2} \qquad (1.29)$$

If the minimum distance between any pair of clusters is too low, then S_{XB} will be very high. Hence, normally a good partition corresponds to a low value of S_{XB}.

Both validity indexes were derived for the FCM but they are still applicable for the H-FCM algorithm. A simple modification is required in the expression for S_{XB} to replace the squared distance $\left\| \cdot \right\|^2$ with the dissimilarity function of Eq. (1.14). No change is required for *PE* since it is only a function of the membership matrix.

External Performance Measures: Precision, Recall, F-Measure

Precision and *recall* are two popular measures for evaluating the performance of information retrieval systems [1,24]. They represent the fraction of relevant documents out of those retrieved in response to a query and the fraction of retrieved documents out of the relevant ones, respectively. Similar measures have been proposed for classification systems [17]. The purpose of such systems is to classify data elements given a known set of classes. In this context, *precision* represents the fraction of elements assigned to a pre-defined class that indeed belong to the class; and *recall* represents the fraction of elements that belong to a pre-defined class that were actually assigned to the class. *Precision* and *recall* can be equally used to evaluate clustering algorithms, which are in fact unsupervised classification systems, when a clustering benchmark exists. For a given cluster α and a reference cluster β, we define *precision* ($P_{\alpha\beta}$) and *recall* ($R_{\alpha\beta}$) as follows:

$$P_{\alpha\beta} = \frac{\text{number of elements from reference cluster } \beta \text{ in cluster } \alpha}{\text{total number of elements in cluster } \alpha} \tag{1.31}$$

$$R_{\alpha\beta} = \frac{\text{number of elements from reference cluster } \beta \text{ in cluster } \alpha}{\text{total number of elements in reference cluster } \beta} \tag{1.32}$$

These two measures can be combined into a single performance measure – the *F-measure* [18,24] – that is defined as:

$$F^{\gamma}{}_{\alpha\beta} = \frac{(\gamma^2 + 1) \cdot P_{\alpha\beta} \cdot R_{\alpha\beta}}{\gamma^2 \cdot P_{\alpha\beta} + R_{\alpha\beta}} \tag{1.33}$$

where γ is a parameter that controls the relative weight of *precision* and *recall* (for equal contribution, $\gamma=1$ is used). The individual $P_{\alpha\beta}$, $R_{\alpha\beta}$ and $F_{\alpha\beta}$ measures are averaged to obtain overall performance measures P, R and F.

For fuzzy clusters, the maximum membership criterion can be applied to count the number of data elements in each cluster for the $P_{\alpha\beta}$ and $R_{\alpha\beta}$ calculation. In the case of maximum fuzziness, *i.e.* all elements with equal membership in all the clusters, *precision* and *recall* will be $P_{\alpha\beta}= N_{\beta}/N$ and $R_{\alpha\beta}=1$, $\forall_{\alpha,\beta}$ (where N_{β} is the total number of documents in reference cluster β and N the collection size).

4. Fuzzy Clustering Experiments

The main objectives of the document clustering experiments were to investigate the suitability of fuzzy clustering for discovering good document relationships and also to carry out a performance analysis of the H-FCM compared with the original FCM. In this section the experimental results are reported and analyzed.

4.1 Data Set Description

Since the CANDLE database was not yet available for these clustering trials, a familiar document collection was selected, a subset of IETF's[3] RFCs (Request For Comments) documents. These documents describe protocols and policies for the Internet and the ones chosen were those describing Internet standards.

Each document was automatically indexed for keyword frequency extraction. Stemming was performed and stop words were removed (see section 1.3.1). Document vectors like in Eq. (1.1) were generated and organized as rows of a $(N \times k)$ matrix, where $N=73$ was the collection size and $k=12690$ was the total number of indexing terms. For the experimental trials two matrices were created, one for each of the term weighting schemes described in section 1.3.3 (*tf* and *tf$\times$idf*).

A clustering benchmark was manually created based on our knowledge of the documents' content and also on the indexing information found in [30]. The RFCs were distributed into six fairly homogeneous clusters although a few documents were very generic and could have been attributed to multiple clusters.

4.2 Experimental Results

Performance Comparison between FCM and H-FCM

The goal of the first experiment was to investigate whether the *Hyperspherical Fuzzy c-Means* (H-FCM) algorithm was able to generate a good partition of the document collection and how it compared to the original *Fuzzy c-Means* (FCM) algorithm. The number of clusters was set to $c=6$ (as the benchmark indicated), the convergence threshold to $\varepsilon=10^{-4}$ and the maximum number of iterations to $t_{max}=300$. FCM and H-FCM were applied to both the *tf* data and the *tf$\times$idf* data (in the H-FCM case, the data vectors were normalized to unit length). The experiment was repeated for increasing values of the fuzzification parameter – $m \in [1.10,1.50]$.

The results are analyzed based on internal and external performance measures (Figs. 4 to 6 and 7 to 9, respectively). The graphs show that in general the *tf* data leads to slightly better results than the *tf$\times$idf* data, both in the FCM case and in the H-FCM case. With the *tf* data matrix, lower *Partition Entropy* and lower values of the *Xie-Beni index* were obtained. Furthermore, both algorithms converge slightly faster with this data. Regarding the quality of the clusters, there are no significant differences in the *Precision, Recall* and F^1 measures of each algorithm.

---

[3] IETF - Internet Engineering Task Force: http://www.ietf.org/

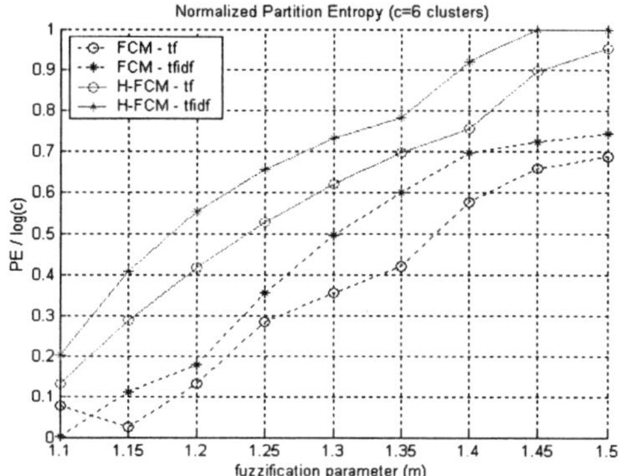

Fig. 4. Internal Performance Measure: normalized *Partition Entropy*

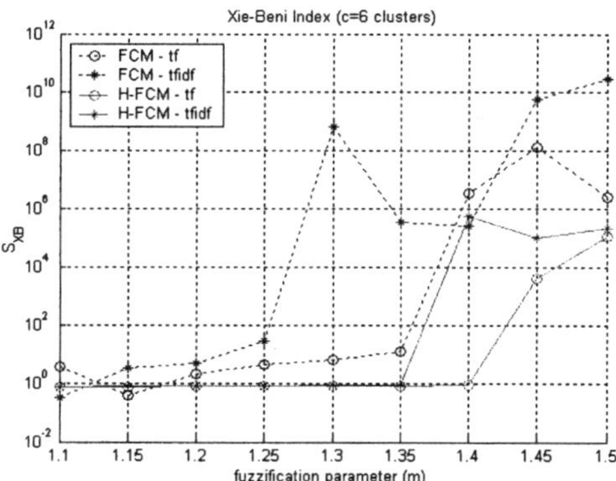

Fig. 5. Internal Performance Measure: *Xie-Beni index*

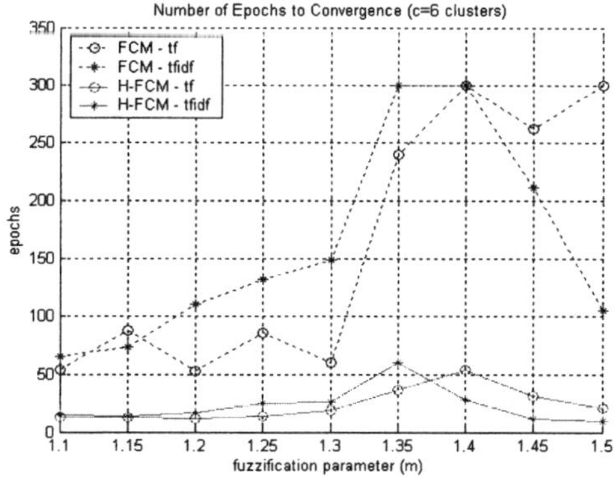

Fig. 6. Internal Performance Measure: number of *epochs* until convergence

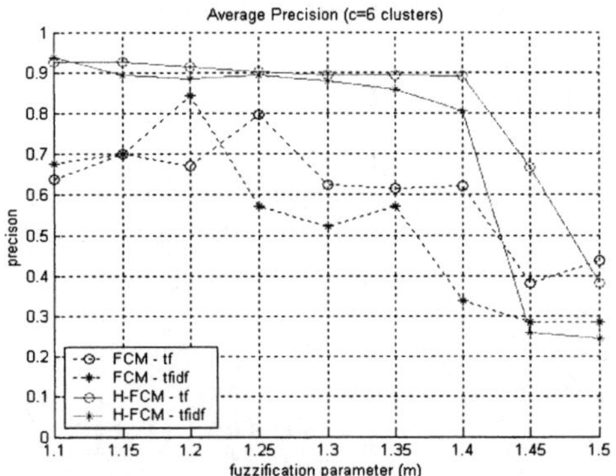

Fig. 7. External Performance Measure: average *Precision*

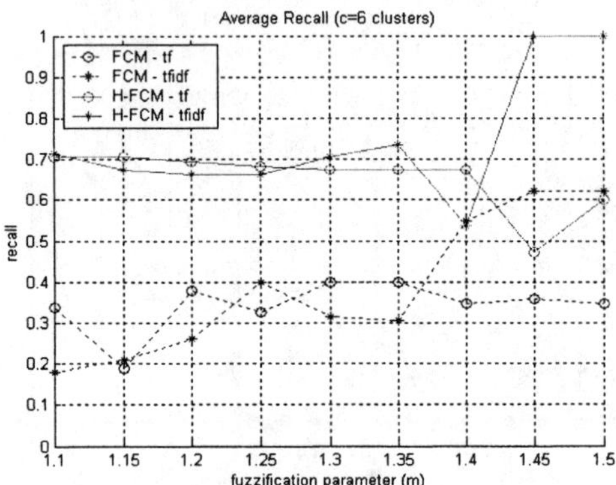

Fig. 8. External Performance Measure: average *Recall*

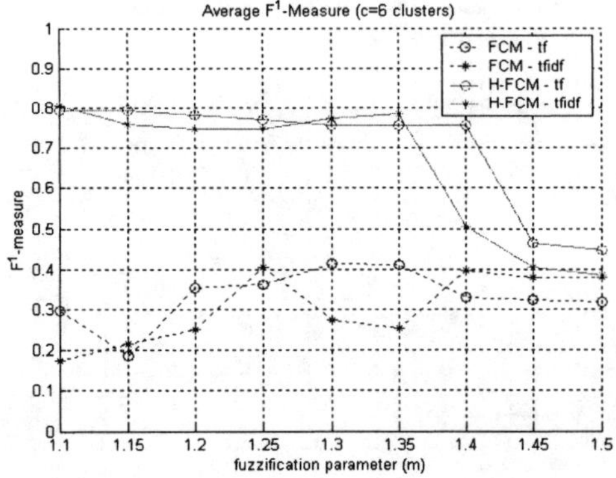

Fig. 9. External Performance Measure: average F^1-Measure

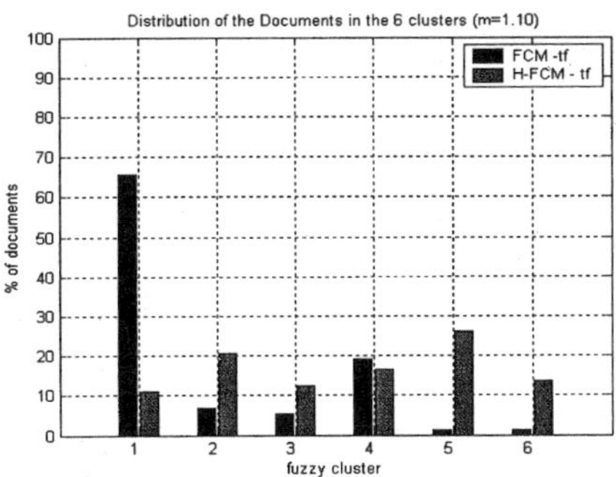

Fig. 10. Percentage of documents attributed to each of the 6 clusters (when m=1.10)

Comparing the performance of both algorithms, the results show that H-FCM performs significantly better than the original FCM. From the *PE* graph (Fig. 4), it may seem that the H-FCM results are worse than the FCM ones because the former produces fuzzier document clusters for a fixed m. However, the S_{XB} graph (Fig. 5) indicates that the H-FCM clusters are more compact and more clearly separated; and that this is true for a wider range of the fuzzification parameter ($m \in [1.10, 1.40]$). A significant advantage of the H-FCM algorithm is its execution time. It is evident from Fig. 6 that H-FCM converges much faster than FCM. In both cases, the number of iterations for convergence increases with m until maximum *Partition Entropy* is reached, point after which it starts decreasing again. Furthermore, the external performance graphs prove that the quality of the six clusters (considering the benchmark) is significantly better with H-FCM. Both *Precision* (Fig. 7) and *Recall* (Fig. 8.) are quite high in the same range of m values where S_{XB} is low. Conversely, FCM exhibits low *Recall* which means that very few documents of the reference clusters are attributed to the corresponding fuzzy clusters. In fact, with m set to 1.10 (close to the hard case) around ~66% of the documents were assigned to a single cluster by the FCM algorithm, as the graph in Fig. 10 shows.

FCM vs. H-FCM with Normalized Document Vectors

In section 1.3.4, the relationship between the squared Euclidean norm and the dissimilarity function was established. It was shown that they were equivalent for unit length documents vectors. The goal of this experiment is to demonstrate that despite the equivalence, FCM and H-FCM produce different results since in the FCM case the cluster centers are not normalized.

In Fig. 11, the *Partition Entropy* plots show that FCM with normalized data vectors reaches maximum fuzziness even for very low values of m, which means that it fails to find any structure in the document collection. Although for $m<1.15$ the quality of the clusters is comparable with the H-FCM case (see Fig. 13), FCM is still much slower than the H-FCM algorithm (see Fig. 12). The results prove that the dissimilarity function is in fact more suitable than the Euclidean norm for document clustering.

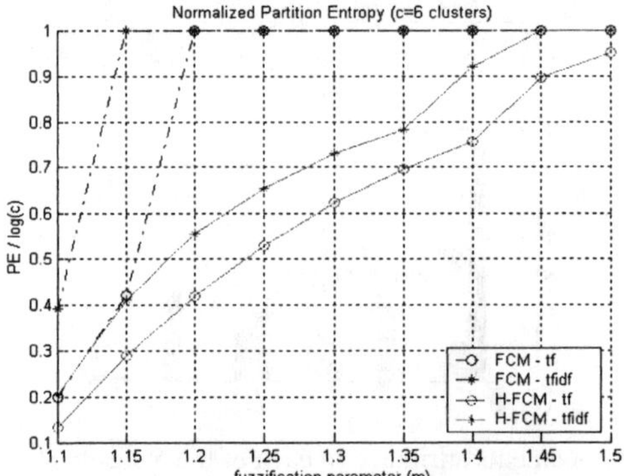

Fig. 11. Normalized *Partition Entropy* (with normalized document vectors)

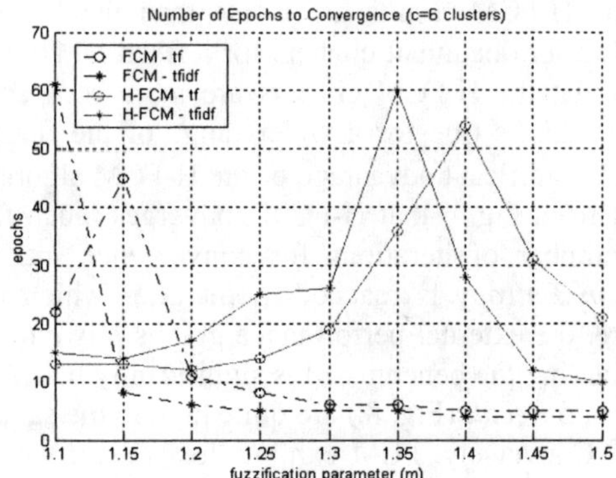

Fig. 12. Number of *epochs* until convergence (with normalized document vectors)

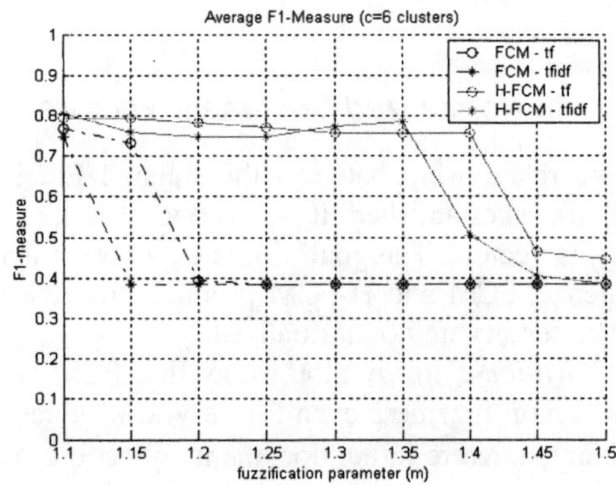

Fig. 13. Average F^l-Measure (with normalized document vectors)

Pre-processing Effects on the H-FCM Results

The aim of this last experiment was to analyze the effects of pre-processing in the H-FCM results. After automatically indexing the document collection, several thresholds for the *entropy* and *specificity* filters (discussed in section 1.3.2) were set. For each threshold, a *tf* matrix was generated and the H-FCM algorithm was applied. Indexing terms were discarded according to the following criteria:

- terms with *entropy* above a given threshold $\tau \cdot H_{max}$ (where H_{max} is the theoretical maximum *entropy*, that occurs for terms with equal probability in every document),
- terms that appeared in a high percentage of documents (indexing terms with *specificity* below a given threshold)
- and terms that appeared in a small percentage of documents (with *specificity* above a given threshold).

Each of these filters was considered separately in the experiment.

Fig. 14 shows the average F^l-Measure obtained using H-FCM for different values of the fuzzification parameter. The graph contains the results of the clustering algorithm after *entropy* pre-processing. With this filter the dimensionality of the document vectors was reduced to k=12627, 12203, 12040, 11837, 11406, 11133, 10803 and 10417 terms, for values of τ equal to 0.75, 0.60, 0.55, 0.50, 0.40, 0.35, 0.30 and 0.25, respectively. It can be observed that for low values of m the quality of the clusters increases slightly for *entropy* thresholds above $0.55 \cdot H_{max}$ but as more terms are removed the results degrade considerably. This means that the presence of terms with very high *entropy* does not have significant impact on the clustering results and that terms with medium *entropy* should not be eliminated, since it degrades the performance of the algorithm.

Figs. 15 and 16 show the average F^l-Measure obtained using H-FCM for different values of the fuzzification parameter after pre-processing with the *specificity* filter. In the first case, terms present in more than 57%, 44%, 34%, 21%, 16% and 12% of the documents were discarded, reducing the dimensionality of the document vectors to k=12754, 12690, 12582, 12253, 12094 and 11889, respectively. From the plots in Fig. 15, it can be observed that keeping terms which are very common does not degrade the clustering results. In the second case, terms which appeared only in 1, 2, 3, 4, 6, 7 and 8 documents were removed, reducing k to 4298, 2906, 2250, 1900, 1631, 1300 and 1192, respectively. From the plots in Fig. 16 it can be observed that removing terms which are very specific does not practically change the F^l-Measure. Such filter produces obvious benefits regarding memory and CPU requirements since a significant dimensionality reduction is achieved (by a factor of 10 with the current document collection).

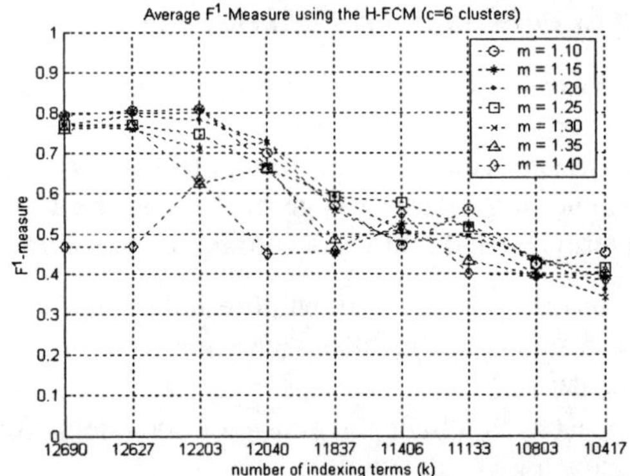

Fig. 14. Average F^1-Measure vs. number of indexing terms – effects of the *entropy* filter (exclusion of terms with *entropy* <u>above</u> a fixed threshold)

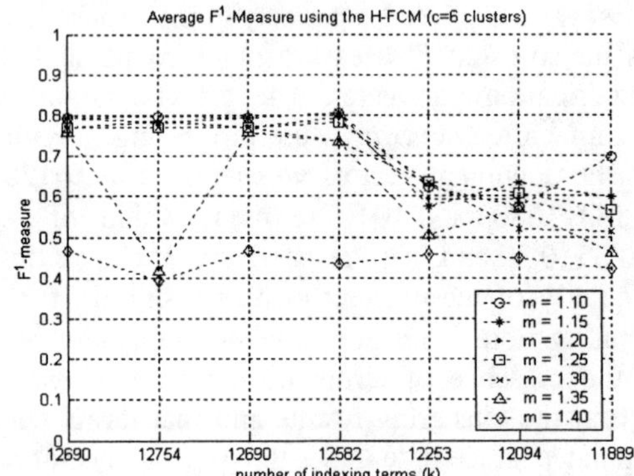

Fig. 15. Average F^1-Measure vs. number of indexing terms – effects of the *specificity* filter (exclusion of terms with *specificity* <u>below</u> a fixed threshold)

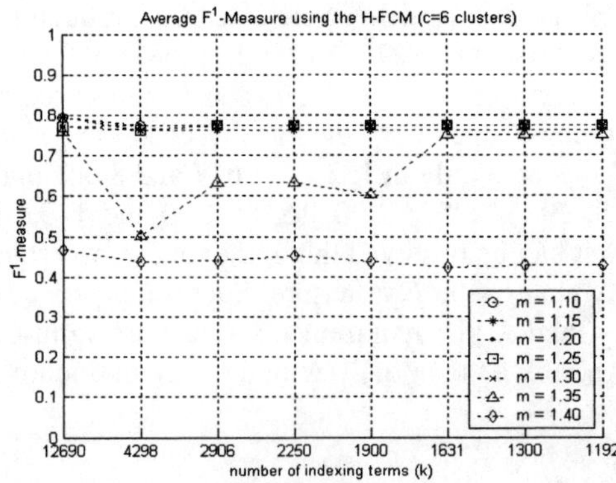

Fig. 16. Average F^1-Measure vs. number of indexing terms – effects of the *specificity* filter (exclusion of terms with *specificity* <u>above</u> a fixed threshold)

5. Adaptive Knowledge-based *e*-Content Navigation

Adaptive content navigation in *e*-Learning applications has been introduced in section 1.1. Here, the integration of the dynamic knowledge representation framework in such adaptive systems is discussed.

In general, adaptive navigation systems enable personalized access to hyper-linked information. Adaptation in hypermedia systems can be provided at two different levels: at the presentation-level and at the link-level [10]. Presentation-level adaptation deals with issues like how to display the content of a page to a particular user, which information should be shown, which should be available on request, and so on. Link-level adaptation deals with the discovery and display of relevant links to the individual user. Although both kinds of adaptation should be present in *e*-Learning applications that aim at providing flexible learning environments, link-level adaptation is particularly important to guide students throughout their learning path. There are two principal approaches to dynamically define the links. One is to log the user's actions so that the system can suggest links based on past information. The other approach keeps a record of the user's current knowledge and interests in a profile and then search for pages that match the individuals' needs.

There are several adaptive educational hypermedia systems that implement link-level adaptation [6,7,22,29]. The basic mechanism for adapting content to each student is based on the representation of both domain knowledge (domain model) and student knowledge level (student model) and uses the notion of pre-requisites and outcomes. Pre-requisites are basically the set of concepts a user needs to know to access a document (or online course) and the outcomes are the set of concepts he/she is expected to acquire after reading (or completing) it. The system then analyses the student's current background (stored in the profile as weighted concepts) to determine which links should be made available.

The domain model is usually static and manually created by experts in the area. By replacing the static model with the dynamic knowledge space obtained with fuzzy clustering, the same adaptation mechanisms can be applied. The use of document clustering for adaptively linking resources has been proposed in [12]. The approach described in this paper employs a hard clustering algorithm and adapts links in context of the user's interests and of the documents' contents. The advantage of having fuzzy clusters instead is that links can be sorted by degree of relevance, computed from the fuzzy memberships, and unobvious links may be revealed.

6. Conclusions

Fuzzy clustering has been proposed for dynamic knowledge representation to support flexible content exploration in *e*-Learning systems. A modified version of the *Fuzzy c-Means* (FCM) clustering algorithm was derived – the *Hyperspherical Fuzzy c-Means* (H-FCM) – to replace the Euclidean norm by a dissimilarity function common in traditional information retrieval systems. The experiments carried out with a test document collection showed that the dissimilarity function is a better measure of document proximity than the Euclidean norm. The results indicate that the FCM algorithm produces poor clusters either with or without normalizing the document vectors to unit length. In contrast, H-FCM is able to discover good document clusters for different levels of fuzziness. It was observed that increasing the fuzziness parameter m to a certain level did not affect significantly the quality of the H-FCM clusters. This means that although the membership of documents in other clusters increases with m, documents still have maximum membership in the right cluster. Therefore, fuzziness uncovers the possibility to associate different clusters representing different concepts. It was also observed that the H-FCM algorithm is much faster to converge than the FCM algorithm, which is an important feature considering its applicability to larger document collections.

Regarding the encoding of document vectors no advantages were found in using the *tf×idf* scheme, as with the *tf* scheme slightly better results were obtained both with FCM and H-FCM. This means that the clustering algorithms actually performed slightly better when the test documents were encoded independently of the entire document collection. This result suggests that in systems like CANDLE, where new learning material is frequently being created and added to the system, it may not be necessary to re-encode every document each time new content is stored in the database. Furthermore, with sequential or incremental algorithms re-clustering of the whole database may not be required each time a document is added.

The experiments with the RFC collection using different pre-processing filters showed that the H-FCM is robust to pre-processing. Eliminating terms which are very specific to very few documents has almost no impact in the clustering results while reducing significantly the dimensionality of the document vectors and consequently reducing memory and CPU requirements. Likewise, filtering out terms that appear in many documents or that have high *entropy* does degrade the quality of the clustering results. Both the *entropy* and *specificity* filters require information about the entire document collection and therefore, every time the collection grows pre-processing and clustering have to be re-done since terms discarded previously may then be deemed important. Nevertheless, the results showed that the H-FCM algorithm performs equally well without discarding any terms.

7. Acknowledgement

This work has been supported by the *Portuguese Foundation for Science and Technology* (FCT - Fundação para a Ciência e a Tecnologia) through the PRAXIS XXI doctoral scholarship programme.

8. References

1. Baeza-Yates R, Ribeiro-Neto B (1999). Modern Information Retrieval. Addison Wesley, ACM Press, New York

2. Berners-Lee T, Hendler J, Lassila O (2001). The Semantic Web - A new form of Web content that is meaningful to computers will unleash a revolution of new possibilities. Scientific American, vol 284, no 5, pp 34-43, May 2001

3. Bertsekas DP (1995). Nonlinear Programming. Athena Scientific, Belmont, Massachusetts

4. Bezdek JC (1981). Pattern Recognition with Fuzzy Objective Function Algorithms. Plenum Press, New York

5. Bray T, Paoli T, Sperberg-McQueen CM, Maler E (2000). XML - eXtensible Markup Language v1.0. W3C Consortium, Oct. 2000. Available at: http://www.w3.org/TR//2000/REC-xml-20001006

6. Brusilovsky P, Eklund J, Schwarz E (1998). Web-based education for all: A tool for developing adaptive courseware. Computer Networks and ISDN Systems, vol 30, no 1-7, pp 291-300, Apr. 1998

7. Calvi L, De Bra P (1997). Improving the usability of hypertext courseware through adaptive linking. Proceedings of the 8th ACM Conference on Hypertext and Hypermedia, HT'1997, pp 224-225, Apr. 1997

8. Cherry, SM (2002). Weaving a web of ideas. IEEE Spectrum, vol 39, no 9, pp 65 -69, Sep. 2002

9. Cutting DR, Karger DR, Pedersen JO, Tukey JW (1992). Scatter/Gather: a cluster-based approach to browsing large document collections. Proceedings of the 15th Annual International ACM SIGIR Conference on Research and Development in Information Retrieval, SIGIR'92, pp 318-329, Jun. 1992

10. De Bra P, Brusilovsky P, Houben GJ (1999). Adaptive hypermedia: from systems to framework. ACM Computing Surveys, vol 31, no 4es, Dec. 1999.

11. Decker S, Melnik S, van Harmelen F, Fensel D, Klein M, Broekstra J, Erdmann M, Horrocks I (2000). The semantic web: the roles of XML and RDF. IEEE Internet Computing, vol 4, no 5, pp 63-73, May 2000

12. El-Beltagy SR, Hall W, De Roure D, Carr L (2001). Linking in context. Proceedings of the 12th ACM Conference on Hypertext and Hypermedia , HT'2001, pp 151-160, Aug. 2001

13. IEEE Learning Object Metadata Working Group (2000). Draft Standard for Learning Object Metadata. IEEE P1484.12/D5, Nov. 2000. Available at: http://ltsc.ieee.org/doc/wg12/LOM_ WD5.pdf.

14. IEEE Learning Technology Standards Committee (LTSC): http://ltsc.ieee.org/

15. Klawonn F, Keller A (1999). Fuzzy clustering based on modified distance measures. Proceedings of the Third International Symposium on Intelligent Data Analysis, IDA'99, LNCS 1642, pp 291-301, Aug. 1999

16. Kraft DH, Chen J, Mikulcic A (2000). Combining fuzzy clustering and fuzzy inference in information retrieval. Proceedings of the 9th IEEE International Conference on Fuzzy Systems, FUZZ IEEE 2000, vol 1, pp 375-380, May 2000

17. Lewis DD (1991). Evaluating text categorization. Proceedings of the Speech and Natural Language Workshop, pp 312-318, Feb. 1991

18. Lewis DD and Gale WA (1994). A sequential algorithm for training text classifiers. Proceedings of the Seventeenth Annual International ACM-SIGIR Conference on Research and Development in Information Retrieval, SIGIR 94, pp 3-12, Aug. 1994

19. MacQueen J (1967). Some methods for classification and analysis of multivariate observations. Proceedings of the Fifth Berkeley Symposium on Mathematics, Statistics and Probability, vol 1, pp 281-296

20. Mendes MES, Sacks L (2001). Dynamic knowledge representation for e-Learning applications. In: *Proceedings of the 2001 BISC International Workshop on Fuzzy Logic and the Internet*, FLINT 2001, Memorandum No. UCB/ERL M01/28, pp 176-181, U. C. Berkeley, Aug. 2001

21. Miyamoto, S (2001). Fuzzy multisets and fuzzy clustering of documents. Proceedings of the 10th IEEE International Conference on Fuzzy Systems, FUZZ IEEE 2001, vol 2, pp 1191-1194, Dec. 2001

22. Pilar da Silva D, van Durm R, Duval E, Olivié H (1997). A simple model for adaptive courseware navigation. Proceedings of INFWET '97, Nov. 1997

23. Porter M (1980). An algorithm for suffix stripping. *Program*, vol 14, no 3, pp 130-137, Jul. 1980

24. van Rijsbergen CJ (1979). Information Retrieval. 2nd Edition, Butterworth, London

25. Sacks L, Earle A, Prnjat O, Jarrett W, Mendes M (2002). Supporting variable pedagogical models in network based learning environments. Proceedings of the 2nd IEE Annual Symposium on Engineering Education, vol 1, pp 22/1-22/6, Jan. 2002

26. Salton G (1975). A Theory of Indexing. Society for Industrial and Applied Mathematics, Philadelphia

27. Salton G, Allan J, Buckley C (1994). Automatic structuring and retrieval of large text files. Communications of the ACM, vol 37, no 2, pp 97-108, Feb. 1994

28. Shannon CE (1948). A mathematical theory of communication. The Bell System Technical Journal, vol 27, pp 379-423 and 623-656, Jul. and Oct. 1948

29. Weber G, Specht M (1997). User modelling and adaptive navigation support in WWW-based tutoring systems. Proceedings of the 6th International Conference on User Modelling, UM'97, pp 289-300, Jun. 1997

30. Wheeler L. IETF RFC Index. Available at: http://www.garlic.com/~lynn/rfcietf.htm

31. Willett P (1988). Recent trends in hierarchical document clustering: a critical review. Information Processing and Management, vol 24, no 5, pp 577-597, 1988

32. Xie XL, Beni GA (1991). A validity measure for fuzzy clustering. IEEE Transactions on Pattern Analysis and Machine Intelligence, vol 13, no 8, pp 841-847, Aug. 1991

33. Zadeh, LA (1965). Fuzzy Sets. Information and Control, vol 8, pp 338-353, Jun. 1965

34. Zamir O, Etzioni O (1999). Grouper: a dynamic clustering interface to Web search results. Computer Networks, vol 31, no 11-16, pp 1361-1374, May 1999

Content Based Vector Coder for Efficient Information Retrieval

Shuyu Yang and Sunanda Mitra
Department of Electrical and Computer Engineering, Texas Tech University
Lubbock, TX 79409-3102
USA
Shu.yang@ttu.edu, Sunanda.Mitra@coe.ttu.edu

Abstract: Retrieval of relevant information and its efficient transmission over the Internet to worldwide users are of utmost interest in many applications such as telemedicine, video conferencing, distance education, to name a few. Content-based source encoding is, however, essential in enhancing information retrieval. Despite some significant work done in this area, indexing and retrieval of medical image data still pose a challenging problem since distinct features are not always present in such data sets. We present a novel hybrid multi-scale vector quantizer (HMVQ) whose codebook is generated by neuro-fuzzy clustering of salient information features in the wavelet domain. Our codec incorporates multi-scale feature extraction, vector quantization codebook training and detail-preserving residual scalar quantization. The performance of this new vector encoder, namely, HMVQ, surpasses that of the well-known scalar coder, the Set Partitioning in Hierarchical Trees (SPIHT) in the fidelity of reconstructed data at all bit rates. Our results also demonstrate that the performance of such encoder is equivalent to an optimized statistical approach while, providing a drastic reduction in execution time. Efficiency in computational cost is of great significance while considering future advances in visual communications using multi-view 3-D auto-stereoscopic systems.

1 Introduction

Content-based source encoding can be achieved by designing appropriate vector codec allowing high fidelity retrieval of the input information. Vector quantization (VQ) has been theoretically proven to be a more efficient coding method than scalar quantization [3]. The design of an efficient VQ encoder involves global codebook generation by selecting a good clustering algorithm and using appropriate features extracted from the training data set. In image vector coding, codebooks obtained from a set of training images tend to retain common features of the samples, thus, when used to decode a particular image, special features of the image will be lost, which usually leads to a blurred reconstructed image. Here we are presenting a content-based vector coder whose codebooks are generated using a neuro-fuzzy clustering algorithm, adaptive fuzzy leader clustering (AFLC) [21,30] together with a new feature extraction method. Usually, the fidelity of a reconstructed compressed image depends on the nature of the codebook used for decoding. Thus, it is highly desirable to choose an efficient clustering algorithm for codebook generation. At the same time, selecting appropriate features from a set of training images is equally critical.

Through multi-level wavelet transform, an image can be decorrelated in both space and frequency, resulting in a hierarchical structure in which most of the image information being packed into a few large-magnitude coefficients at higher levels, while a large population of the coefficients concentrates around zero [1], leaving much room for compression. Successful wavelet based scalar quantization methods such the Embedded Zerotree Wavelet (EZW) [24] and SPIHT [23] exploit this dependency between inter-scale coefficients by retaining and quantizing coefficients according to their significance. Although SPIHT has been regarded as one of the best scalar quantization schemes, the intra-scale dependency among wavelet coefficients are not explicitly exploited. On the other hand, traditional wavelet-based VQ schemes tend to take advantage of the intra-scale dependency by grouping adjacent coefficients in the same level into vectors [1, 16], without taking the relationship among inter-scale coefficients into consideration. Here we will present a new vector forming method that exploits both dependencies.

Besides generating wavelet feature vectors in a new way, we compare two efficient clustering algorithms that can be used for codebook training. The AFLC algorithm, based on neuro-fuzzy concepts, has been proven to be quite successful in many clustering tasks such as image vector quantization, image segmentation and noise removal [8,16]. Deterministic annealing (DA) [22] is entirely based on a statistical approach capable of yielding global optimum. We compare the clustering accuracies of both methods in terms of the distortion of reconstructed image from codebooks generated from these two methods respectively. After the image is vector quantized, the error residual, which contains image-specific information rather than the common image features found in the codebook, is scalar quantized using SPIHT. Such a combination allows the detail contents of the image to be preserved, rendering a higher fidelity reconstructed image than coded otherwise.

In section 2, the AFLC and DA algorithms will be briefly described. Our new feature extraction method will be explained in section 3. The coding method and results are presented in section 4. Section 5 presents the conclusions.

2 AFLC and DA

2.1 AFLC

AFLC is an integrated neural-fuzzy clustering algorithm that can be used to learn cluster structure embedded in complex data sets in a self-organizing manner. The algorithm can be described as a two-layered structure (Figure 1). The first layer uses a self-organizing neural network similar to ART1 [5-7] to find hard clusters. Let C be the current number of centroids and $\mathbf{v}_i$ (i=1, …, C) be the centroids. When a new sample $\mathbf{x}_k$ comes in, it is normalized and then initially classified into the cluster whose centroid has the largest dot product with the sample vector:

$$y_{i*} = \max_i \{ y_i \} = \max_i \left\{ \sum_{j=1}^{n} b_{ij} \bar{x}_{kj} \right\} \qquad i = 1,...,C, \tag{1}$$

where $b_{ij} = \dfrac{v_{ij}}{\|\mathbf{v}_i\|}$, $\bar{x}_{kj} = \dfrac{x_{kj}}{\|\mathbf{x}_k\|}$, and i^* is the index of the winning cluster.

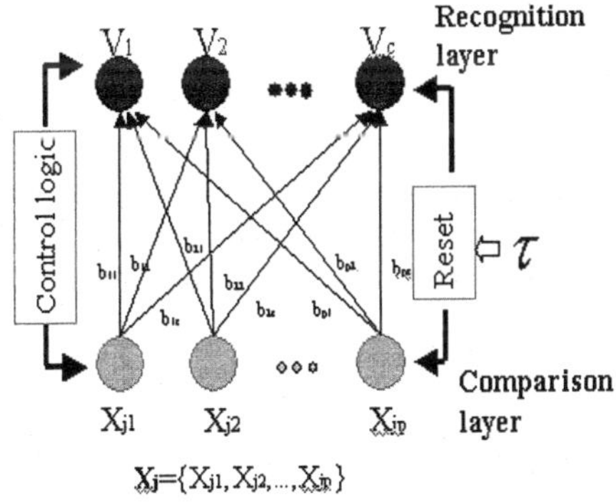

Fig. 1. AFLC algorithm structure

This initial classification is verified through a vigilance test in the second layer, the recognition layer. The second layer serves as a verification process. By verifying the initial sample recognition through a vigilance test, the algorithm is able to dynamically create new clusters according to the data distribution when the verification fails, or optimize and update the system when the initial sample recognition is confirmed. The vigilance test consists of calculating a ratio between the distance of the sample to the winning cluster and the average distance of all the samples in this cluster to the cluster centroid. When the test fails, a new cluster is created, otherwise the system is optimized and clusters are updated. The vigilance test consists of calculating a ratio between the distance of the sample to the winning cluster and the average distance of all the samples in this cluster to the cluster centroid,

$$R = \left\| \mathbf{x}_j - \mathbf{v}_i \right\| \Big/ \frac{1}{N_i} \sum_{k=1}^{N_i} \left\| \mathbf{x}_k - \mathbf{v}_i \right\|. \tag{2}$$

When this ratio is higher than a user-defined threshold, the test fails and a new cluster is created, taking the sample as the initial centroid. Otherwise, the sample is officially classified into the winning cluster, and then its centroid and the fuzzy membership values are updated using fuzzy C-means [4] (FCM) equations:

$$\mathbf{v}_i = \sum_{j=1}^{p} (u_{ij})^m \mathbf{x}_j \Big/ \sum_{j=1}^{p} (u_{ij})^m \qquad i = 1,2,...,c \tag{3}$$

where

$$u_{ij} = \frac{(1/\left\| \mathbf{x}_j - \mathbf{v}_i \right\|^2)^{1/m-1}}{\sum_{k=1}^{c} (1/\left\| \mathbf{x}_j - \mathbf{v}_k \right\|^2)^{1/m-1}}, \qquad i = 1,2,...,c; j = 1,2,...,p$$

The number of clusters finally generated depends on the user-specified threshold.

2.2 The Deterministic Annealing (DA) Algorithm

Deterministic annealing is an optimization algorithm based on principles of information theory and statistical mechanics. Specifically, it minimizes the expected distortion of a given system while maximizing its randomness (Shannon's entropy). Let D be the average distortion,

$$D = \sum_x \sum_y p(x,y)d(x,y), \tag{4}$$

where p(x,y) is the joint probability distribution of input x and codeword y, p(y|x) is the association probability that relates x to y, d(x,y) is the distortion measure. Shannon's entropy of the system is given by

$$H(X,Y) = -\sum_x \sum_y p(x,y)\log p(x,y). \tag{5}$$

The optimization is achieved by minimizing the Lagrangian F=D-TH, where T is the Lagrange multiplier.

When used for image vector quantization, mass-constrained DA is preferred, where the constraint $\sum_i p(y_i) = 1$ is added into the Lagrangian, resulting in the optimization of F' with respect to the centroids y_i and the encoding rule $p(y_i|x)$:

$$F' = D - T[H + \sum_i p(y_i) - 1]$$ (6)

The encoding rule and the centroids are thus given by:

$$y_i = \sum_x xp(x)p(y_i \mid x) \Big/ p(y_i)$$ (7)

where $p(y_i \mid x) = \dfrac{p(y_i)e^{-(x-y_i)^2/T}}{\sum_{j=1}^{K} p(y_j)e^{-(x-y_j)^2/T}}$, and $p(y_i) = \sum_x p(x)p(y_i \mid x).$

When the variance of a cluster reaches certain limiting value ruled by the covariance matrix of the cluster, splitting of the cluster happens. The Lagrangian parameter at which a splitting occurs is called critical temperature T_c, from the terminology of statistical physics. It is calculated by $T_c=2\lambda_{max}$, where λ_{max} is the maximum eigenvalue of the covariance matrix $C_{x|y}$ of the posterior distribution $p(x|y)$ of the cluster corresponding to codeword y:

$$C_{x|y} = \sum_x p(x \mid y)(x - y)(x - y)^t$$ (8)

As the Lagrange multiplier decreases, the number of clusters increases. Implementation of DA can be found in its original paper in [22].

2.3 The Performances of AFLC and DA

When considering the use of AFLC or DA for codebook training, a comparative evaluation of the two is necessary. However, unlike DA, which is derived strictly from a statistical framework, a mathematical analysis of AFLC is difficult because of its complicated structure. Therefore, to get around the problem, it would be appropriate to evaluate their performances based on the reconstructed image quality (relating to the clustering accuracy) and the time it takes for both to generate the same amount of clusters under the same condition.

Our experiment consists of using 11154 samples extracted from a set of 26 magnetic resonance (MR) images (which will be described in detail in section 4). Figure 2 and Figure 3 shows the performances of both algorithms in terms of codebook generation time and reconstructed image peak signal to noise ratio (PSNR) (VQ only). It is noteworthy that the time is plotted in logarithmic scale so that the large difference of the two can be illustrated. We can clearly see that the reconstructed image quality is comparable while the speed is drastically different. For example, for the MR images used, DA takes 156 hours as opposed to 2.97 hours by AFLC to generate around 800 codewords. Because of its statistical computational complexity involving updating the association probabilities and centroids, the clustering speed of DA slows down considerably when the population and dimension of samples as well as the number of existing clusters increase.

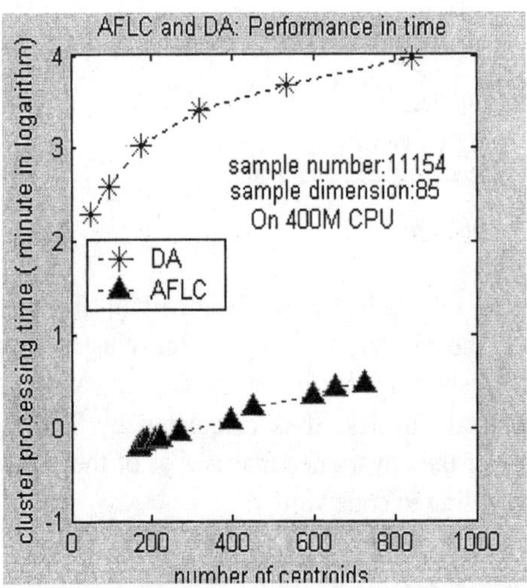

Fig. 2. Comparison between AFLC and DA in clustering speed

AFLC provides a good compromise between clustering accuracy and speed. However, what plagues AFLC is just the same problem that troubles algorithms with ART-1 structure—the sample-order- dependency. Just as the first sample is considered the first initial centroid, the centroids of the following new clusters created in the recognition layer of AFLC are initialized with the sample that does not fit in with any existing clusters. In other words, the algorithm "sees" and only considers the samples in the order they are presented. To eliminate the sample-order-dependency, proper initialization of the centroids is essential. Instead of assigning randomly incoming samples as initial centroids, the algorithm can be forced to "see" the entire sample population when a decision is made to increase the number of clusters and assign a new centroid. An algorithm aiming to improve AFLC using such a method is under development in our laboratory.

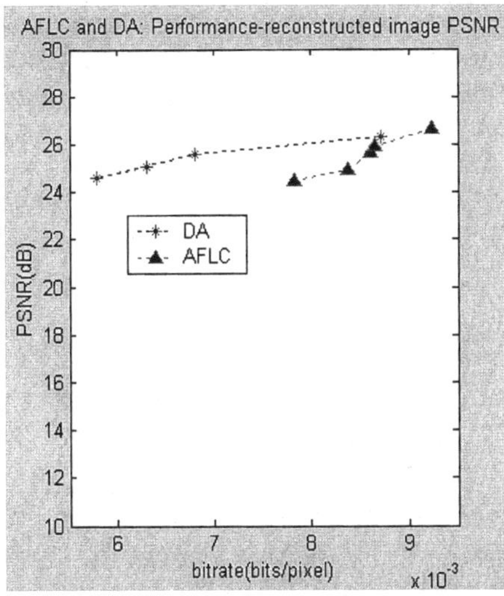

Fig. 3. Comparison of DA and AFLC in reconstructed image quality

3 Feature Extraction

Vectors formed in the spatial domain by partitioning the spatial image into non-overlapping blocks have a random distribution that cannot be embedded into the optimization of the clustering algorithm. For the DA clustering used above, we approximate the distribution function of the training sample vectors with uniform distribution, which is not necessarily true for any type of images. Since vector samples with a general distribution is always preferred over undefined distributions, it is of interest to look for a way to extract feature vectors from the image so that a generalized distribution can be applied.

The study of wavelet transformed coefficients of images show that wavelet coefficients do have such desirable property. In [1], it is shown that the wavelet coefficients can be fitted to a generalized Gaussian distribution. Further more, for multi-level wavelet transformed coefficients, it can be observed that if a wavelet coefficient at certain scale is insignificant with respect to a given threshold, then all the coefficients of the same orientation in the same spatial location at finer scales are insignificant with a high probability. This hypothesis has been successfully applied to code wavelet coefficients at successive decomposition levels using coefficient's significance maps [26, 28]. Although such zerotree based scalar coding takes advantage of the redundancy among different scales, redundancy inside the scales are not exploited. In contrast to scalar coding of wavelet coefficients, intra-scale redundancy can be easily exploited by VQ, since VQ removes redundancy inside the vector. One of the most popular but a simple way of achieving this is by blocking individual wavelet coefficients inside one scale as feature vectors, as is done in [1].

3.1 Vector Extraction

Traditionally, vectors in the wavelet domain are generated by grouping neighboring wavelet coefficients within the same subband and orientation. Square blocks as well as a combination of square blocks and orientation-dependent grouping are usually used [1, 9,12, 27,29]. After a partitioning method is chosen, bit allocation based on rate-distortion optimization is performed in order to lower the distortion [1, 19]. It decides the dimension of the vectors in different subbands and orientations as well as the codebook size for each dimension. As a result, multi-resolution codebooks are usually needed, which consist of sub-codebooks of different dimensions and sizes. Although the use of sub-codebooks makes vector-codeword matching process faster, the resulting vector dimension and codebook size become image-size dependent. It is difficult to use such vector extraction method for training and generating universal codebooks.

On the other hand, motivated by the success of the hierarchical scalar coding of wavelet transform coefficients, such as the embedded zerotree wavelet (EZW) algorithm [24] and set partitioning in hierarchical trees (SPIHT) [23], several attempts have been made on adopting similar methodology to discard insignificant vectors (or zerotrees) as a preprocessing step before the actual vector quantization is performed using traditional vector extraction method [9,15,18]. In [18], the set-partitioning approach in SPIHT is used to partially order vectors of wavelet coefficients by their vector magnitudes, followed by a multi-stage or tree-structured vector quantization for successive refinement. In [15], 21-dimensional vectors are generated by cascading vectors from lower scale to higher scales in the same orientation: in a 3-level wavelet transform, 1, 4, and 16 coefficients from the 3rd ,2nd, and 1st level bands of the same orientation are sequenced. If the magnitudes of all the elements of such a vector are less than a threshold, the vector is considered to be a zerotree and not coded. After all zerostrees are designated, the remaining coefficients are re-organized into lower dimensional vectors, and then vector quantized.

3.2 Multi-Scale Vector Extraction

Vector quantizer designs consisting zerotree elimination reduce the number of vectors to be coded by using extra bits to code the location; it also helps to include more detail information into the codebook without increasing its size. Nevertheless, the actual vector quantization is performed by using traditional vector extraction methods, in which only inter-scale redundancy is exploited.

Our approach of vector extraction resembles that of [15] but different in the way it is used as explained below. Firstly, instead of using the multi-scale vectors just for insignificant coefficient rejection, we use the entire multiscale vectors as sample vectors for codebook training. Secondly, the dimension of the vector is not limited to 21. Depending on the level of wavelet transform and the complexity of the quantizer, it can be varied. Our new way of forming sample vectors takes both dependencies into consideration. Vectors are formed by stacking blocks of wavelet coefficients at different scales at the same orientation location. Since the scale size decreases as the decomposition level goes up, block size at lower level is twice the size of that of its adjacent higher level. The same procedure is used to extract feature vectors for all three orientations. This is illustrated in Figure 4 how a multi-scale vector of dimension 85 is formed from a four level wavelet transform. The dimension of the vector is fixed once the decomposition level is fixed.

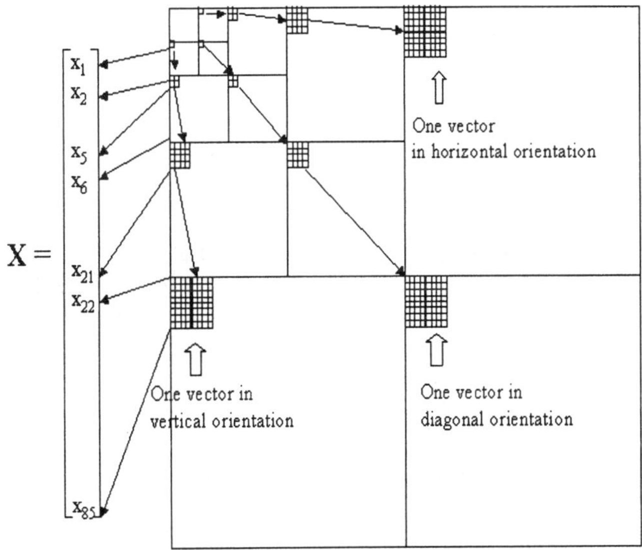

Fig. 4. Multi-scale vector extraction from 4-level wavelet coefficients

The advantages of using such multi-scale vector are the following:

1. It is image size independent. The vector dimension is fixed once the wavelet transform level is decided. The block size in a finer scale is twice of the block size in the subband immediately above. If a block in the coarser subband contains shape features of the image, for example, edges or curves, then detail information of this feature exists in the finer subband, which together with the coarser version, is included in the same vector vector. In this way, we are able to

capture image features, from the coarser version to finer version, within one vector.

2. The hierarchical structure of the wavelet coefficient magnitudes across subbands is embedded in the vector, making it as one of the common features. Thus, when vectors are trained into a codebook, the codebook incorporates both image features and wavelet coefficient properties. This common feature is illustrated in Figure 5, where a number of vectors from different images are plotted together.

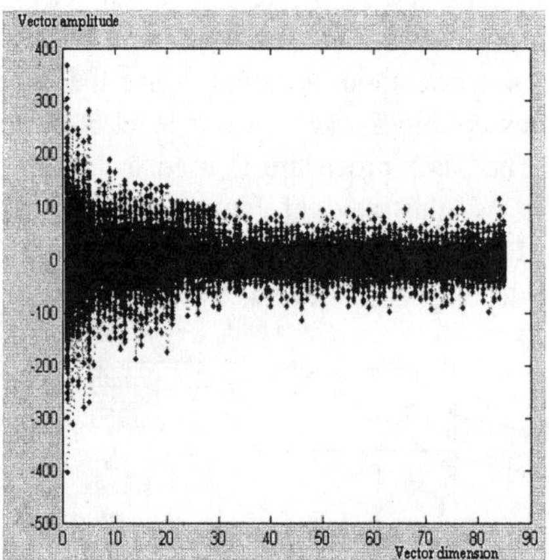

Fig. 5. Multi-scale vectors

3. Both intra-scale and inter-scale redundancy among wavelet coefficients can be efficiently exploited, since the vector contains coefficients inside the subbands and across the subbands. To compare the efficiency of using regular rectangular vectors and the multi-scale vectors, we compute the MSE of the reconstructed training samples with codebooks generated from the same wavelet coefficients. LBG is used to train the sample vectors and generate codebooks of different sizes. According to [14], the training set contains at least 10,000 vectors per codeword. We generate the training samples from two sets of T1 MR images [13,17], applying a 2-level wavelet transform to generate vectors of dimension 5 using both the multi-scale method and traditional method. To consider the square block grouping method, we also use the traditional way to generate vectors of dimension 5 and 6. Mean square error (MSE) is calculated for the whole training set. (MSE of samples from outside the training set will be slightly larger). Figure 6 shows the result. We can see that the multi-scale vector method renders a much lower MSE than those from traditional methods. For example, a MSE reduction of 0.13 and 0.8 for a moderate codebook size of 225, corresponding to an 0.9dB and 0.53 dB enhancement in PSNR.

In addition, the use of a sorted multi-scale codebook also improves lossless coding of the vector indices. In Figure 7, we compare the histogram of vector indices from unsorted and sorted codebooks of different sizes. It is obvious that the lossless coding efficiency using differential or adaptive arithmetic coding will be higher for the sorted codebook index histogram than that of the unsorted one.

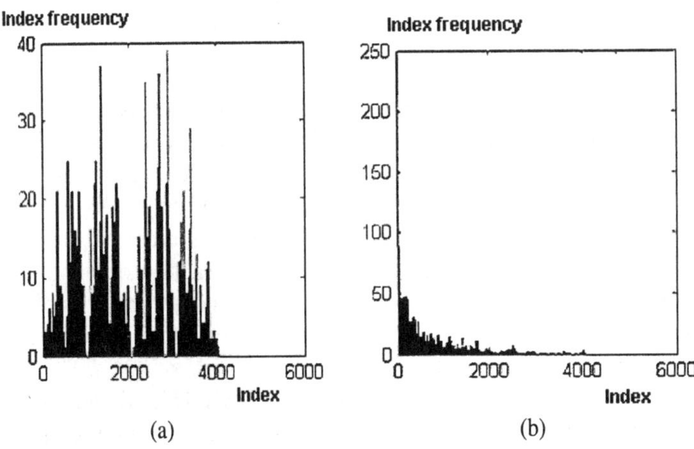

Fig. 6. Vector quantization index distribution. (a) index of unsorted codebook (b) index of sorted codebook

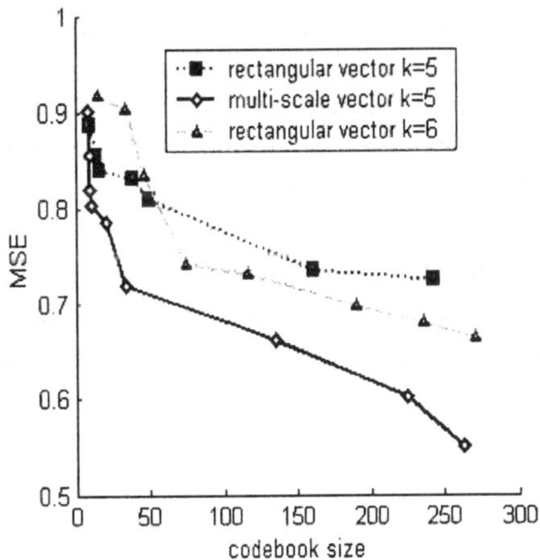

Fig. 7. Comparison between multi-scale vector and traditional vector

Depending on the level of decomposition, the multi-scale vector dimension can be 5, 21, and 85, etc, corresponding to 2, 3, and 4 level decomposition. As the dimension of the vector increases, computational complexity increases, and the codebook becomes larger.

4 Coding Method

When the multi-scale vectors are used for codebook training using the Euclidean distance as distortion measure, distortions from each coefficient of the vector are equally weighted, thus, the contribution to the distortion depends on the coefficient itself, instead of on the coefficient order. This, in fact, codes the coefficients according to their importance (large magnitude), not to their occurring order, which is the same principle that has been used in EZW and SPIHT. The advantage of using multi-scale vector is that we don't have to spend valuable bits to mark the positions of important coefficients, instead, all location information has been embedded in the vector and in the order of the vector.

Clustering algorithms such as AFLC, DA, and LBG [11,14] used to generate codebooks from a set of training samples tend to blur images because of the 'average' or 'centroid' condition used to generate the codewords. When we use the centroid of a partition to represent all the vectors within that cell, some information is lost. When the vector space is not separated into enough partitions because of the limitation of the codebook size, a lot of detail information can be lost when such codebook is used for image coding. This phenomenon is most obvious at high compression ratios (low bit rates). To address this problem, structured codebooks such as tree-structured codebooks [11,12], classified codebooks and multi-resolution codebooks [2] are often designed to increase the codebook size (to include more details) while maintaining reasonable   codebook search time.   However, the blurring effect comes essentially from the nature of codebook design methodology. For a universal codebook design, it is important to capture the common features of images into the codebook. It is not possible to include everything in it since the dimension of the codeword and the size of the codebook have to be reasonable for practical applications. Therefore, it is important to find other approaches to preserve some particular information in individual images when they are vector quantized with a universal codebook.

4.1 The New Hybrid Coding Scheme

To accomplish such a goal, we use a second step residual coding, in which the difference between the original vector and the nearest codeword is scalar quantized. The residual represents the detail information lost during vector quantization. When the codebook is well designed, the residual contains only a few large magnitude elements and the number of such elements is small. In this case, we only have to code a few large magnitude ones to compensate for the detail lost during vector coding. This is especially important at low bit rates. We have used SPIHT as our scalar quantizer for residual coding for two reasons. Firstly, the residual still maintains the hierarchical structural property of the wavelet coefficients. Secondly, with just a few important coefficients in the residual, we can use SPIHT to code them efficiently without sacrificing too many bits to location designation.

Figure 8 depicts a block diagram of the encoding and coding process of the hybrid multi-scale coding scheme. Since wavelet coefficients obey generalized Gaussian distribution, the vectors formed in this way can be approximated by a multi-dimensional generalized Gaussian distribution.

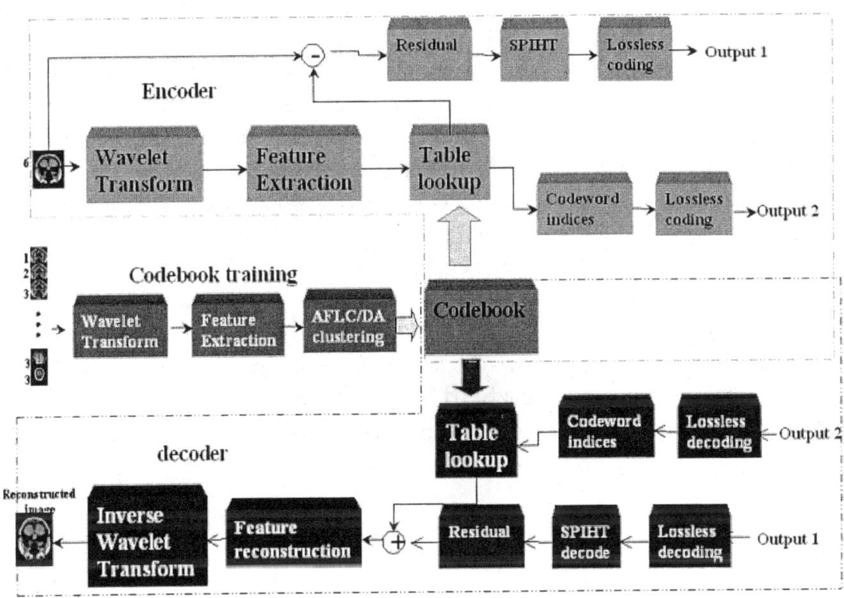

Fig. 8. . Block diagram of HMVQ coding scheme

4.2 Results

There is tremendous flexibility in using the multi-scale vector extraction to generate codebooks of different dimensions. However, as the dimension increases, computational complexity increases. Codebooks specifically designed for different types of images can also be created to improve reconstructed image quality. In this section, we will demonstrate the effectiveness of using vector quantization in combination with residual scalar quantization to compensate for the detail information lost during vector quantization, especially at low bit rates for various types of images. In example 1, images are compressed using a codebook trained from the same kind of images, namely, the MR images. In example 2, a general codebook generated from randomly selected images of different categories is used to decode the lena image. Example 3 extends HMVQ to color images. In all the above cases, the HMVQ reconstructed images shows better quality than well-known scalar codecs such as SPIHT or JPEG 2000.

Example 1: HMVQ For MRI Images

The training data we used is a group of slices (slice 1 to slice 31) from a 3D simulated MR image of a human brain [17]. Thus, the training images are reasonably different because of the span from top of the brain to the lower part of the brain despite belonging to the same class. Figure 9 shows some of the images from the training set. A few slices inside the group, for example, slice 6 and slice 12, etc are randomly chosen and excluded from the training set and later used as test images.

After the codebooks are generated, VQ is performed on the test image (slice 6, shown in Figure 10 (a)). The residual obtained by subtracting the vector-quantized coefficients from the original wavelet is then scalar-quantized with SPIHT. Figure 11 compares the reconstructed image peak signal to noise ratio (PSNR) of HMVQ coder and that of using SPIHT alone. Our coder surpasses SPIHT in all bit rates but by far at low bit rates. Figure 10(b)-(g) shows the reconstructed images at various bit rates. It is to be noted that at low bit rates the SPIHT reconstructed images appeared to be much blurrier than the content-based encoder even though PSNRs are comparable.

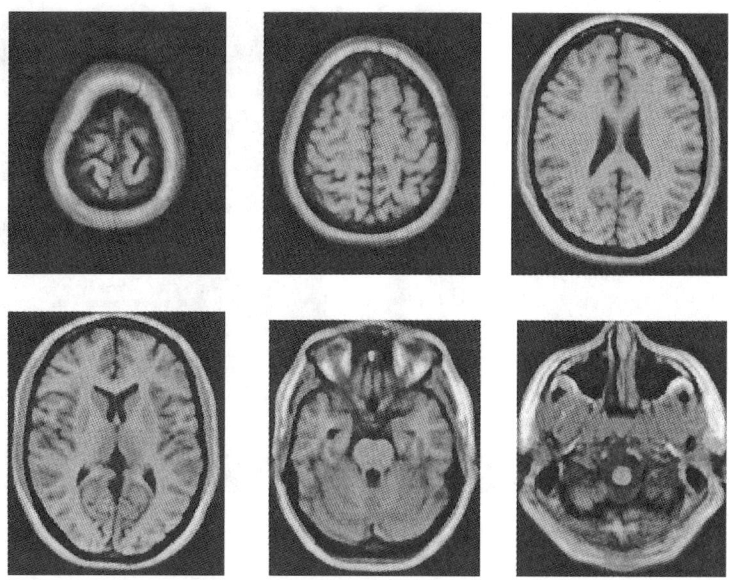

Fig. 9. Some MRI images from the training set

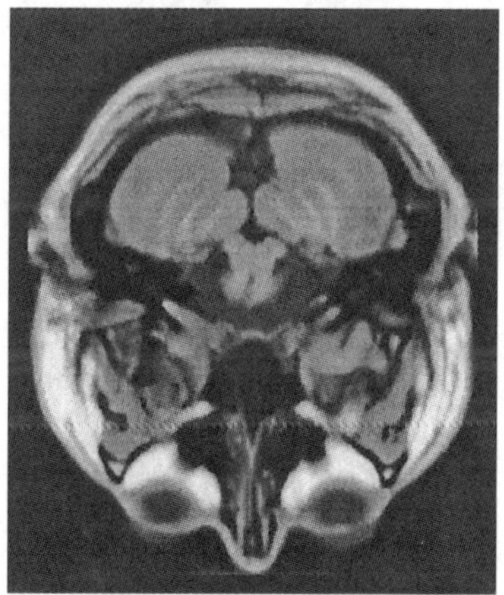

(a) Original test image (MRI slice 6)

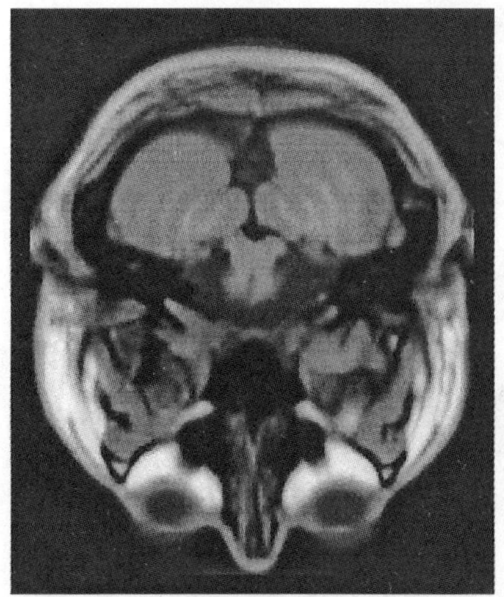

(b) HMVQ coded
Bit rate: 0.36bpp
PSNR: 40.87dB

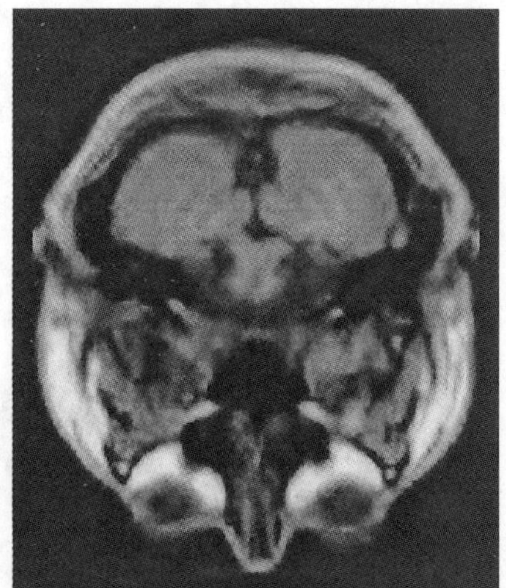

(c) HMVQ coded
Bit rate: 0.095 bpp
PSNR: 32.51 dB

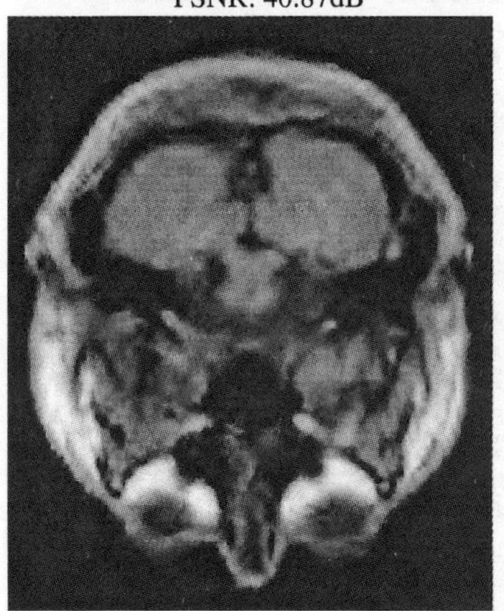

(d) HMVQ coded
Bit rate: 0.048 bpp
PSNR: 29.81 dB

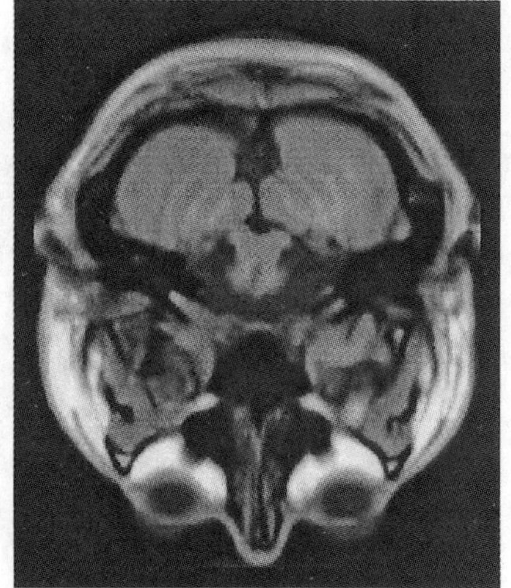

(e) SPIHT coded
Bit rate: 0.37 bpp
PSNR: 40.86 dB

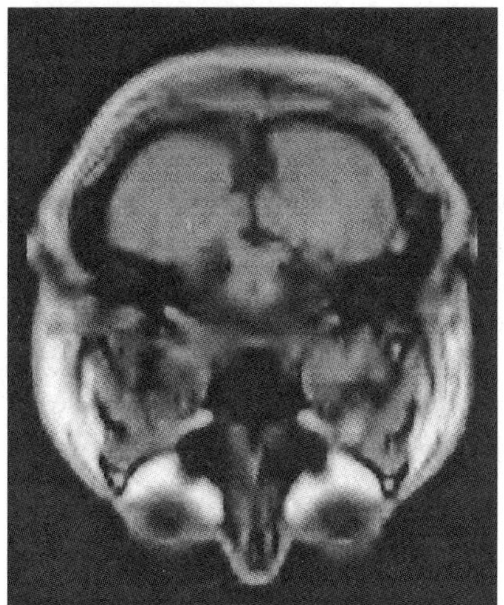
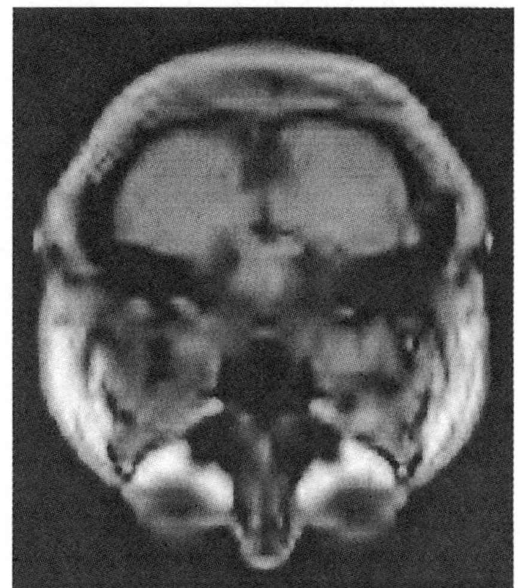

(f) SPIHT coded
Bit rate: 0.1266 bpp
PSNR: 32.53 dB

(g) SPIHT coded
Bit rate: 0.07 bpp
PSNR: 28.87 dB

Fig. 10. MRI reconstructed images from HMVQ and SPIHT

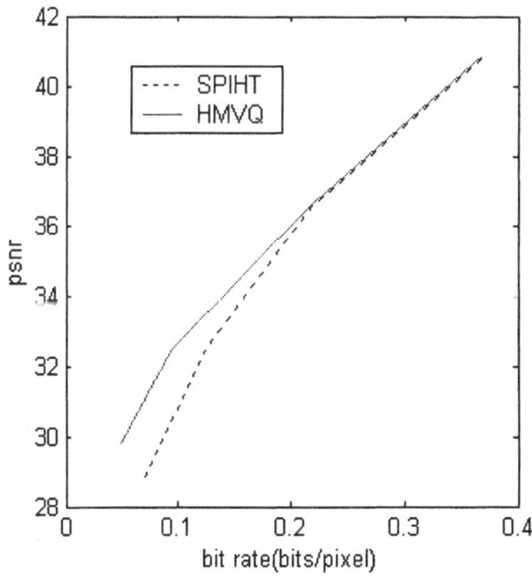

Fig. 11. Reconstructed image PSNR for HMVQ and SPIHT

Example 2: HMVQ for Images from Different Categories

HMVQ seems to perform well in the last case, in which training images belong to the same class. Here we want to test the robustness of the algorithm for images of larger variety. In the following experiment, we generate vectors of dimension 85 from a set of 28 images, most of them come from the USC standard image database, and some of them are pulled from the author's own database. A codebook size of 256 is used in this experiment. Figure 12 shows some images of the training set. They contain many more differences than the previous example. Our test image is the "lena" image (8bpp), which is outside the training set. "lena" is selected as the test image because it is the most frequently used test image in the image compression community. It contains complicated features that are ideal to test the performance of a compression algorithm. For example, "lena" has smooth areas, such as the shoulder. It also has rough and fine features, such as the feather and the rings on the hat. Using "lena" also allows us to compare our results with those of others. (Also it should be noted that there might be small differences in the electronic version of the image, which might introduce minor differences in the result, which are negligible.) Reconstructed images are given in Figure 13. As we can see, at low bit rates such as around 0.03 bpp (Figure 13 (e) and Figure 13 (f)) and 0.05bpp(Figure 13 (a) and Figure 13 (b)), reconstructed images from HMVQ demonstrate more detail information than SPIHT, even though the distortion of the image is perceptually disturbing at such low bit rates. On the other hand, at higher bit rates such as 0.2bpp (Figure 13 (c) and Figure 13 (d)), although there is a PSNR difference, the images are pleasant and the differences are subtle to the eyes. Again, at low bit rates, we are able to obtain more than 2dB increase over SPIHT. The result of this experiment is summarized in Figure 14. Also plotted in Figure 14 is the PSNR of the reconstructed "lena" using traditional rectangular vector extraction and multi-resolution codebook optimized with. To obtain the result, uses two codebooks, each having 256 codewords. We can see that that our PSNR is far better although we are using just one codebook of 256 codewords.

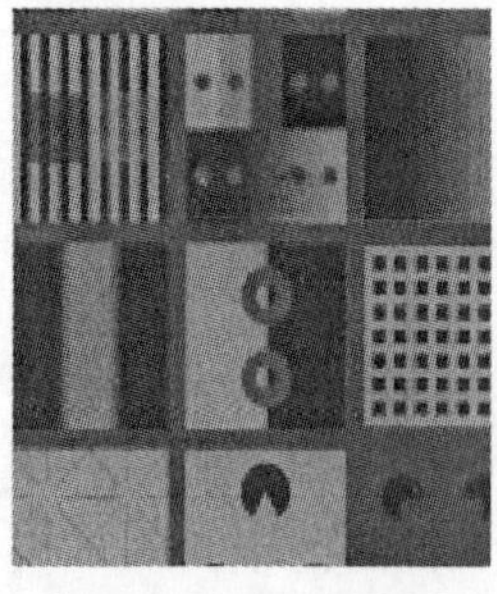  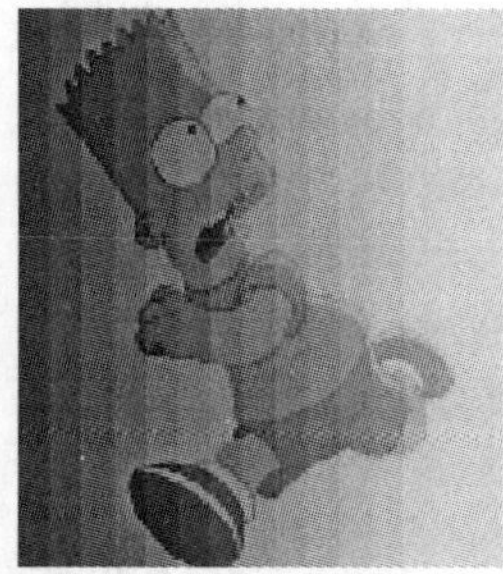

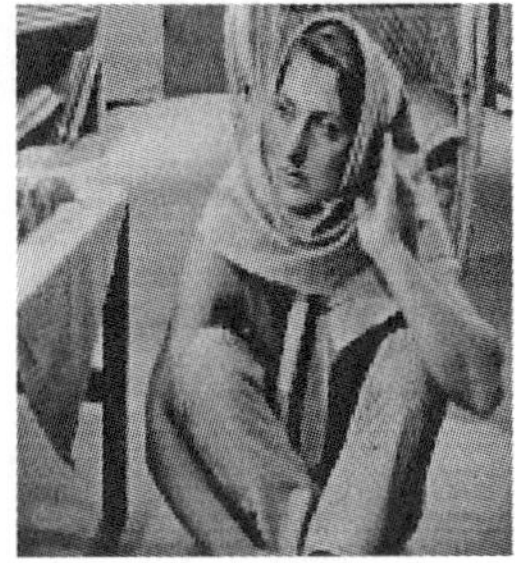

Fig. 12. Some images belonging to different categories from the training set

(a) Lena: HMVQ coded
Bit rate: 0.049bpp, PSNR: 27.48

(b) Lena: SPIHT coded
Bit rate: 0.06bpp, PSNR:26.17

(c) Lena: HMVQ coded
Bit rate: 0.2bpp, PSNR: 32.11

(d) Lena: SPIHT coded
Bit rate: 0.22bpp, PSNR: 32.24

(e) Lena: HMVQ coded
Bit rate: 0.030 bpp, PSNR: 23.57

(f) Lena: SPIHT coded
Bit rate: 0.035bpp, PSNR: 23.48

Fig. 13. Lena reconstructed images from HMVQ and SPIHT

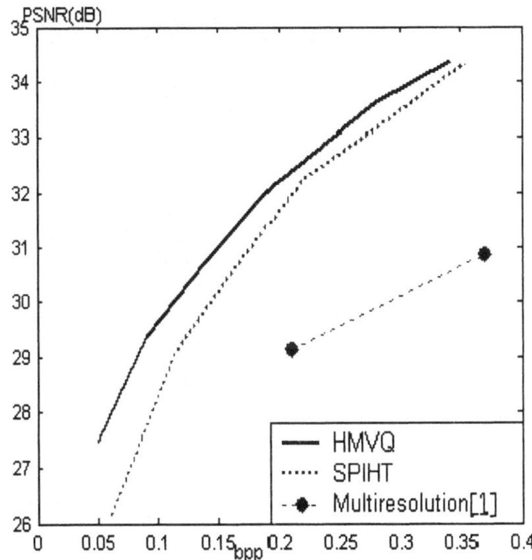

Fig. 14. PSNR of HMVQ and SPIHT for "Lena" image

To illustrated edge information in reconstructed images from HMVQ and SPIHT, Figure 15 shows the edges in a section of the images from Figure 13 (a), Figure 13 (b) and the original lena image. The Canny algorithm is used for edge detection. It is clear that the HMVQ decompressed image retains more edge information than that from SPIHT.

Example 3: Color Image Compression Using HMVQ

Application of HMVQ can also be extended to color image compression. The need can arise from high compression of color medical images from storage space reduction or faster transmission over the network. In this example, HMVQ reconstructed image is compared with that from JPEG2000 [25] at the compression ratio around 100 for a 1024 by 1024 cervigram image. Figure 16 shows a section within the region of interest in a cervigram image from each reconstructed image as well as the original image. Figure 16 (b) clearly shows higher perceptual quality and PSNR than Figure 16(c).

(a) Edges of the original lena image

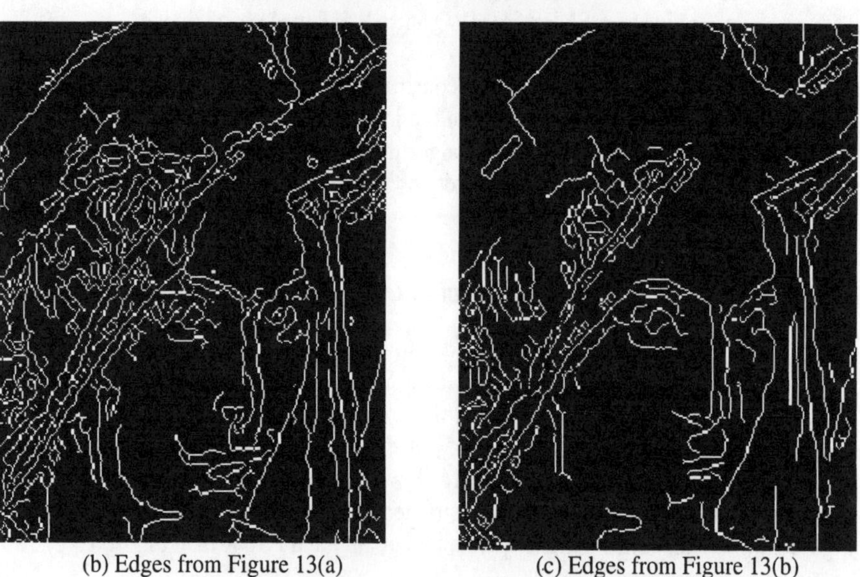

(b) Edges from Figure 13(a)   (c) Edges from Figure 13(b)

Fig. 15. Comparison of HMVQ and SPIHT on edge preservation

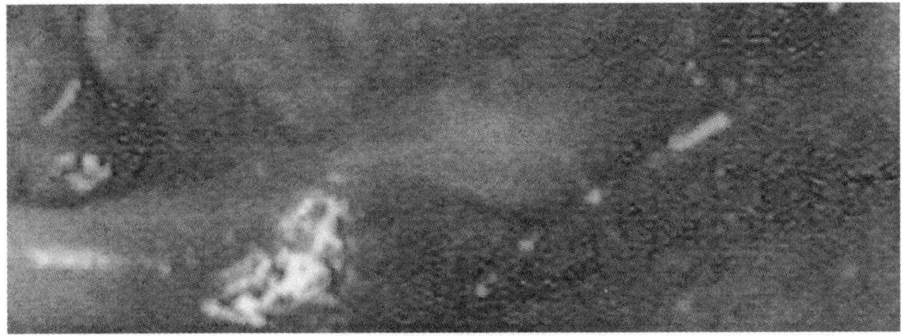

(a) Original section

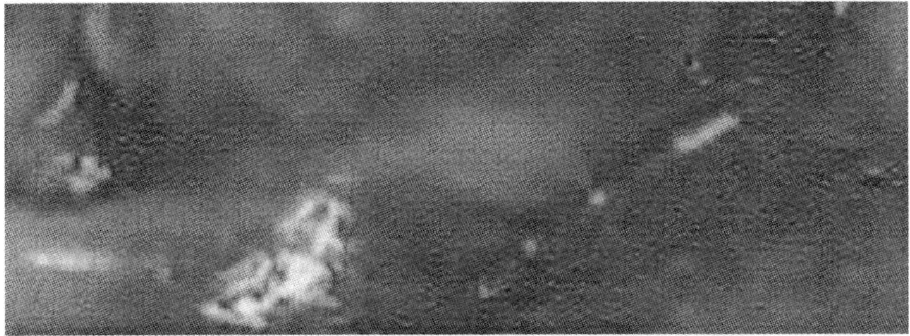

(b) HMVQ reconstructed image (CR=94, PSNR=35.67dB)

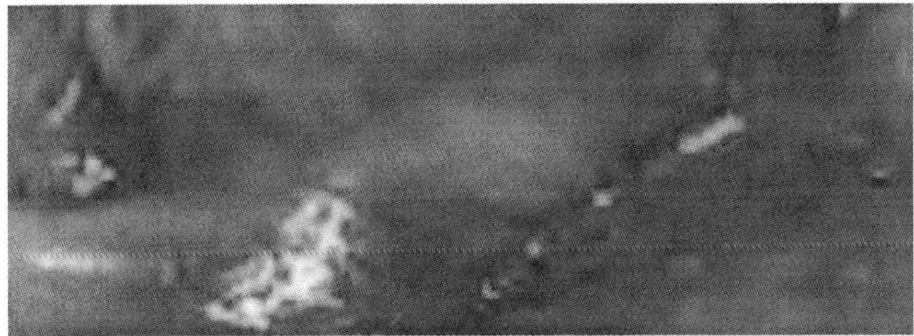

(c) JPEG2000 reconstructed image (CR=94, PSNR=30.68 dB)

Fig. 16. Color image compression using HMVQ and JPEG2000

5 Conclusions

Recent advances in image communications in many medical and industrial applications necessitate management of vast amounts of data for archival and transmission. Design of efficient image coding has, therefore, become a vigorous research area. Loss of information is not tolerated in most medical image handling thus requiring high fidelity lossy image coding since lossless coding is not sufficient in managing vast image data without resorting to image specific designs [16].

In this work, we employed two clustering algorithms (AFLC and DA) for codebook training and presented a new vector quantization scheme using a hybrid multi-scale vector quantization (HMVQ) as a content-based vector coder to preserve important image features in the wavelet domain. We also demonstrate that this coder yields much better performance than the state-of-the-art scalar quantization scheme, namely, SPIHT, at low bit rates. demonstrate its effectiveness, specifically, for medical image coding when compared with the state-of-the-art scalar quantization scheme such as the SPIHT and JPEG2000. The performance of HMVQ is particularly noticeable in retaining edge information due to the addition of residual coding. These preliminary results showing superior performance of HMVQ over SPIHT and JPEG2000 can be further improved by studying the properties of the vector indices and using a more structured codebook. Investigation of analyzing zerotree effect on codebooks and the study of the statistical properties of the codebook and codeword indices are currently in progress.

This method can be extended to encode other types of medical images such as ultrasound images, x-ray images etc., and universal codebooks can be easily generated. This shows a promising application of this encoder to medical image compression in which a codebook can be generated for specific types of medical images (for example, X-ray images). The computational complexity of DA culminates in excessively long computational time required for generating universal codebook. The use of other adaptive methods [21] for partitioning vectors drastically reduces the computation time and can be used for other practical applications such as multimedia applications [22].

We also would like to point out that the codebook we use now is a non-structured codebook. Based on the distribution of the feature vectors, we can impose certain structure on the codebook to make the coding more efficient. In addition, a better clustering algorithm that eliminates the sample-order dependency of AFLC can be used for codebook generation. These issues are being investigated currently.

We are currently developing a wavelet codec system for applying HMVQ specifically for indexing and retrieval of digitized images of vertebral radiographs taken as a part of the Second National Health and Nutrition Examination Survey (NHANES-II) organized by the National Center for Health Statistics (NCHS) and archived at the National Library of Medicine (NLM). The primary objectives of this work are to develop and implement this system that will efficiently and with high visual quality allow the retrieval of these digitized x-ray images from the NHANES II survey, and will operate compatibly with the existing Web-based Medical Information Retrieval System (WebMIRS) developed at NLM [20]. Work toward the extraction of such features by clustering algorithms is potentially of value not only for the immediate problem of image delivery and display over the World Wide Web, but also for the retrieval of images by content in future image database systems.

Acknowledgments

This research has been partially supported by funds from the Advanced Research Program (ARP) (Grant No. 003644-176-ARP), the Advanced Technology Program (ATP) (Grant # 003644-0280-ATP) of the state of Texas, and the NSF grant EIA-9980296.

The authors are grateful to Rodney Long from the National Library of Medicine for providing us with the cervigram images from the Guanacaste Project, Costa Rica, sponsored by the National Cancer Institute of USA.

References

1. Antonini M, Barlaud M, Mathieu P, Daubechies I (1992) Image coding using wavelet transform. IEEE Trans. on Image Processing 1(2):205-220
2. Antonini M, Gaidon T, Mathieu P, Barlaud M (1994) Wavelet transform and image coding. In: Baulaud M (ed) Wavelet in image communication, Elsevier, Amsterdam
3. Berger T (1971) Rate Distortion Theory. Englewood Cliffs, Prentice-Hall,NJ
4. Bezdek J (1981) Pattern recognition with fuzzy objective function algorithms, Plenum Press NY
5. Carpenter GA, Grossberg S (1987) A massively parallel architecture for a self-organizing neural pattern recognition machine. J Computer vision, graphics and image processing 37:54-115
6. Carpenter GA, Grossberg S (1987) Art-2: self organization of stable category recognition codes for analog input patterns. J Appl. Opt. 26: 4919-4930
7. Carpenter GA, Grossberg S (1990) Art-3: hierarchical search using chemical transmitters in self-organizing pattern recognition architectures. J Neural networks, 3:129-152
8. Castellanos R, Castillo H, Mitra S (1999) Performance of nonlinear methods in medical image restoration. SPIE proceedings on nonlinear image processing 3646
9. Cosman PC, Perlmutter SM, Perlmutter KO (1995) Tree-structured vector quantization with significance map for wavelet image coding. In: Proceeding of data compression conference, Snowbird Utah

10. Daubechies I (1992), Ten lectures on wavelets in CBMS conference on wavelets, Society for Industrial and Applied Mathematics 61

11. Gersho A, Gray RM (1992) Vector quantization and signal compression. Kluwer Boston MA

12. Gray RM, Neuhoff DL (1998) Quantization. IEEE Transactions on information theory 44(6): 2325-2383

13. Johnson KA, Becker JA, (2001, July). The whole brain atlas, normal brain, Atlas of normal structure and blood flow. [Online] Available: http://www.med.harvard.edu/AANLIB/cases/caseM /mr1_t/

14. Linde YL, Buzo A, Gray RM (1980) An algorithm for vector quantizer design. IEEE Trans. Commun. 28: 84-95

15. Lyons DF, Neuhoff DL, Hui D (1993) Reduced storage tree-structured vector quantization. In: Proc. IEEE Conf. Acoustics, Speech, Signal Proc. 5:602-605 Minneapolis

16. Mitra S, Yang S (1998) High fidelity adaptive vector quantization at very low bit rates for progressive transmission of radiographic images. J Electronic Imaging 11(4) Suppl. 2:24-30

17. Montréal Neurological Institute, McGill University (2001) BrainWeb: simulated brain database. Montréal Neurological Institute, McGill University, (2001, May). BrainWeb: Simulated Brain Database. [Online] Available: http://www. bic. mni. mcgill. ca/brainweb

18. Mukherjee D, Mitra SK (1998) Vector set partitioning with classified successive refinement VQ for embedded wavelet image coding. In: Proc. IEEE international symposium on circuits & systems: 25-28, Monterey CA

19. Nasrabadi N, King R (1988) Image coding using vector quantization: a review. IEEE Trans. commun. 36(8): 957-971

20. National Library of Medicine (2002) World wide web medical information retrieval system. [Online] Available: http://archive.nlm.nih.gov/proj/webmirs/

21. Newton SC, Pemmaraju S, Mitra S (1992) Adaptive fuzzy leader clustering of complex data sets in pattern recognition. IEEE Trans. Neural Networks 3:794-800

22. Rose K (1998) Deterministic annealing for clustering, compression, classification, regression, and related optimization problems. In: Proc. of IEEE 86(11)

23. Said A, Pearlman WA (1996) A new,fast and effcient image codec based on set partitioning in hierarchical trees. IEEE Trans. Circuits and systems for video technology 6(3):243-250

24. Shapiro JM (1993) Embedded image coding using zerotrees of wavelet coefficients. IEEE Trans. signal processing, 41(12): 3445-3462

25. Skodras A, Christopoulos C, Ebrahimi T (2001) The JPEG2000 still image compression standard. IEEE signal processing magazine Sept: 36-58

26. Strang G, Nguyen T (1996) Wavelets and filter banks. Wellesley-Cambridge Press, Wellesley MA

27. Van Dyck RE, Rajala SA (1994) Subband/VQ coding of color images with perceptually optimal bit allocation. IEEE Trans. circuits and systems for viedo techn. 4(1): 68-82

28. Vetterli M, Kovacevic J (1995) Wavelets and subband coding, Prentice Hall, Englewood Cliffs NJ

29. Yang S, Mitra S (2001) Rate distortion in image coding from embedded optimization constraints in vector quantization. The International Joint INNS-IEEE Conference on Neural Networks, Washington DC
30. Lotfi A. Zadeh, (1973) "Outline of a new approach to the analysis of complex systems and decision processes" IEEE Trans. On Systems, Man, and Cybernetics, SMC-3 (1): 28-44

A Reference Software Model for Intelligent Information Search

Ivan L. M. Ricarte, Fernando Gomide

School of Electrical and Computer Engineering (FEEC), State University of Campinas (Unicamp)

Campinas, São Paulo, 13083-970, Brazil

Abstract: This chapter provides a tutorial review of the current state of the art in the area of Web search and addresses information retrieval models that induce a reference software model for intelligent search systems. For these purposes, we review current information Web search models and methods from the point of view of information retrieval systems. Next, we present a reference software model which abstracts the search and retrieval process. This abstraction is important to identify the points of adaptation to integrate soft computing techniques into the information search and retrieval. We discuss the contributions that machine learning, artificial and computational intelligence brought to improve information retrieval models to enhance information search effectiveness, and to develop intelligent information search. The purpose of the model is to capture the relationships between computational intelligence and information search systems as a means to promote development and implementation of innovative, intelligent information search systems.

Introduction

A decade ago, computer technology was evolving towards cheaper and faster hardware with software components breaking the limits of the feasible with sophisticated interfaces, data and knowledge bases, and information processing systems and engines. In the mean time, the phenomenon of the World Wide Web has surprised most the technical and social world. The Web is revolutionizing the way people access information and acts. It opened up new possibilities in information dissemination and retrieval, education, business, entertainment, government, and industry. However, the Web brings difficulties to classical information processing and retrieval methodologies. Whereas relatively static collections of documents are indexed by traditional information retrieval systems, a dynamic and rapidly growing set of resources is offered by the Web.

Considerable research effort is being developed to fill the gap between traditional retrieval techniques and search needs for the Web – improved search requires more effective information retrieval. Ideally, effective information retrieval means high recall and high precision. Recall is the ratio of the number of relevant documents retrieved to the total number of relevant documents in the collection; precision is the ratio of the number of relevant documents retrieved to the total number of documents retrieved. Ideally, we would like to achieve both high recall and high precision. In practice, we must accept a compromise.

A traditional information retrieval system must perform two main tasks, building a retrieval database from its set of documents and accessing this database to retrieve relevant documents for the user (Fig. 1). The first task involves extracting from each document the set of its representative terms, which are associated with the document in the database through a set of auxiliary structures, such as inverted indexes and postings files (Salton and McGill 1983). The second task starts with a user query, which is used by the system to access relevant information from the retrieval database. This information is then returned to the user, preferably ranked by order of relevance.

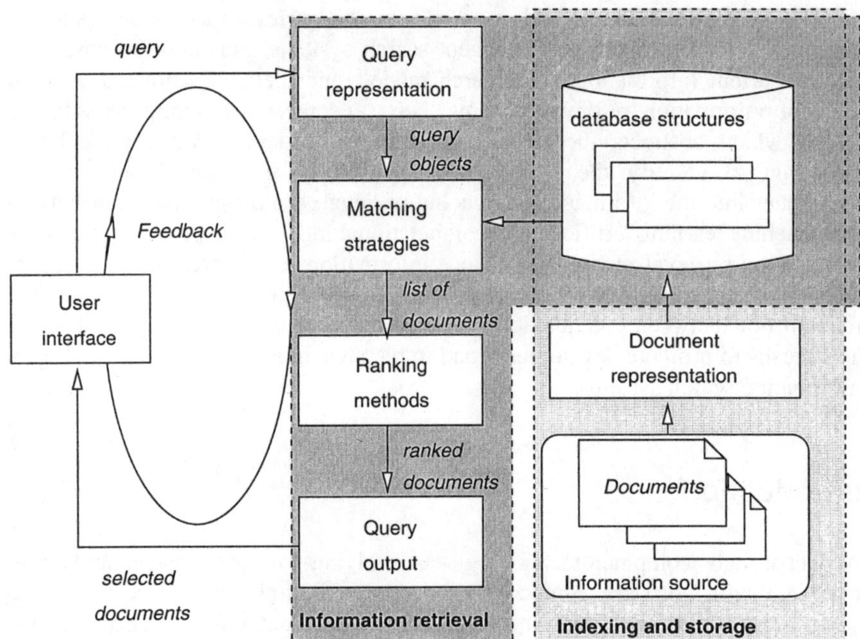

Fig. 1. Information retrieval system model

Recall and precision strongly depend on the indexing, query, matching, ranking and feedback techniques used by the information system. There are three basic models for information retrieval (Baeza-Yates and Ribeiro-Neto 1999). *Set theoretic models*, which characterize each document by a set of relevant terms, provide the basis for retrieval techniques based on a simple Boolean model. Models based on fuzzy sets have been proposed to generalize classical set operations in queries, document representation, and ranking (Radecky 1979; Harman 1992). *Algebraic models* represent each document in the collection as a vector in a high-dimensional vector space, in which dimensions are associated to terms in the vocabulary of interest. Thesauri (Joyce and Needham 1997) and latent semantic indexing (Dumais et al. 1988) are strategies to reduce the vector space dimension. *Probabilistic models* associate with each retrieved document a probability of relevance, thus capturing the uncertainty about the actual user interest on it. Through interactions with the user, the system can refine these probabilities calculation and provide better ranked selections. A related technique is *relevance feedback* (Salton and Buckley 1997), a traditional query reformulation strategy in which users indicate which documents are closer to the retrieval expectations in the initial results for their queries. Hybrid models combine techniques from these basic models.

When considering the problem of retrieving information from the Web, however, these techniques address only part of the problem, since they are imbedded within classical information models. Efforts should instead focus on the information retrieval model itself, especially its constituents and computational paradigms that support them.

The aim of this chapter is twofold: To provide a tutorial review of the current state of the art in the area of Web search, and to address information retrieval models that induce a software reference model to develop intelligent search techniques. We first review current information Web search models and methods from the point of view of information retrieval systems. From abstracting the main issues identified in this review, we present a generic software reference model. Next we discuss the potential contributions that machine learning, artificial and computational intelligence brought to improve information retrieval models to enhance information search effectiveness and to develop intelligent information search. We establish relationships between such contributions and the software reference model as a means to promote the development of innovative, intelligent information search systems. Remarks concerning relevant related issues not emphasized here and suggestions of items deserving further exploration conclude the chapter.

Information Search in the Internet

The Web is a large distributed dynamical digital information space, the most extensive and popular hypertext system in use today. The need to make sense of the swelling mass of data and misinformation that fills the Web brings a crucial information search problem. Efficient search, along with browsers, domain name servers, and markup languages, becomes an essential ingredient to turn the Web useful.

However, searching in the Internet is still far from achieving its full potential. Current search methods and tools for the Internet retrieve too many documents, of which only a small fraction is relevant to the user query. Furthermore, the most relevant documents do not necessarily appear at the top of the query output order. The design and development of current-generation search methods and tools have focused on query-processing speed and database size, but information retrieval is still based on techniques from the past forty years applied to the Web-based text. The focus should instead shift to provide a short, ranked list of meaningful documents. That is, search should shift from lexicographical spaces to conceptual spaces of documents because this is where sense may be found in a compact and meaningful form from the point of view of the user query.

There are two major ways to search for documents in the Web. One way is to use a Web agent, frequently called a robot, wanderer, worm, walker, spider, crawler, or knowbot in the literature. These software programs receive a user query and systematically explore the Web to locate documents, evaluate their relevance, and return a rank-ordered list of documents to the user. The main advantage of this approach is that the result reflects the current state of the Web, and the probability of returning an invalid reference is very low. However, this approach is impractical due to the considerable dimension of the Web space – current estimates indicate that the Web size is close to 1 billion resources[1].

The second way, the alternative currently adopted by most if not all search tools available, is to search a precompiled index built and periodically updated by Web traversing agents. The index is a searchable database that gives reference pointers to Web documents. Clearly, we note two different, albeit related flows of search tasks: the first to mediate the user with the database during information retrieval, and the second to interface Web traversing agents with the database to store and update documents (Fig. 2). Popular usage currently name systems that perform the first task as *search engines* and the ones that perform the second task as *search agents*.

Currently, most Web search schemes function similarly. The document-collecting program, the search agent, explores the hyperlinked resources of the Web looking for pages to index. References to these pages are stored in a database or repository. Finally, a retrieval program takes a user query and creates a list of links to Web documents that matches the query. Therefore, Web search systems are built upon extensions of the information system model depicted in Fig. 1. Clearly, if we assume the Web itself as an information source and add a search agent module to traverse the information source to catch documents, extract the information required to represent and to store them in a database, then a Web search system model emerges as shown in Fig. 2. However, information search in the Internet has features that expand the problem complexity.

---

[1]Some statistics can be seen on http: //www.searchengineshowdown.com.

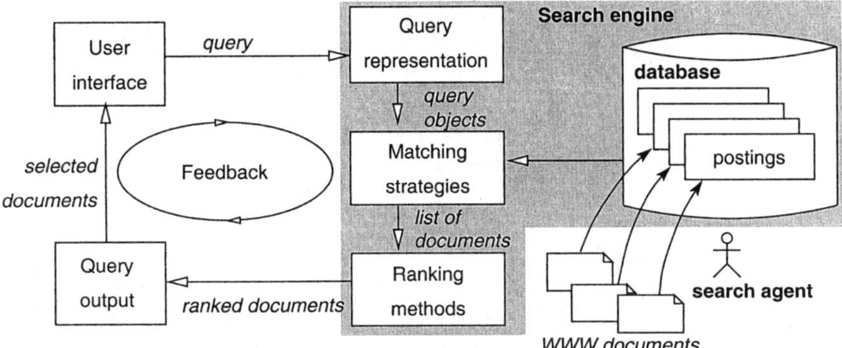

Fig. 2. Web information search model

The Web, unlike traditional information retrieval applications, has to deal with many distinct representation formats. Hypertext is structured using nodes and hyperlinks to provide nonsequential access to information. In the Web, nodes may contain documents, graphics, audio, video, and other media. For simplicity, since most Web documents are still stored as files using the Hypertext Markup Language (HTML), here we assume nodes with documents only. However, the Web was built for human use and although everything on the Web is machine readable, it is not necessarily machine understandable. Thus, it is difficult to automate anything on the Web and, because of the huge volume of information, practically impossible to handle manually.

The Web structure, with hyperlinks connecting nodes to provide a mechanism for nonlinear organization of information, is similar to that of a directed graph, so it can be traversed using graph-traversal algorithms. All search agents use essentially the same scheme to traverse the Web to fetch pages. They create a queue of pages to be explored, with at least one Web page in the queue, and choose a page from the queue to explore. Next, they fetch the chosen page and extract all the links to other pages and add any unexplored page links to the queue. The fetched page is processed to extract representation parameters such as title, headers, keywords or other representation information and store them in a database. Next, a new page is selected from the top of the queue. Search algorithms differs only in the way new elements are added to the queue, with the depth-first, breadth-first, and best-first strategies (Pearl 1984) being among the most used.

Extracting relevant information from a page to create a document descriptor is another problem. The Hypertext Markup Language provides constructs to minimally describe a document appearance, structure, and linking to other documents. The two main sections of any HTML document are represented by the HEAD and BODY elements. The head section contains elements which represent information about the document, such as its title (a TITLE element), structural relationship with other documents (LINK elements), and any property expressed as a pair name-value (using META elements). Although some search agents are able to explore information contained in this section for indexing purposes, most Web documents are created without such elements. Thus, indexing is mostly performed by extracting plain text from the body section of HTML documents. Since Web documents have not been designed to be understandable by machines, the only real form currently feasible to index information is through full-text analysis.

The open and dynamic nature of the Web also brings additional complexity to the problem. It is usual, as the result of a Web search, to receive many references to a same document – an identical copy stored in another site, a previous version still hanging around. Some documents that are referenced in the search result, since they were present in the prebuilt index, may not exist anymore in the referenced site – see Kobayashi and Takeda (2000) for an overview of Web indexing limitations.

Traditional Web search engines are based on users typing in keywords or expressions to query for the information they want to receive. These schemes completely hide the organization and content of the index from the user. There are other schemes that feature hierarchically organized subject catalog or directory of the Web, which is visible to the users as they browse and query (Gudivada et al. 1997; Mauldin 1997). Most directories rely on manually built indexes, although some may use agents to update their databases (Greenberg and Garber 2001).

Query terms are translated into query objects whose form depends on the document representation method used. Query objects are matched against document representation to check their similarity or adjacency of their postings, which are tuples consisting of at least a term, a document identifier, and the weight of that term in that document. Matching strategies may take in account distinct clustering techniques to retrieve similar documents. Besides content similarity, related documents, linked by HTML LINK elements or by hypertext links, can be clustered together. Hypertext links also provide another dimension to the search, since the retrieved document may contain the desired information or may contain links to pages with the information.

Matching documents are ranked according to a relevance measure. Relevance measure considers the number of query terms contained in the document, the frequency of the query terms used in the documents, and proximity of the query terms to each other in the document. The position where the terms occur in the document and the degree to which query terms match individual words is also considered in most of the systems. Web ranking strategies may also introduce some measures for page relevance based on number of accesses or on number of hypertext links originating from other pages.

Processing power for efficient search and information retrieval is not a major bottleneck, given the current technological standards with powerful workstations and advanced parallel processing techniques. But retrieval effectiveness is a different matter. Indexing and query terms that are too specific yields higher precision at the expense of recall. Indexing and query terms that are too broad yields higher recall at the expense of precision (Gudivada et al. 1997). Current search tools retrieve too many documents, of which only a small fraction are relevant to the user query. Clearly, this is a result of two main issues. The first refers to the limitations of the information retrieval and search models in use today, especially the ones behind query and document representation, matching strategies and ranking methods, their main building blocks. This is where doors are open for methods inspired in computational intelligence paradigms. The second concerns the lexicographical view of the models, which uses the word structure of documents content only. Enhanced models should strive to include conceptual views of documents to enhance search effectiveness.

A Reference Framework

Computational intelligence methods open new doors to information retrieval – so many that they must be explored sistematically. Information search and retrieval encompass a wide range of techniques and domains; when combined with methods from artificial and computational intelligence, the number of alternatives and possible concrete implementations becomes huge. In this section we propose a model that captures the generic aspects of this type of system, along with indicators to adaptation points in the model to incorporate techniques used to improve the information search and retrieval process.

To present this model, we adopt the primitives and vocabulary from object orientation, here graphically represented using diagrams from the Unified Modeling Language (Booch et al. 1999), adopted as standard by the Object Modeling Group (OMG). We present a set of interrelated abstract classes – a framework – along with method specifications that constitute the framework adaptation points. In this type of model, we concentrate on what has to be done rather than on how it is done. Nevertheless, the framework establishes clearly in which methods each aspect is contemplated and how the implemented classes should collaborate to perform the application tasks.

In an actual implementation, concrete classes, derived from the framework abstract classes, represent entities from the application domain. These concrete classes must implement the methods defined in the framework specification using a specific technique. In some cases, a single class will be enough to provide the required functionality. Usually, one framework class may be an abstraction of a complex subsystem, composed by many classes and external services, such as a Web server or a database management system.

We present the framework in four parts, one for the common model for documents and the remaining corresponding to the three main subsystems that can be extracted from Fig. 2, namely, search, retrieval, and user interface. This section closes with considerations regarding some implementation options.

Document model

The main point regarding the modeling of a document in this framework is that the system has no control about the document itself, which is in the Web. Thus, we have two concepts to represent documents, the WebDocument and Document classes (Fig. 3).

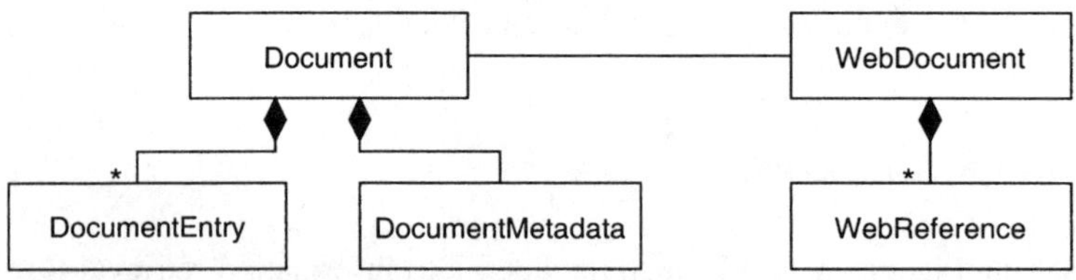

Fig. 3. Model for a document in the framework

Objects of the WebDocument class represent resources in the Web, which may have an inaccurate representation in the system as objects from the Document class. This innacuracy comes from the dynamic nature of the Web. Content changes may alter the document classification, which should be reflected by changes in the associated DocumentMetadata and DocumentEntry objects. The Web document could also be simply removed from the Web, which should be reflected by removing the Document object and associated entries and metadata from the internal database.

Another aspect which is specific to the Web is that some documents have several distinct references to it. The search agent should ideally recognize this situation and create simply another WebReference object for the same Web document, rather than creating another Document object for the same content.

Search model

The goal of the search subsystem is to create a classified collection of documents from the document domain (the Web). To achieve this goal, the search agent has scan the Web and register its findings into the retrieval database.

Using the vocabulary introduced in Fig. 3, a search agent is continuosly performing the following activities:

1. Get a WebReference from the set of unexplored references.
2. Fetch this WebReference to get the corresponding WebDocument.
3. Extract the set of WebReferences contained in this WebDocument and add it to the set of unexplored references.
4. Classify the WebDocument to generate a set of DocumentEntry.
5. Associate the set of DocumentEntry to a new Document object related to this WebDocument.
6. Generate (and associate) the DocumentMetadata for this Document.
7. Add the Document to the collection of classified documents.

There are many conditions that may prevent the search agent to perform these activities from the beginning to the end. Such conditions are discussed next.

No reference to explore. In this case, the search agent should rest and wait for someone to change this condition. This "someone" can be another agent (see below) or even human users submitting new documents to be classified, as it happens in Web directories.

The reference cannot be reached. The search agent should, in this case, move the reference to a set of "broken" references and proceed to another reference. Another agent will get a reference from this set and retry periodically to fetch it. This retry agent must be able to differentiate permanent failures (*e.g.*, a "file not found" error) from transient failures (*e.g.*, "server busy" messages). In case of permanent failure, the reference is simply eliminated. When the reference can be reached again, the retry agent put it back in the set of unexplored references.

The reference has already been explored. The search agent should, in this case, move the reference to a set of "to be verified" references and proceed to another reference. This verification step must be performed to capture eventual changes in Web content. Another agent, a verification agent, explores this set of pending references. For each reference, the WebDocument header information (*e.g.*, "last update") is checked against the Document information. If the WebDocument is newer than the Document, the verification agent proceeds to classify the new document version and update the Document entry as required.

The Web document was already classified. If the reference appears to be new but the document analysis results to a document entry already classified, then the search agent registers this WebReference as a mirror for this WebDocument and proceeds to the next unexplored reference.

These activities related to the search of documents in the Web yields a partial specification for methods associated to each class in the document framework. For example, it is clear that, given a WebReference object, someone (the search agent, the retry agent, the verification agent) should be able to fetch from this reference the corresponding WebDocument. This is translated as a method `fetch` which receives no argument (since the reference is already the object being activated by this method) and returns a reference to a WebDocument object, created upon succesfull retrieval of the resource. This initial specification is presented in Fig. 4. For clarity, basic methods (object creation, properties get and set) are omitted.

This figure also includes the DocumentBase class, which is the collection of Document objects as classified by the search agent or reviewed by the verification agent. Similarly, the agent must update the collection of document entries, EntryBase, and the collection of document metadata, MetadataBase.

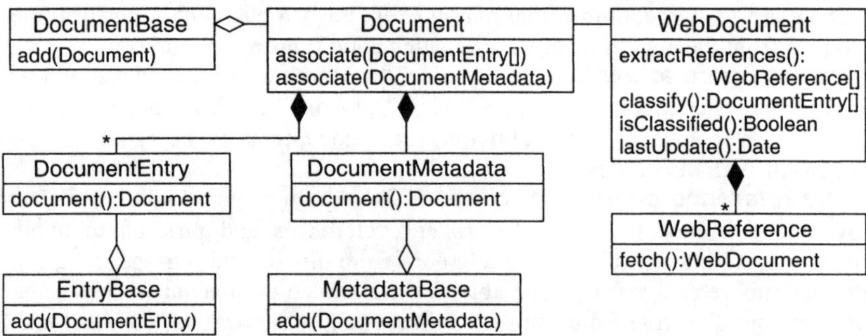

Fig. 4. Specification for the search model in the framework

We can also specify at this point the structures used by the agents to keep references. There are three sets of references, represented by classes UnexploredReferences, PendingReferences, and BrokenReferences. For each class, a ReferenceIterator is associated. The role of the iterator is to provide a mechanism to scan a collection whose specification is independent of the collection internal structure (Gamma et al. 1994). In this case, this mechanism is provided through the methods to verify that the collection has unscanned elements, hasMore, get the next element, next, and to remove an element from the collection, remove. As the three classes present the same behavior and associations, these common aspects are captured by an abstract class ReferenceCollection (Fig. 5).

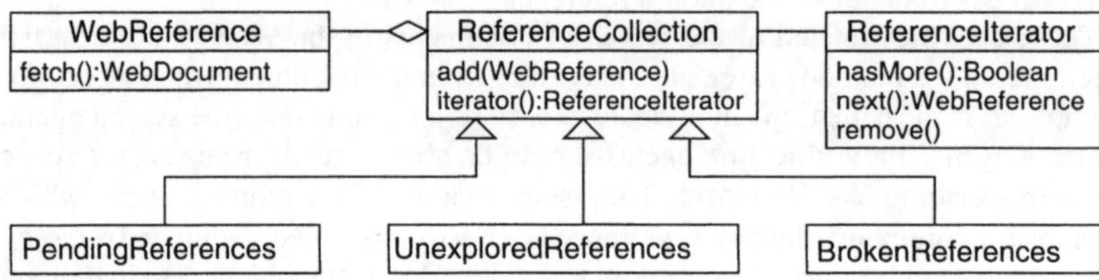

Fig. 5. Collections of references for the search model

Although some methods have straightforward implementation, others may be quite complex. Take as an example the classify method for the WebDocument class (Fig. 4). This method encapsulates the strategy used to cluster or classify documents, generating from their available data the information to be stored for posterior retrieval. This strategy could be as simple as "extract most frequent words, disregarding those in the given stop list" or could include elaborate approaches from computational intelligence to provide more effective retrieval.

320

User interface model

The retrieval process starts when the user specifies a query expressing what she is looking for. From her input, the retrieval application must be able to create an internal representation that can be used to match this input against the set of information about the stored documents. If the user's input is a text string, then there must be a method to create such UserQuery object receiving the text string as argument.

Besides creating a new UserQuery object from a raw user's input, some facility must be provided to create a UserQuery departing from another existing UserQuery object. This facility is essential when the user wants to refine her query by providing some feedback for the current query. These methods related to the user query are presented in Fig. 6.

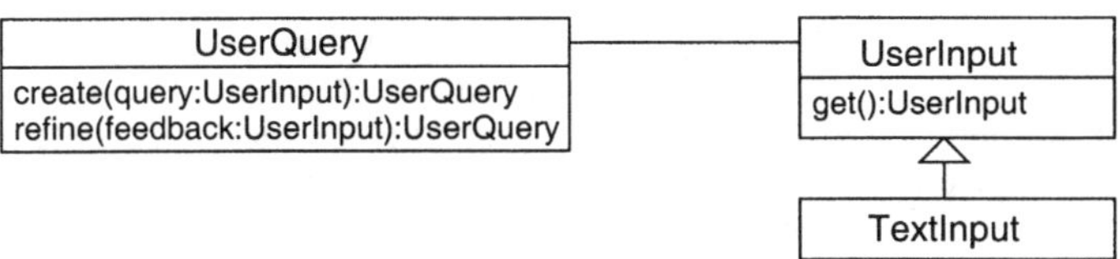

Fig. 6. Classes related to the user query

Instead of specifying the argument for the creation of a user query as being a string, a reference to an abstract class, UserInput, is used. If another type of input is to be provided, such as with a voice or a virtual reality interface, then it suffices to derive a concrete class from UserInput with the appropriate implementation of the get method. In this way, we isolate the internal retrieval process from the type of user interface.

Retrieval model

In this model, the goal of the retrieval process can be established as extracting from the collection of classified documents (Fig. 4) a subset of Document objects that matches to some degree a given UserQuery object (Fig. 6). In the following, the classes that provide support to achieve this goal are presented.

As retrieval is done based on the information extracted from the document during the search and storing process, the user query has to be matched against collections of DocumentMetadata and DocumentEntry objects. Such collections, respectively MetadataBase and EntryBase (see Fig. 4), must be built as the DocumentBase is being filled. Both have match methods which receive objects derived from the UserQuery object as arguments.

A user query, ideally expressed in natural language, may contain information that is related to the document metadata (author, publication date, title) and information regarding the document content (subject, concepts, words). Thus, from a given UserQuery, we must be able to extract a pattern to be matched against the MetadataBase, here called a set of MetadataPattern, and another to be checked against the EntryBase, the set of EntryPattern. Both sets of patterns are generated by the `extractMetadata` and `extractEntry` methods from the UserQuery class.

The `match` methods receive these objects with a pattern as arguments and return a corresponding set of matched documents. A MatchedDocumentSet object is a transient object that relates a Document to a UserQuery with some matching measure. The retrieval application receives the several sets of MatchedDocumentSet objects, one for each pattern, operates on these sets as required in the user query (unions, intersections or differences) and generates a single MatchedDocumentSet object, from which the query output is presented. As the final result is a collection of objects, it can also be scanned using an iterator, RankedIterator. The scanning order defined by this iterator is related to the ranking criteria adopted for the collection.

The framework for retrieval components is presented in Fig. 7. Some abstract classes are introduced to capture the general behavior associated to the manipulation of information about a document. Therefore, should a new type of information regarding each document be contemplated in the framework, these abstract classes determine the general behavior that the new classes must provide.

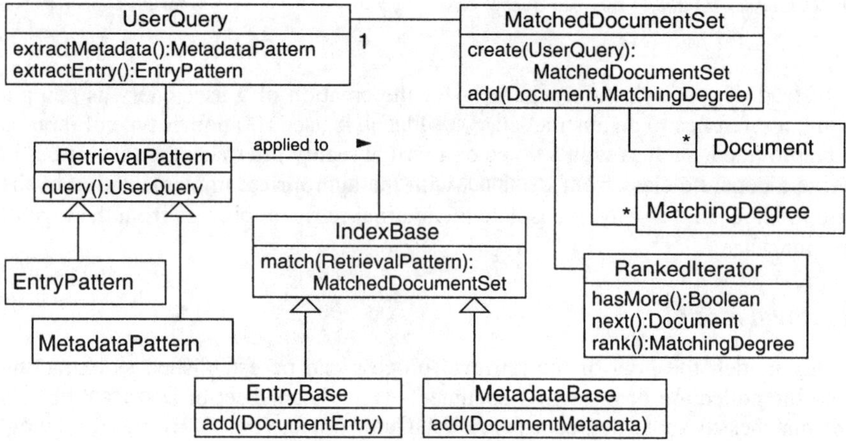

Fig. 7. Model for retrieval components

The MatchedDocumentSet object is the result generated for a given UserQuery, which is registered in the moment the object is created. Each document that matches a RetrievalPattern associated to this query is added to this set, along with information about its matching degree. When the retrieval is complete, the resulting set can be scanned in order of relevance by the associated RankedIterator, from which the documents and corresponding degree of matching are obtained.

Considerations about implementation

The complete framework for intelligent Internet search and retrieval is obtained from joining the classes and methods presented in the previous figures. A whole vision of this framework is presented in Fig. 8.

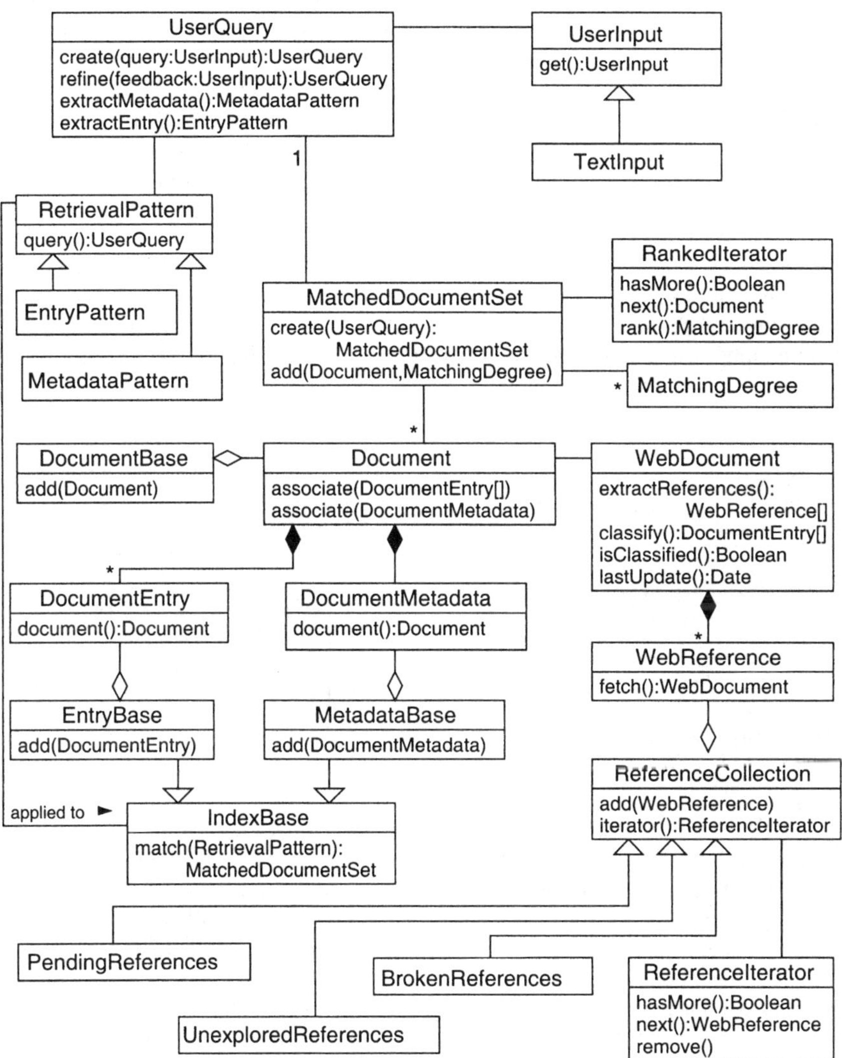

Fig. 8. Intelligent Internet search and retrieval framework

As the framework enables the simultaneous activity of several agents, it is appropriate to implement it using techniques for concurrent and distributed software. It already defines the set of interfaces required to specify distributed object architectures, such as those in the Java Remote Method Invocation architecture or in the OMG Common Object Request Broker Architecture (CORBA). Several classes in the presented framework are persistent classes, that is, the corresponding objects must have their information preserved between distinct sessions or executions of the application. The typical solution for this problem is to use a database management system.

Ideally, an object-oriented database management system would be used. In this type of system, simply defining a class that represents a collection of objects as persistent shoud suffice to store that information in the database. Retrieving information from the object-oriented database would also be transparent, without any difference from accessing a conventional (in main memory) object.

In general, however, a relational database management system is adopted, since the relational technology is well established and there are many good available implementations for this type of system. There is an impact in adopting this solution to store an object collection, which is to provide a mapping between the application class and a database accessor class. Fig. 9 illustrates this mapping for the collection of document entries.

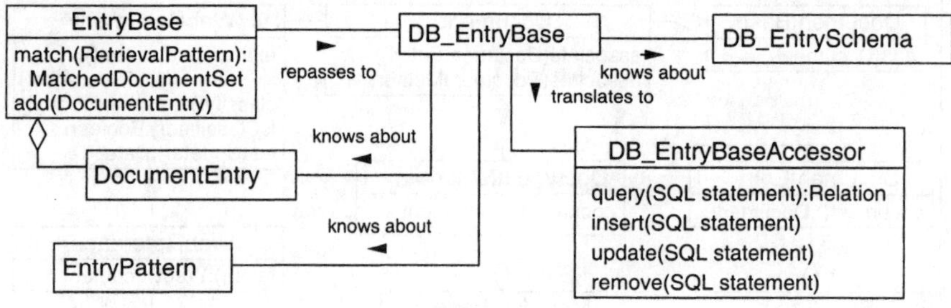

Fig. 9. Database components

EntryBase and DocumentEntry are the framework classes, and the goal is to access the collection of document entries registered in an EntryBase object through a relational database. The DB_EntryBase class is an adapter class, which receives messages from the EntryBase object and translates these to calls to the underlying database system, here accessed through an object of the DB_EntryBaseAccessor. To translate the messages correctly, the DB_EntryBase class should know about the internal structure of the framework objects, as well as about the database schema adopted to represent the document entries. With this information, an object of this class can translate these messages to SQL statements and use them, through a DB_EntryBaseAccessor object, to operate on the relational database.

A similar structure should be provided to the DocumentBase and MetadataBase.

Intelligent Information Search

The Web information search and retrieval framework provides a general model upon which distinct system implementations can be derived. Decisions related to how its methods will operate open the opportunity to incorporate techniques to improve search and retrieval effectiveness.

Many approaches have been suggested to improve search. Among them we note relevance feedback, filtering and routing techniques (Gudivada et al. 1997; Chang and Hsu 1999), context and page analysis (Lawrence and Giles 1998), semantic similarity (Green 1999), and case-based reasoning (Vollrath et al. 1998). However, these techniques address only part of the problem because they are imbedded within classical information models. Efforts should instead focus on the information retrieval model itself, especially its constituents and computational paradigms that support them.

Information retrieval using probabilistic methods has produced a substantial number of results over the past decades. In the eighties, artificial intelligence and fuzzy set theory also made an impressive contribution to intelligent information retrieval and indexing. More recently, attention has shifted to inductive learning techniques including genetic algorithms, symbolic learning, and neural networks. Most information retrieval systems still rely on conventional inverted index and query techniques, but a number of experimental systems and computational intelligence paradigms are being developed.

Knowledge-based information retrieval systems attempt to capture domain knowledge, search strategies and query refinement heuristics from information specialists. Most of these systems have been developed based on the manual knowledge acquisition process, but data mining and knowledge discovering techniques indicate a major source for automatic knowledge elicitation (Chen 1995). Some of the knowledge-based systems are computer-delegated in that decision making is delegated to the system, and others are computer-assisted wherein user and computer form a partnership (Buckland and Florian 1991). Computer-assisted systems have shown to be more useful and several systems of this type have been developed during the last decade (Chen 1995). Many of them embody forms of semantic networks to represent domain knowledge, accept natural languages queries, and include a knowledge base of search strategies and term classification similar to a thesaurus.

Systems and methods of information retrieval that are based upon the theory of fuzzy sets have been recognized as more realistic than the classical methods (Miyamoto 1990). Extensions of Boolean models to fuzzy set models redefine logical operators appropriately to include partial membership and to process user query in a manner similar to the Boolean model. Earlier information systems based on fuzzy set models are reported to be nearly as capable to discriminate among the retrieved outputs as systems based on Boolean model. Both are preferable to other models in terms of computational requirements and algorithmic complexity of indexing and computing query-document similarities (Gudivada et al. 1997). Fuzzy set models provide the means to represent document characteristics and allows user queries in the form of natural language propositions. Similarly to classical knowledge-based systems, various forms of thesaurus have been added to improve retrieval performance. Algorithms to generate pseudo-thesaurus based on cooccurrences and fuzzy set operations, and its use in information retrieval through fuzzy associations have been developed (Miyamoto 1990). More recently, the use of fuzzy relational thesauri in information retrieval and expert systems has been introduced as a classificatory problem solving approach (Larsen and Yager 1993). In general, fuzzy information retrieval systems consider each piece of knowledge as a pattern represented as a set of attributes, objects, and values. The values of the attributes can be either quantitative or qualitative and are represented by possibility distributions in the attribute domain. Fuzzy connectives allow the user to elaborate complex queries. Each query is fuzzy-matched with the characteristic patterns of each document in the database. Matching degrees provides information to rank document relevance and query output.

We note that earlier knowledge-based and fuzzy information retrieval models follow the same vein as classical models. Except for the underlying set theoretical formalism, knowledge bases to include various forms of thesaurus and expert search knowledge, and forms of natural language-based query, they share the same structure.

Unlike manual knowledge acquisition process and linguistic-based natural language processing technique used in knowledge-based systems design, learning systems rely on algorithms to extract knowledge or to identify patterns in examples of data. Various statistics-based algorithms have been used extensively for quantitative data analysis. Computational intelligence techniques, namely neural network-based approaches, evolution-based genetic algorithms and fuzzy set theory, provide drastically different schemes for data analysis and knowledge discovery.

Neural networks seem to fit well with conventional models, especially the vector space model and the probabilistic model. From a broader perspective of connectionist models, vector space model, cosine measures of similarity, automatic clustering and thesaurus can be combined into a network representation (Doszkocs et al. 1990). Neural nets can been used to cluster documents and to develop interconnected, weight/labeled networks of keyword for conceptbased information retrieval (Chen 1995). Generally speaking, neural networks provide a convenient knowledge representation for information retrieval applications in which nodes typically represent objects such as keywords, authors, citations and their weighted associations of relevance.

Often compared with neural networks and symbolic learning methods, the self-adaptiveness property of genetic algorithms is also appealing for information retrieval systems. For instance, genetic algorithms may find the keywords that best describe a set of user provided documents (Chen 1995). In this case, a keyword represents a gene, a user-selected document represents a chromosome, and a set of user-selected documents represents the initial population. The fitness of each document is based on its relevance to the documents in the user-selected set, as measured by the Jaccuard's score. The higher the Jaccuard's score, the stronger the relevance. Genetic algorithms have also been used to extract keywords and to tune keyword weights (Horng and Yeh 2000). Genetic algorithms and genetic fuzzy systems are ones of the most fertile computational intelligence tools for Web search engines. This is to be contrasted with the knowledge and neural network-based approaches discussed above, which has shown modest presence in the Web world. Genetic algorithms can dynamically take starting homepages selected by the user and search for the most closely related homepages in the Web based on their links and on keyword indexing (Hsinchun et al. 1998). More advanced systems use genetic information retrieval agents to filter Internet documents to learn the user information needs. A population of chromosomes with fixed length represents user preferences. Each chromosome can be associated with a fitness value to be considered the system belief in the hypothesis that the chromosome, as a query, represents the user information needs. In a chromosome, every gene characterizes documents by a keyword and an associated occurrence frequency, represented by a fuzzy subset of the set of positive integers. Based on the user's evaluation of the document retrieved by the chromosome, compared to the score computed by the systems, the fitness of the chromosomes are adjusted (Matin-Bautista and Vila 1999). A similar, collaborative approach to develop personalized intelligent assistants uses a metagenetic algorithm to evolve populations of keywords and logic operators (Horng and Yeh 2000). In this approach, a primary genetic algorithm creates a population of sets of unique keywords selected at random from a dictionary. A secondary genetic algorithm then creates a population of sets of logic operators for each of the primary genetic algorithm members.

Concluding Remarks

Currently, major interest concentrates on structuring and turning the Web more amenable to computer understanding (Cherry 2002). The aim is to include semantic processing to capture the intended meaning of user words and queries. Semantics continues to be a key, but difficult issue in any human-centered computer and complex systems. Semantic processing provides a way to enlarge syntactic processing through relationships between terms and real things, and ontology. The Semantic Web is the most visible attempt in this direction (Fensel and Musen 2001). Structuring induces a mean to exploit taxonomies and to analyze links to create implicit links between pieces of information.

Despite the value the Semantic Web might bring, conventional Web search completed with soft computing technologies will continue to play a significant role in the near future. For instance, scatter and gather approaches to cluster concepts is a mechanism to dynamically process single topic words to discover particularly desirable sets of words (Fensel and Musen 2001). An alternative approach for dynamic knowledge processing is to use participatory learning (Yager 1990) to create clusters of words (Silva and Gomide 2002) and relate them in conceptual networks. This seems to be a key issue to improve natural language query. Natural language query provides search engines a way to infer labels and field names of each token in the query string. Hence, it can parse the entire query string, label its tokens, and hold an interactive conversation about the query to confirm its expectations and to approximate as much as possible to the intended meaning. Therefore, participatory learning turns natural language query feasible to construct intelligent search engines. In addition to participatory learning, reinforcement learning has shown to be successful in specialized domains (Siverman and Al-Akharas 2001). Fuzzy semantic typing (Subasic and Huettner 2001) is an example on how to deal with ambiguity of words in natural language processing. New information search systems (Fig. 10) should process queries in a user-directed fashion and have learning capabilities to capture the dynamic nature of the domain knowledge involved and to model the user requirements.

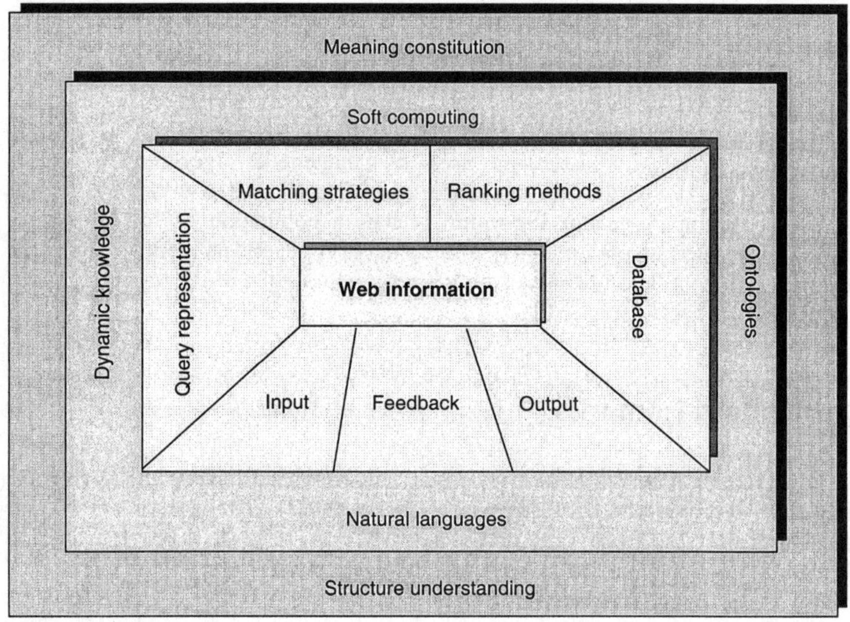

Fig. 10. Intelligent Web information search and retrieval

References

Baeza-Yates R, Ribeiro-Neto B (1999) Modern information retrieval. ACM Press, New York

Booch G, Rumbaugh J, Jacobson I (1999) The Unified Modeling Language user guide. Addison-Wesley, Boston

Buckland M, Florian D (1991) Expertise, task complexity, and artificial intelligence: A conceptual framework. Journal of the American Society for Information Science 42:635–643

Chang C, Hsu C (1999) Enabling concept-based relevance feedback for information retrieval on the WWW. IEEE Transactions on Data and Knowledge Engineering 11:595–609

Chen H (1995) Machine learning for information retrieval: Neural networks, symbolic learning, and genetic algorithms. Journal of the American Society for Information Science 43:194–216

Cherry S (2002) Weaving a web of ideas. IEEE Spectrum 9:65–69

Doszkocs T, Reggia J, Lin X (1990) Connectionist models and information retrieval. Annual Review of Information Science na Technology 25:209–270

Dumais ST, Furnas GW, Landauer TK, Deerwester S, Harshman R (1988) Using latent semantic analysis to improve access to textual information. In: O'Hare JJ (ed) Conference proceedings on human factors in computing systems. ACM Press, New York, pp 281–285

Fensel D, Musen M (2001) The Semantic Web: A brain for humankind. IEEE Intelligent Systems 2:24–25

Gamma E, Helm R, Johnson R, Vlissides J (1994) Design patterns: Elements of reusable object-oriented software. Addison-Wesley, Reading

Green S (1999) Building hypertext links by computing semantic similarity. IEEE Transactions on Data and Knowledge Engineering 11:713–730

Greenberg I, Garber L (2001) Searching for new search technologies. IEEE Computer 32:142–144

Gudivada V, Ragahavan V, Gorski W, Kasanagottu R (1997) Information retrieval on the World Wide Web. IEEE Internet Computing 1(5):58–68

Harman D (1992) Ranking algorithms. In: Frakes W, Baeza-Yates R (eds) Information retrieval data structures and algorithms. Prentice-Hall, Upper Saddle-River, pp 363–392

Horng J, Yeh C (2000) Applying genetic algorithms to query optimization in document retrieval. Information Processing and Management 36:737–759

Hsinchun C, Ming C, Ramsey M, Yang C (1998) An intelligent personal spider (agent) for dynamic Internet/intranet searching. Decision Support Systems 23:41–58

Joyce T, Needham RM (1997) The thesaurus approach to information retrieval. In: Sparck-Jones K, Willett P (eds) Readings in information retrieval. Morgan Kauffman, San Francisco, pp 15–20

Kobayashi M, Takeda K (2000) Information retrieval on the Web. ACM Computing Surveys 32(2):144–173

Larsen H, Yager R (1993) The use of fuzzy relational thesauri for classificatory problem solving in information retrieval and expert systems. IEEE Transactions on Systems, Man, and Cybernetics 23:31–41

Lawrence S, Giles C (1998) Context and page analysis for improved Web search. IEEE Internet Computing 2:38–46

Matin-Bautista M, Vila M (1999) A fuzzy genetic algorithm approach to an adaptive information retrieval agent. Journal of the American Society for Information Science 50:760–771

Mauldin M (1997) Lycos: Design choices in an Internet search service. IEEE Expert 12:811

Miyamoto S (1990) Fuzzy sets in information retrieval and cluster analysis. Kluwer, Boston

Pearl J (1984) Heuristics: Intelligent search strategies for computer problem solving. Addison Wesley, Reading

Radecki T (1979) Fuzzy set theoretical approach to document retrieval. Information Processing and Management 15:247–259

Salton G, Buckley C (1997) Improving retrieval performance by relevance feedback. In: Sparck-Jones K, Willett P (eds) Readings in information retrieval. Morgan Kauffman, San Francisco, pp 355–364

Salton G, McGill MJ (1983) Introduction to modern information retrieval. McGraw-Hill International, Tokyo

Silva L, Gomide F (2002) Participatory learning in fuzzy clustering of data. Unicamp-FEEC-DCA Internal Report, pp 1–12

Siverman B, Al-Akharas K (2001) Do what I mean: Online shopping with a natural language search agent. IEEE Intelligent Systems 16:48–53

Subasic P, Huettner A (2001) Affect analysis of text using fuzzy semantic typing. IEEE Transactions of Fuzzy Systems 9:483–496

Vollrath I, Wilke W, Bergmann R (1998) Case-based reasoning support for online catalog sales. IEEE Internet Computing 2:47–54

Yager R (1990) A model of participatory learning. IEEE Transactions on Systems, Man and Cybernetics 20:1229–1234

Soft Computing and Personalized Information Provision

Marcus Thint, Simon Case, Ben Azvine
thintm@info.bt.com, {simon.case / ben.azvine}@bt.com
BTexact Technologies
Adastral Park
Martlesham Heath
IP5 3RE
UK

Abstract: In the recent past, the development and deployment of software agents as digital assistants for managing electronic information has steadily increased. While corporate intranets and the world wide web provide the appropriate infrastructure for agent-based applications, automating the search, access, and filtering of information specifically relevant to the individual, from diverse and distributed resources poses a difficult challenge. We discuss the issues and design considerations for personalised information management frameworks in networked environments, and present one implementation solution that ensures privacy, promotes trust, and achieves efficient information sharing among the software agents and the users. We also present selected agent services in the framework that employs soft computing to improve agent autonomy, quality of service, and human-computer interface.

1. Introduction

Organizations typically store information in many different places and in many different formats. As the needs of the organization change, new repositories are created and new data formats are used. However, each repository typically has its own means of access and method of storage, requiring different tools and the know-how to access the information. In this way, new projects and initiatives tend to force information into information islands. This is also true in the Internet environment where individuals or companies post information in uncoordinated, independent fashion, in numerous formats standardized or otherwise.

Hence, the result is that users (of information) can be adversely affected by the heterogeneous sources. They become unaware of all relevant information sources, do not have the time to search all those sources, or do not have the resources to access the relevant sources. Moreover, the cost of missing important pieces of information is rising, especially for time-sensitive applications. Clearly, tools for finding quality information and filtering relevant content for the individual are becoming not only desirable, but a necessity.

Software agents are becoming essential components of productivity management tools for corporate and personal use. Solutions that integrate flexible platforms, advanced middleware, software agents, and personalised services have emerged[1-3], and will soon evolve into mature products. There are, however, many practical challenges faced by software agent developers including: interfacing to legacy systems; developing efficient distributed systems; building agent competence and user trust, and privacy issues. This paper discusses the issues and design considerations for personalised information management services. It begins with a brief overview of software agents, followed by key issues pertaining to personal agents systems in particular. Then, we describe how those issues are addressed in one personal assistant framework (iPAF) including the applications of soft computing techniques to augment agent intelligence.

1.1 Software Agents

Software agents are essentially software modules with *some* intelligence that can perform tasks delegated to them at a high level. The agents are sufficiently autonomous to decide how and when they should perform the task, and some agents have the ability to negotiate the task or even reject the task. The diverse types of software agent are covered in depth in Nwana's paper [4]. The personal assistant agents championed by Maes [5] have received much attention in the recent past because their utility is easily understood and they are desired by many computer users. The idea is that these personal assistants would take over some of the more mundane, repetitive tasks in the office environment. The agents would "look over your shoulder" and monitor how you performed certain tasks, when they were confident how you would react in certain situations, they take over or assist in that task. These could be agents to look after your e-mail: filtering out the low priority or junk mail; agents that managed your diary; or agents to draw your attention to relevant or interesting news. These types of agents have one thing in common — they are all personalised for a particular user and adapt to the individual's behavior.

Another form of software agents are those that collaborate to solve problems. Each agent possesses sufficient "intelligence" to schedule its work and delegate or negotiate with other agents to perform (other/sub) tasks on their behalf. For example, these agents could represent different specialists in a business, each called upon to perform work at the right time as part of a business process. In order to collaborate with others, these agents need to know about other agents in the society, and how they can be contacted, and negotiated with on task issues. In the personal assistant framework discussed in this paper (iPAF), we employ personal agents that also collaborate to share information and contribute to user modeling.

1.2 Agent Systems

Common issues that have to be addressed in all agent systems include task handling, agent systems development, and agent competence. Moreover, issues that are specifically important to personal agent systems include trust, privacy, and security as discussed in Section 2.

Software agents perform tasks on behalf of their "users", whether a human or another agent. This requires agents to understand task requests, translating them into the actions required to carry out their tasks, and interact appropriately with other software systems (databases, information management systems, legacy systems). This is accomplished via an explicit command via a user interface, an implicit designed task (e.g. generate a daily personalised newspaper), or via agent-to-agent message exchange. The inter-agent interface may be system-specific protocols, distributed object-based (e.g. CORBA), or standard agent languages such as KQML (Knowledge Query and Manipulation Language) [6], or ACL from FIPA (the Foundation for Intelligent Physical Agents) [7]. The latter languages provide the infrastructure for communications between agents, and thus allows for collaboration between agents of different systems. If an agent performs tasks that necessitate interfacing to external or legacy systems , they must be given access to, or possess methods to request access to those systems. Unfortunately, if some sources do not have a well-defined API, agents may even have to resort to screen scraping (sending command sequences to obtain information targeted for screen/human user and extracting data as necessary).

For agent systems design, developers could create agents from scratch, or employ tools such as Zeus [8], JADE [9], MAST[10], ABE[11], AdventNet [12], and others. Background knowledge in software agents, strengths and limitations of such development environments and methodologies are important to design, develop and debug multi-agent systems.

In order to be "useful", agents must possess a specialized skill or knowledge and some degree of autonomy to successfully engage and complete their tasks. Preferably, agents should adapt in the environment where they operate and improve service or their performance over time. Agents that cooperate and negotiate with one another will need to learn which agents are prompt and helpful, so as to devise the better negotiation strategy and increase competence over time. Agents that act on behalf of a single user need to take into account the particulars of that user's needs in order to be effective.

2. Personal Agent Systems

Personal agents are an important component in the management of electronic information, and will play an increasing role in our corporate and personal lives. Agents that proactively engage in tasks on behalf of the user to find, filter, and present relevant information to the user in the most appropriate manner are the desired goal of personal agent systems. Personal agents have been constructed to assist users with tasks such as information retrieval, information filtering, email handling and meeting scheduling [4,13].

For personal agents to provide an effective, value-added services to a user, they must know their user's preferences and habits. Furthermore, accurate user profiling is a critical to the personalisation of agent-based services. Profile construction may be manual or automatic [25-26], although fully automatic profiling remains an active research topic. With knowledge about the user, personal agents can specifically tailor how they interact with the user. Since personal agents operate in environments where they have access to personal and confidential information, security, trust, and privacy are of great importance. Other desirable features for personal agents systems are also discussed below.

2.1 Trust and Privacy

A personal agent may be tasked to prioritise e-mail, find information about a specific interest topic, or negotiate for goods and services for its "owner". The knowledge and information related to these tasks may be of a private and personal nature - not something openly shared by the owner. Moreover, some agents may be authorised to share limited information about its owner with other agents to obtain information or services, so they must also preserve confidentiality and pass on only the data pre-approved for restricted sharing. In order for the use of personal agents to proliferate, owners must develop trust in their digital assistants, which requires owners' confidence in the competency and security of their personal agents.

Many of the underlying privacy issues are being addressed by the cryptographic infrastructures being put in place, for example the use of high encryption, public and private keys [14], together with some certification authority (e.g. Verisign). These can provide basic services such as: encryption of data during transmission; authentication of the sender or receiver of commands or information; digital signatures to prove that the message has not been tampered with on route; time-stamping to prove when it was received; and non-repudiation mechanisms to ensure that if a person or system received the message, they cannot deny it was received. Moreover, personal information and preferences should only be shared in a very controlled manner. Initiatives such as the Open Profiling Standard [15] go some way towards this. The standard only allows information to be shared directly between two parties — a client (the user) and server of a web site (the product or service provider), and the user is made aware that information has been requested.

Notwithstanding the above, as members of an e-community, users will wonder: Who can access the contents of my profile? How much (which fields) can they see? How do I *really* know what information is being shared? Despite all good intentions by the developers, how do I know an agent did not violate privacy due to a programming bug?, etc... These are all valid concerns, and the system designers can provide certain features to alleviate most of these concerns as summarized in Section 3. The un-intentional programming bug is a bane to all software, and not unique to the personal agent environment (although risks are perceived as much higher). Although this issue may never be fully resolved, it can be overcome gradually, through reputation of the system developers, system performance, and the absence of adverse incidents and privacy violations from select vendors.

2.2 Security Issues

The security regarding personal agent systems affects two areas: storage of personal information, and communication of such information between personal agents, and other users. Personal profile may be obtained directly from the user, or through the use of machine learning techniques to automatically extract the relevant profile. Whatever the method for creating the information, its storage and use should be secure. An appropriate method of securely storing personal information would be key based data encryption, such that the data could not be easily deciphered by malicious agents or humans. This would require an agent to have the proper key (e.g. assigned by the system administrator or its owner/user) and the ability to encrypt and decrypt information dynamically. Typically, only one profile manager agent would have the use of the key and be accountable for the secured access, both to retrieve and store information. An alternative is to store personal profiles in a commercial database (e.g. Oracle) using secured modes of data entry and access.

A user may have several personal agents performing different tasks. These service agents must make requests to the profile manager agent (via an established API) to obtain personal user information. Within an agent framework, all service agents could be required to use a specified protocol/password when requesting user profile information to ensure that they are among the approved consumer of such information. In turn, the profiler agent must ensure not to divulge data that has been marked private by the owner. Hence, this is a section of code that must be carefully checked to guarantee that all necessary privacy level checks are performed without fail.

2.3 Other Desirables for Personal Agent Systems

A Common Profile: A personal agent framework should employ a common profile that can be shared among all the users' agents. Each personal agent should not have to construct a profile through queries or interaction with the user; the profile should be collectively built and used by all agents. A common profile improves the accuracy, efficiency, and security of personal data. The need for a standard profile is also argued well by the Open Profiling Standard committee, which seeks to provide a standard, controlled access to a personal profile for web based applications.

Access to Diverse Sources: The profile content should reflect, and be constructed from diverse information sources as possible, so that the agents can provide a more complete, well rounded service. For example, agents should not only be able to provide appropriate documents about a particular interest, but also the right people (contacts), events and timely reminders. Hence, it is highly desirable that personal agents have access to many different systems/sources that the user interacts with, to be able to build up a useful profile. Previous research has also shown that to generate and maintain an accurate user profile, information from as many different sources should be used [16].

Adaptability: Typically, the profile of active users would not be static; people's interests, work assignments, and curiosities change, and ideally, the personal agents must be able to adapt to those changes. The alternative to this is to place the onus on the user to inform his/her agent when there are any new/changed interests or preferences. Although this is a possible short term solution, it is not a desirable operation mode, since it would quickly become tedious for the user to keep remembering to update his/her profile. Quick, superficial surveys provide insufficient details, and repeated lengthy surveys would irritate the user. Hence, agents must become sufficiently intelligent to detect and adapt to changes in users' needs.

Information Sharing: Just as it is beneficial for people to share information among each other, it is often more efficient or advantageous for software agents to exchange information to complete their tasks. Thus, in addition to sharing profile contents, any other information (e.g. which documents the user has read/written) should also be shared among the framework's agents. Having information that an agent cannot acquire independently allows that agent to extend its services – even if indirectly by passing the user onto other agents for complementary services.

Agent Collaboration: Personal agents may need to collaborate with each other to perform tasks for their users. A novice personal agent may need to ask other personal agents about resources or procedures to perform a certain task. An agent may also delegate or request a sub-task to other agents for load balancing or efficiency reasons. It is important that a multi-agent framework is sufficiently flexible to allow for these types of collaboration, even if all features are not used in the initial deployment.

3. A Personal Assistant Framework

Initially, personal agents have been developed in isolation from one another, their purpose being to assist the user with one specific task, e.g. to search for documents or products on the web. However, users typically require a collection of several services to realise significant benefits in their daily routines, especially in a corporate environment. Stand-alone agents cause inconvenience to the user by forcing the user to supply each agent with their interests or preferences for that service. Furthermore, stand-alone systems do not allow for collaboration between their agents and those of other services. A personal agent framework (iPAF) developed at BTexact Technologies (formerly BT Laboratories) provides a unified environment in which several personal agents are integrated. This framework has evolved over a number of years and relevant prior publications can be found in [17-20]. Hence only certain highlights and latest enhancements are summarized below, with supplemental discussion focused on the applications of soft computing to augment the intelligence of personal agents.

iPAF is both a framework for integrating a variety of information management services, and a suite of personal agents for filtering and locating information for a given member of an organization. At the core of the platform is a set of member profiles. Each profile contains a description of the interests and preferences of a member and is accessed by a suite of personal assistants which work on behalf of the user. Although each agent performs a specialized task, all assistants use the common profile and make suggestions to improve it – thus providing a single point of contact (Profile Manager agent) and minimal time on maintenance for the user. Other assistants in the framework include a personal newspaper agent (myPaper), a people finder agent (iContact), an authoring assistant that provides context-sensitive reminders/references (iRemind), intelligent diary assistant for flexible schedule management (iDiary), email assistant that learns to prioritise messages(iMail), personalised document alert agent (DocAlert), and intelligent directory enquiry assistant with natural language interface (IDEA) for semi-structured resources (e.g. yellow pages).

3.1 Architectural Benefits

iPAF utilizes a clear separation between individual information sources and the framework itself to allow the use of multiple heterogeneous sources without troubling the user with the details. The software assistants access information sources through the iPAF framework. Each information source has its own wrapper agent which communicates with the iPAF platform and is registered to the platform via a service directory. The directory agent accepts requests from the application agents, routes them to the correct information sources, and relays the replies. The personal agents are thus decoupled from the information sources. This facilitates the introduction of new sources; creates virtual aggregated information sources for the user and contains failure on the part of individual information sources. An overview of an iPAF system can be seen in Figure 1.

Within iPAF framework, benefits are derived by both the service agents and the users. Each agent does not need to build its own user model for personal information, as this is encapsulated in the profiler agent. The framework specifies interfaces between the agents and profile information to expedite the addition of new agents into the existing framework, and to facilitate the use of the profile information generated by any of the existing agents. Hence, each agent merely invokes the interface functions to the personal information component. The use of a standard (at least within a unified framework like iPAF) interfaces to personal information and data sources hides the details of that component and provides modularity and flexibility. Thus, approaches to profile generation or acquisition can be altered without requiring modification to the user interface (UI). Although the UI for each agent service differs due to the varying nature of functionalities provided, the user needs only to become familiar with the usage of one platform to benefit from the services of many personal assistants.

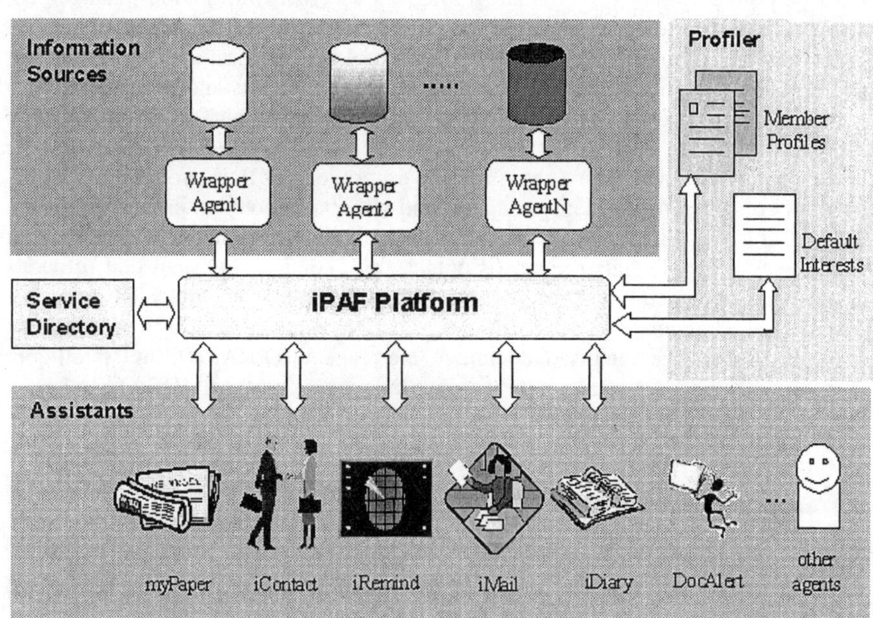

Figure 1: iPAF Framework Components

The framework is also flexible and extendable – exiting services can be easily disabled or removed, and new ones added. When new data sources are added, either the content provider or the system administrator must first create a wrapper program to handle the specific access protocols for that source, but thereafter, it needs only to be added to the service directory.

3.2 Satisfying the Design Criteria

The iPAF framework addresses the privacy and security issues in the following ways to gain trust of its users. It is deployed as an e-community that requires authenticated registration (e.g. via a binding email), and password-protected login procedures to access it. We recognize that additional level of user identification could be requested, e.g. using SecureID [27] tags, but this is not currently implemented in iPAF – it is merely a trade-off between extra security and convenience for the users; all users are accustomed to simple password protected accounts and have thus far expressed high comfort level with this mode of operation.

The registration process requires that a given installation hosts only known, limited number of identifiable (verifiable) individuals. Users can be subsequently identified by a unique ID or email address, and registered users are provided with a system password plus their chosen private password. Complete documentation and help is provided on-line to fully explain the features of the system. Users clearly understand what information is stored in their profile, and they are able to specify a privacy level for each topic in the profile as:

- *Public:* anyone can know about the content for this topic
- *Restricted:* only specified individuals or groups are allowed to know about this content. Likewise, users can ban specific individuals or groups.
- *Private:* No one will know about the user's interest in this topic. Personal agents can find information about this topic for the user, but will not disclose it to other agents or users.

The user's personal profile in iPAF is essentially a list of interest topics, which are described by a set of keywords and phrases. Each interest has attributes associated with it, including privacy levels (specified above), expertise levels, importance/priority levels, and duration factor. The topics can be selected from a default hierarchy setup for a particular e-community, or created from scratch by the user. Profile elements are fully accessible and controllable by the owner so that s/he can examine and modify all the contents at all times. The data is stored in encrypted format, such that it could be deciphered only by agents with the proper key.

Moreover, using the system password, registered users may also create temporary anonymous accounts. This allows a user to simulate a different user and exercise the system to convince him/herself of its integrity. For example, concerns about whether private data from their profile will show up in a search by others, etc. can be tested as an anonymous user – the system only knows that user is a valid registered member, but does not know who in particular it is. Anonymous accounts are intended for such temporary use, and are valid only for one hour.

With regard to updating of personal profiles by agents, rather than making the modifications automatically, the personal assistants present the changes as suggestions to the user. Then, only upon explicit approval by the user, changes are applied. (In the near future, a preference option will be added whereby the user may allow automatic updates of the profile by selected personal agents.) The combination of the above-mentioned design approaches help to build user confidence in their personal agents and the iPAF framework, and alleviate concerns about profile security and integrity of their privately marked profile topics.

In consideration of features mentioned in Section 2.3, iPAF employs a common profile shared among all its service agents. Only the profile manager agent has access to retrieve or update personal information , and that agent provides an interface via which all other agents can request or submit profile data. All agents can contribute to the refinement of the profiles based on specific knowledge they have collected (e.g. documents accessed or people contacted by the user), but their requests are first approved by the user and then handled by the profile manager.

To update the user profile in an accurate manner, iPAF assistants examine user interaction with a variety of sources: news articles read, people contacted, documents referenced/accessed, topics of meetings and personal tasks, subject and people from emails, and other sources such as FAQs, databases, and digital libraries used and searched by user. In addition to those activities, direct (optional) feedback from the user as to which articles s/he found good or bad are included in the analyses if those ratings are supplied.

The information collected about the user is shared among the iPAF assistants (subject to privacy ratings). In some cases tasks are relayed between agents – if, for example, the user wishes to contact people with expertise in a particular news article s/he has just read, the personal newspaper agent will submit the request to the contact finder agent with the appropriate query. In all cases, the personal assistants share user activities (again subject to privacy ratings) in their domain with the adaptive profiling agent, which suggests appropriate updates to the user profile as explained in Section 4. The framework allows agents to deposit information for asynchronous use by itself or by other agents. In this way, information sharing between agents is possible, but it is also controlled in such a manner as to allow modular inclusion of agents within iPAF.

The framework is designed to facilitate straightforward specification and development of personal agents, together with the foundations for collaborative personal agents. The assistants communicate with each other via internal iPAF protocols, but one interface agent on a specified port (dubbed Agent0) is also being added to communicate with external applications via FIPA ACL.

4. Enhancing Framework Intelligence

A useful personal assistant framework is realised by integrating a suite of competent agents and a variety of data sources into a flexible, scalable platform. To increase the intelligence of the framework and ensure that iPAF can provide high quality of service over time, its agents are designed to adapt to the changing needs of their users. Since personal agents services are largely dependent on the user model (user profile) iPAF also supports a "background" agent whose role is to track user models and propose appropriate modifications to refine and improve user profiles. This adaptive profiling agent collects evidence of user activity from the suite of applications as input data, and employs fuzzy logic and approximate reasoning to ascertain changes in user profile. Readers interested in an introduction to fuzzy sets and fuzzy inferencing are referred to [21-22].

4.1 Adapting User Profiles

Fuzzy logic based modeling and approximate reasoning are used in this module, to take advantage of the natural, intuitive knowledge representation and the ability of fuzzy systems to process complex, non linear, and incomplete relations in compact form. Moreover, the fuzzy rule-based approach yields an intuitive, easily understood explanation facility when "fired" rules are examined. The adaptation process is based on the usage/activity on different iPAF services, optional user relevance ratings of documents accessed, and detection of newly developing interests. The rating is currently implemented as radio buttons (single choice) among "high", "satisfactory", and "poor". In this way both implicit (via usage logs, non-invasive) and explicit user feedback are employed as inputs to the adaptive profiling decision module.

Although the iPAF profile is sufficiently rich in attributes and structure, certain components such as expertise level (*curious, novice, competent, good, expert*), importance level (*low, medium, high*) and duration factor (*short, medium, long*) are user specified and assumed "correct" by iPAF agents. That is, those self-proclaimed attributes are normally not disputed nor substantiated, except by the adaptive profiling agent that verifies such ratings based on the actual user behavior (interaction with iPAF services). The assessment of a user's expertise level on a subject requires the ability to assess the expertise level of the documents the user has read/written (which is under development), thus current implementation is limited to the adjustment of importance and duration levels. Figure 2 below depicts a high-level block diagram of the adaptive profiling subsystem of iPAF, and its inputs and outputs are detailed in Figure 3 below.

The usage log data (further detailed below) are used to compute the frequency of the actions as well as to extract other data (e.g. category and keywords added, document contents). A design parameter in this pre-processing stage is the time-interval under consideration. Three metrics are currently being used: number of actions per week, per day, and per hour, to enable computation of fast adaptive actions as well as detect slower shifts in interests.

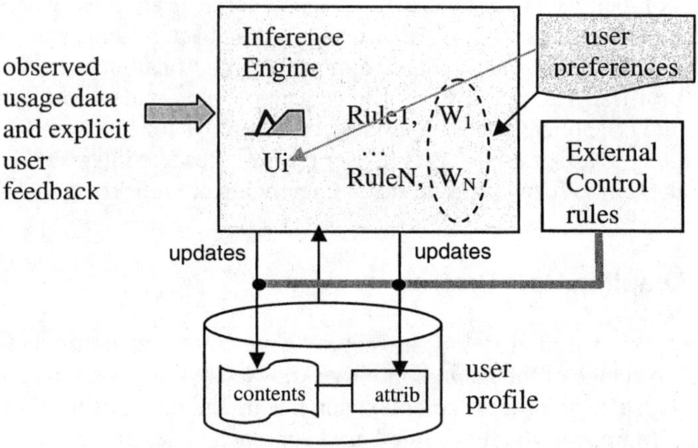

Figure 2: Block Diagram of Adaptive Profiling Module (see text above for discussion)

Note, however, that although concepts of "short" and "long" term can be readily represented by fuzzy sets, their domain values differ between users – for some users, short term profile may imply interests lasting just a few days (e.g. for temporary assignment) and for others, it may span weeks or months. This system allows for personalisation of the time concept – in the preference setting page, user can specify both terms as "about N days" where N is user defined. Internally, fuzzy sets are employed to represent that selection for further processing, and the user's definition affects the domain limits for the fuzzy sets - their universe of discourse. Moreover, the user can specify which attributes may or may not be updated – weights for the rules corresponding to the blocked attributes are set to zero to prevent automatic adaptation. (Typically, fuzzy rules do not employ rule weights in this regard, but this extension is implemented in this module to facilitate the effect of turning some rules on/off).

Since the iPAF framework has numerous services, it is possible to obtain actual usage data logs from the applications as follows: Profile Manager can provide information about which interest(s) are viewed, added, modified, or deleted; iContact can provide people (and the related interest topic) sought; myPaper can provide the articles read, special topics searched, updates of keywords; iRemind can provide links (and thus full document) accessed from the list of offered references; iDiary can provide tasks accessed (added, deleted); DocAlert can inform about any other documents accessed by the user, and iPAF search engine can provide explicit search requests made by the user. The module's inputs and outputs are detailed in Figure 3 below. Note that adding keywords for a specific profile element is also a profile update action, but since these suggestions are provided by other agents (e.g. myPaper), that component is not discussed in this module.

Once the statistics are computed, their values (to corresponding variables) are mapped to fuzzy sets and propagated through a fuzzy reasoning process. For each of the monitored events, three metrics are computed regarding the frequency of the event on hourly, daily, and weekly basis. Each of those metrics (*hourlyFreq<Event>*, *dailyFreq<Event>*, *weeklyFreq<Event>*) have associated fuzzy sets such as *low, medium, high*, with their respective numerical ranges. A second set of design parameters are encountered at this stage, regarding the <u>fuzzy set domain limits</u>. It is understood by fuzzy system designers that these values are initially chosen to reflect reasonable values for the specific problem, and may require tuning during the design-test cycle. (Alternatively, neuro-fuzzy approach for deriving the fuzzy sets and tuning membership functions is another route for investigation. [23, 28].) The fuzzy sets used in this applications may be any linear polyline object, but triangular and shouldered-trapezoids are used in this system.

Currently the outputs for the main module are *importance* and *duration* parameters. Those interest attributes are also denoted by fuzzy terms: *low, medium, high*, and *short, medium, long* respectively. Numerically, they may be represented by singletons or a fuzzy set. The fuzzy set representation also facilitates the encoding and mathematical manipulation of linguistic terms like "frequent", "average", "seldom", and output tuning strengths "slightly", "significantly", etc. Fuzzy rules then establish the mapping between the input and output variables of this module.

The merging and conflict resolution necessary to compute the output "crisp" value is also handled through the fuzzy inference (e.g. max-min, max-dot) and defuzzification (centroid, max-moment, height) processes. Declarative IF-THEN rules are used to compute the output variables. A third design step is the <u>rule creation</u> which is application dependent, but often generic and human readable. For example, a collection of rules like:

IF *readArticleAboutThisInterest* IS ***frequent*** THEN *importance* IS ***increased_slightly***

IF *accessToThisInterest* IS ***negligible*** THEN *duration* IS ***decreased***

IF *numOfLowExpertiseArticles* IS ***high*** THEN *expertise* IS ***decreased_slightly***

are used, where items in *italics* are variable names and items in ***bold italics*** are fuzzy sets. Rules for all possible antecedents are defined in order to reliably predict system performance. During the inference process, different rules fire with varying input strengths and contribute to the overall decision process.

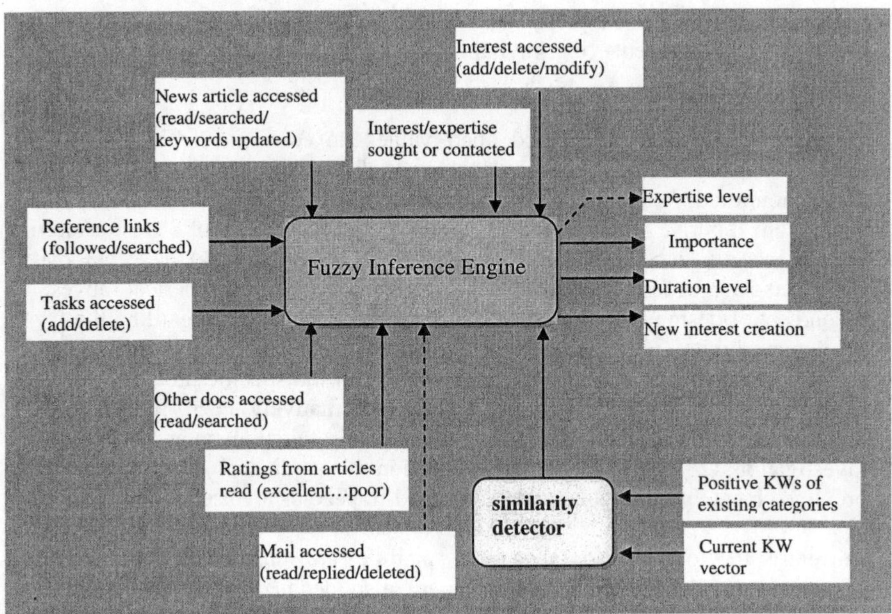

Figure 3: Details of I/O parameters for (dashed lines represent future additions)

During operation, the fuzzy reasoning is achieved similar to a Mamdani-type fuzzy controller [29]. Briefly, multiple antecedents may be combined with AND and OR operators, which are computed with max and min operations, fuzzy sets are either scaled or clipped depending on the type of inference process (e.g. max-min, max-dot). The resultant fuzzy set is an aggregated shape from the contributions of all the fired rules. During defuzzification (computing a single number that best represents the outcome of the fuzzy rule evaluations), centroid or max-moment schemes are used, where the centroid operation provides a "compromise" among the output actions recommended by different rules, and the max-moment approach yields the "strongest" action among the recommendations by different rules.

To detect developing new interests, a sub-processing module initially focuses on the documents explicitly searched and retrieved by the user (via an option in myPaper to search for specific topics). This collection of "new" documents, i.e. documents not already provided to the user based on known interests, are accumulated in a separate user directory. Keywords from those documents are extracted, and a document vector space model is used to perform comparison of current keyword vector with those of existing keyword vectors. A design parameter for this stage is the <u>length and contents</u> of the feature vector. It is currently contains 123 elements: TFIDF value, relative frequency based on number of terms, relative frequency based on the maximum frequency against 40 randomly selected documents from the corpus, plus the number of keywords that matched in the term vector of the corpus, total number of keywords from the new cluster, and cumulated number of documents containing the keywords. The feature values of documents whose indices don't exist with respect to the randomly generated list are set to zero. Note that in any given run, the same [random] index list is used in generating the feature vectors for the existing and new keyword clusters. The Euclidean distance and a nonmetric similarity measure between feature vectors x and y are computed for the feature vectors as **sqrt** $((x-y)^T(x-y))$ and $(x^T y) / (\| x \| \| y \|)$ respectively. The values of the distance and similarity are mapped to fuzzy sets for approximate reasoning about when a new interest category should be created/suggested. Improvements to this module would be to include personal email (via iMail) and web-pages to the cluster of articles that are analyzed, or more radically, entirely different procedures/algorithms such as [16] could be substituted for the document similarity detection.

Note that profile update "actions" in iPAF are implemented as "suggestions" to the user, in accordance with the basic iPAF ideology as described in Section 3.2. The end goal of adaptive profiling is to track changes in user profile, so that application services can be continuously tailored for the individual. Without adaptation, the system may "get it right" in the beginning, but will be unable to maintain the quality of service over the longer term. Hence, profile adaptation improves personalized services by enabling the personal agents to continuously adjust their operation to suit changing user needs in an autonomous manner.

4.2 Soft Concepts in iDiary

Another aspect of enhancing the iPAF framework is to provide an intelligent human-computer interface (e.g. use of natural-language terms), and accommodate flexibility in its operation. The diary assistant (iDiary) allows users to schedule tasks or appointments using soft concepts, and is considered a time management aid. iDiary provides the following features:

- entries to schedule a task or meeting, such as preferred start time, preferred start day, can be defined using fuzzy concepts, such as *"early morning"* or *"around 1:30"* or *"next week"*.
- each entry also contains: description, deadline time and day, and "interruptability", a feature that impacts flexibility of the scheduling algorithm.
- entries can be rescheduled dynamically – for example, if there are no available slots for an event, then the system attempts to reschedule previously scheduled events; rescheduling is either local (to minimize disruption to a schedule) or global.
- agents for different people can communicate together to enable meetings and reschedule if possible. (There is an additional assistant known as the *Co-ordinator assistant* which manages the actions and pro-activity of the other assistants)
- people are disallowed from the socially irresponsible act of giving all times but their most preferred time a preference value of 0.

The user can specify task and meeting preferences with iDiary similar to the way s/he would instruct a human assistant in the same role. Beyond the friendly user interface, iDiary contacts the corresponding agents for other participants (for a meeting), finds common times and attempts to schedule a slot that is suitable to all parties. Machine learning algorithms are also being added to its backend, to recognize repetitive sequences in certain scheduled tasks and meetings, and to automatically plan dependent or subsequent events appropriately. Future work will address (i) enabling the user to specify sequences of actions, and (ii) learning of user preferences for timings for different types of task, and (iii) learning the personal preferences of different users' meaning for fuzzy terms such as *early morning* or *late afternoon*.

5. Discussion and Summary

We presented issues and design criteria for personalized information services in environments like the Internet and corporate intranets. The personal agent framework described in this paper provides a secure, e-community environment in which user profile information is efficiently shared among the service agents while respecting user privacy settings. Its key features include the scalable design that expedites reconfiguration or addition of agent services, decoupling of information resources to facilitate support of heterogeneous contents, a simple browser based user interface, and support of the required features for personal agent system as described in Section 2.

Agents in the iPAF framework provide information of different kinds, ranging from other people who share similar interest areas, through news and documents relevant to your interest profile, to assisting in a flexible schedule management. Beyond the obvious, the true value-added benefits of such a system is that the personal agents effectively convert latent and implicit information (of which the user may not even be aware) into explicit information that the user can use. For example, as the number of information resources and the e-community grows, users can no longer keep up with what information is stored in which resource (including other people) and personal agents become valuable. Users' appreciation of digital personal assistants begin when they are delighted for the first time with a useful link to a resource - which they would have otherwise never found or thought about.

Personal agents require a user model, or personal profile to complete a task effectively. With knowledge about the user, personal agents can specifically tailor their services and interaction with the user. Equally important is the fact that user profiles change over time, especially with respect to the importance and duration attributes of a given interest topic. We have described one approach to adapting user profiles based on fuzzy logic and approximate reasoning. Although this approach requires empirical tests to validate desired behavior, the intuitive and "open" rulebase facilitates debugging and system tuning.

Key issues for the adoption of personal agents by the general public are trust and agent competency, which are both time dependent and inter-related. Trust of another party, whether it is a person, a corporation, or a software agent, will require interaction and evaluation over time. (A prerequisite is that issues of security and privacy are satisfied, which are more tractable with advanced technologies.) Agent competency is also not easy to achieve, and it requires time (for advancements in machine intelligence) and accurate evaluation and disclosure by the system designers to build trust among the users. When both components are satisfied, there will be a bright future for personal agents.

Currently, the iPAF agents all function as designed, and coexist in a stable, integrated framework. Improvements to their quality of service and issues for further research are related – i.e. better semantic understanding of natural language media. Semantic understanding of resource contents and relationships to other resources or concepts have been elusive to date, but efforts are underway to exploit the use of metadata to help increase agent competency as envisioned in the Semantic Web. We expect advances in semantic reasoning to further advance the intelligence of iPAF agents as well.

References

Teamware Group, Pl@za with Knowledge Browser™ (www.teamware.com)

Autonomy, plc., Intelligent Data Operating Layer (IDOL). (www.autonomy.com)

Verity, Inc., Verity K2 Enterprise (www.verity.com)

Nwana HS. "Software agents: An overview." *Knowledge Engineering Review* 11(3) November, 1996.

Maes, P. "Agents that Reduce Work and Information Overload." *Communications of the ACM*, 37 (7) 31-40, 1994.

Finin, T., Fritzson, R., McKay, D. and McEntire, R. "KQML as an Agent Communication Language." *Proceedings, 3rd International Conference on Information and Knowledge Management* (CIKM). New York, November 1994.

The Foundation for Intelligent Physical Agents (FIPA) organization. (www.fipa.org)

Nwana HS, Ndumu DT, Lee LC. "ZEUS: An advanced toolkit for engineering distributed multi-agent systems." *Proceedings of PAAM 98*.

TILAB, Java Agent Development Framework (JADE), (sharon.cselt.it/projects/jade/)

Intelligent System Group, Technical University of Madrid. Multi-Agent Systems Tool (MAST) (www.gsi.dit.upm.es/~mast/mast.html)

alphaWorks and IBM, Agent-Building Environment. (www.alphaworks.ibm.com/tech/abe)

AdventNet Inc., Java Agent Toolkit. (www.adventnet.com/products/javaagent/)

BTexact Technologies, white paper number WP112, July 08, 2002.

searchSecurity.com. PKI introduction. (searchsecurity.techtarget.com/sDefinition/0,,sid14_gci214299,00.html)

Hensley, P., Metral, M., Shardanand, U., Converse, D. & Myers, M (1997). "Proposal for an Open Profiling Standard". Technical Note, w3c, 2 June 1997. (www.w3.org/TR/NOTE-OPSFrameWork.html)

Crabtree I.B. and Soltysiak, S.J. "Identifying and Tracking Changing Interests". *International Journal on Digital Libraries*. 2 (1), 1998. pp. 38-53.

Case, S., Azarmi, N., Thint, M., Ohtani, T. "Enhancing E-Communities with Agent-Based Systems." *IEEE Computer*, July 2001. pp. 64-69.

Soltysiak, S., Ohtani, T., Thint, M., and Takada, Y. An Agent-Based Intelligent Distributed Information Management System for Internet Resources, INET 2000, Yokohama, Japan, Jul. 2000. (www.isoc.org/inet2000/cdproceedings/2f/2f_1.htm)

Crabtree, B., Soltysiak, S., Thint, M. "Adaptive Personal Agents", *Personal Technologies* journal, vol 2, No 3, Springer-Verlag, London,1998.

Ohtani, T., Case, S., Azarmi N., and Thint, M. "An Intelligent System for Managing and Utilizing Information Resources Over the Internet", *International Journal on Artificial Intelligence Tools*, Vol 11, No. 1, 2002. pp. 117-138.

Zadeh, L. "Fuzzy Sets." *Journal of Information and Control*, vol 8, 1965. pp.338-353.

Lee, C. "Fuzzy Logic Controllers", parts I and II. *IEEE Transactions on Systems, Man, and Cybernetics*, vol 20, 1990. pp. 404-435.

Nauk, D., Nauk, U., Kruse, R. "Generating Classification Rules with a Neuro-Fuzzy System NEFCLASS." *Proceedings of NAFIPS 1996*.

Knowledge Based Systems, "Inductive Logic Programming (ILP) " (www.kbs.twi.tudelft.nl/Education/Cyberles/Trondheim/ILP/html/ilp_th_01introd.html)

Pazzani M and Billsus D, 1997. "Learning and revising user profiles: The identification of interesting web sites" Machine Learning 27(3).

Soltysiak, S.J. and Crabtree, I.B. Automatic Learning of User Profiles - Towards the Personalisation of Agent Services. BT Technology Journal, (16), 3, July 1998.

RSA Security, "Two-factor authentication" (www.rsasecurity.com/products/securid/)

Azvine B, Azarmi N, Tsui KC. "Soft computing - a tool for building intelligent systems." BT Technology Journal, 1996 Vol. 14 No.4.

Mamdani, E.H. "Application of Fuzzy Logic to Approximate Reasoning Using Linguistic Synthesis". Proceedings of the 6th International Symposium on Multiple-valued Logic, 1976. pp. 196-202.

iArchive: An Assistant To Help Users Personalise Search Engines

Géry Ducatel, Andreas Nürnberger
gery.ducatel@bt.com; anuernb@cs.berkeley.edu
BTexact Technologies; IS Lab
Adastral Park; Martlesham Heath
IP5 3RE; UK
University of California at Berkeley
EECS; Computer Science Division
Berkeley, CA 94720, USA

Abstract. In this paper we present a method to enhance search engine queries for applications that support the Boolean querying model. The implemented solution provides the casual search engine users with an interactive interface that suggests positive and negative keywords to expand their queries. The underlying idea is to expand a query with user specific positive and negative keywords prior to starting the process of searching. This is achieved by storing some text documents the users are interested in into a personalised repository that is used to perform an analysis of the content of these documents. The analysis consists of finding related keywords based on their occurrence in similar contexts.

1 Introducing iArchive

This document describes a system that personalises a search engine by use of a user profile. This application has been developed as part of an existing platform named iPAF © (Intelligent Personalised Assistant Framework, formerly known as Idioms ©) [1,14,15]. This type of platform acts as a host to a community of users and provides them with on-line services including news sources or corporate databases. The application offers a personalised experience to the users gathering documents that are relevant to the their interests. Within this system, users are delivered with personalised newspapers everyday [13]. The solution proposed here aims to improve the performance of an on-line search engine. This will be achieved by gathering and maintaining user profiles analysing the documents that are relevant to the users. The system builds and maintain user profiles in a two fold process. First the system uses an algorithm discussed in [9,10] to extract contextually related keywords from a set of documents. Secondly, the keywords in the concepts are given attributes: life span and a relevance value. The life span indicates to the system when some words within a concept have not been found relevant for too long and therefore should be removed. The relevance value is a link between two keywords of a concept; this value reflects the strength of the relation between the two keywords. The users have control over these parameters. They can decide if words should have a long or a short life span, and if the strength between keywords should be strong or weak before they are used within their profiles.

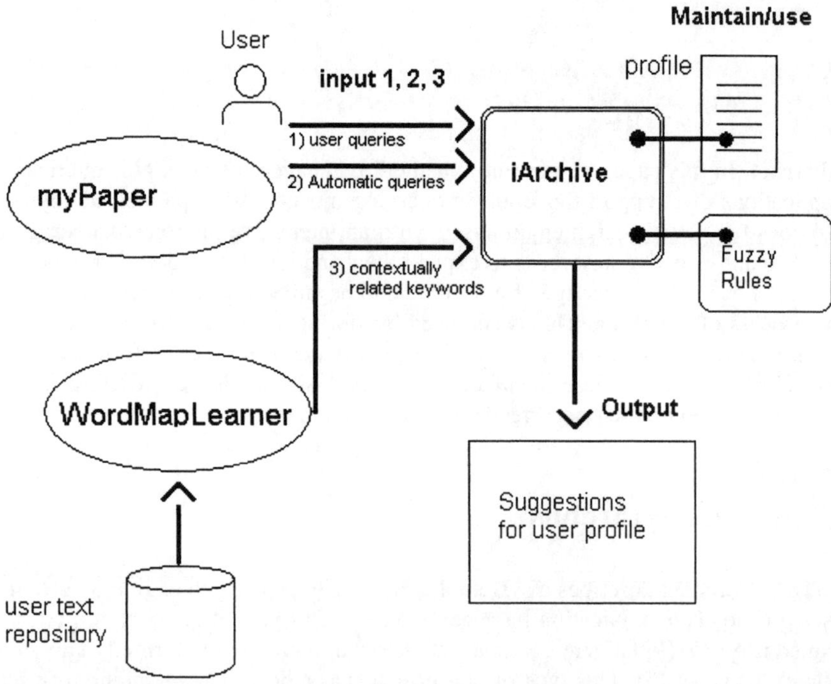

Fig. 1. Overview of the proposed solution (labelled iArchive). The figure shows the different components and their interaction with the solution (input, output and maintenance). The myPaper module is the service that offers personalised newspaper to the iPAF users.

2 User Profiles in Information Retrieval

Building user profiles that change as users' interests evolve is a problem that has seen many very effective solutions in the domain of task scheduling [16,7]. Trying to apply the same techniques to profiling for Information Retrieval is more challenging because of the size of the search space. The solutions proposed in [16] and [7] are based on decision tree algorithms that have input vectors with a number of features below thirty. In Koychev's approach the application does not only rely on a window based approach but their algorithm attempts to freeze an interest in time and save it for future use. When a new interest is found it is checked against "past interests" to see if it corresponds to an old interest, if it does, then the application merges the old interest into the new one; this augments the new interest with information that is relevant to it. The principle is very sound and enables advantageous learning capabilities. Within the scope of Information Retrieval the number of features in a vector is orders of magnitude larger, every keyword that has any relevance must be taken into account and consequently the size of a vector rapidly reaches thousands of features. Therefore, the solution must first prove its scalability to the size of the search space. In order to adapt user profiles to changes in interests there are two main approaches: the window frame and the ageing mechanism. Maintaining interests in a window frame is a solution that is beneficial to discover and maintain a list of recently introduced interests, because they appear fast and distinctively as shown in [2]. However, the drawback of the window frame approach is that it is difficult to maintain past interests. Typically, if an interest changes or disappears, the "old interest concept" is discarded. Algorithms must therefore compromise between the ability to discover new interests fast and the costs of maintaining past interests. This has lead to experiments with optimised "interest forgetting functions" [6], this method is a function that decreases the influence of an interest in time; old interests gradually disappear as their importance is reduced linearly over a period of time. The classification of the interests is a crisp set that discards interests when the linear function of the "gradual forgetting" process comes to term. In order to compensate for the large dimensionality of Information Retrieval a popular technique consists of using user feedback in various forms e.g. relevance feedback [12], or user rating [8]. The disadvantage of requiring feedback from users is not technical but practical. It is often unrealistic to expect users to provide any feedback regardless of how valuable it is to future requests in the system. It is very likely that users do not want to interact with the search engine once it has returned the results because it is perceived (or arguably misperceived) as an annoyance rather than a benefit. The solution proposed here offers the users to rebuild a query that is more valuable based on their initial query and their profile. The difference is that the interaction with the system is to be performed before the documents are retrieved when the users are more receptive to more interaction with the system. For convenience it is perfectly possible to implement the interaction with the system as an optional feature.

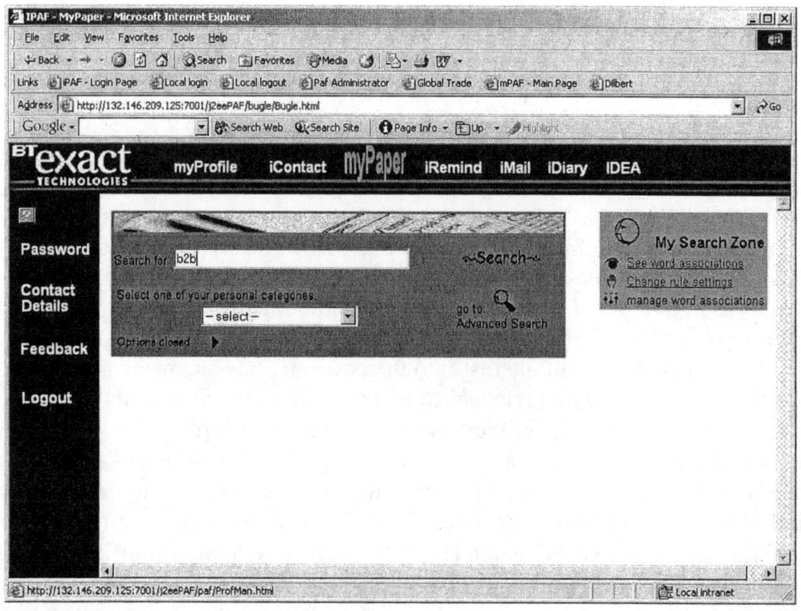

Fig. 2. Screenshot of the application

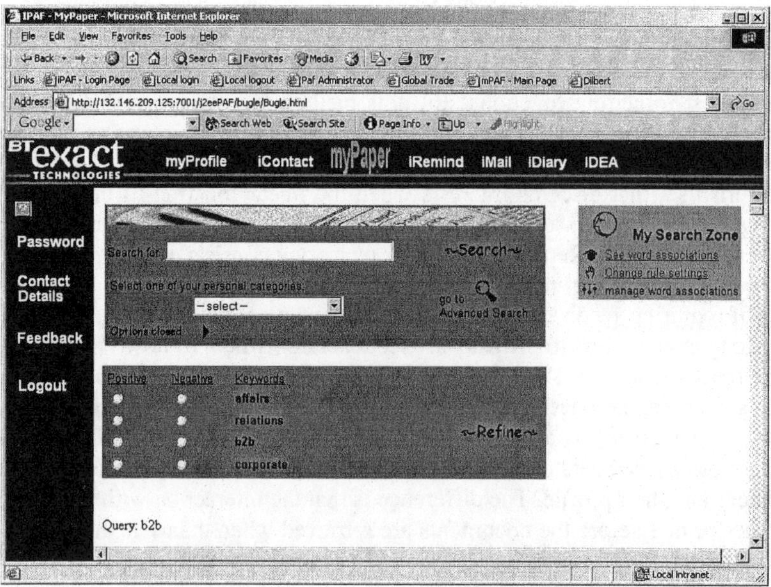

Fig. 3. Screenshot of the application

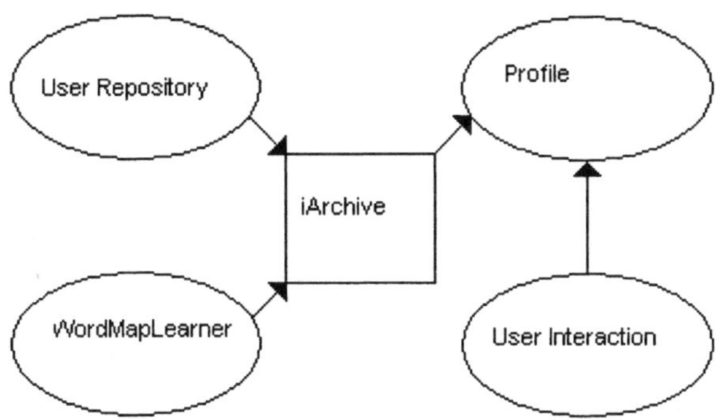

Fig. 4. System's input: user repository, WordMapLearner. System's output is the user profile. The users can change their profile.

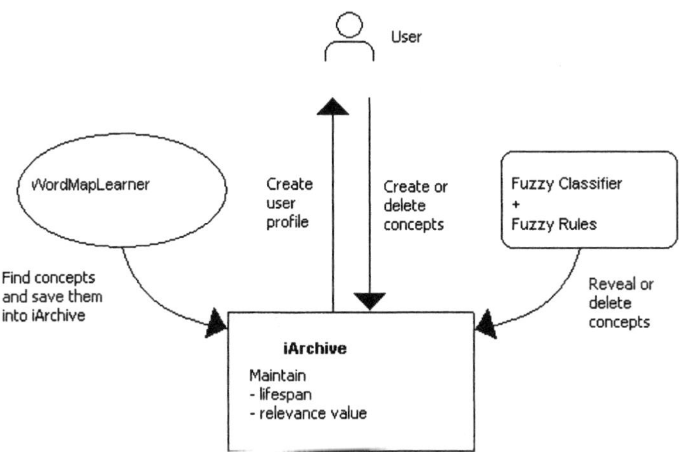

Fig. 5. The modules that contribute to the creation and profile maintenance. The WordMapLearner algorithm provides the iArchive module with concepts. iArchive maintains a date stamp and computes a relevance value for every keyword association. The fuzzy rules reveal the concepts that can be part of the user profile. Finally the users themselves can create or delete the concepts present in the system.

3 A Fuzzy Logic Approach

The overview of the system is shown on Fig 1. The figure shows how the WordMapLearner algorithm uses the user personal document repository. The algorithm can identify contextually related keywords. The extraction of related keywords is achieved by exposing word triples (represented in a numerical format) into a SOM (Self-Organising Map) algorithm [4,5]. Hence, the terms are clustered onto a two-dimensional map where strongly related keywords appear close to one another. The basic idea of this approach is to discover words that are used in similar contexts, i.e. if a, b, x, and y are words that can be found in a text corpus T and the word arrangements axb and ayb are frequent across T, then x and y are co   ι [9-11].

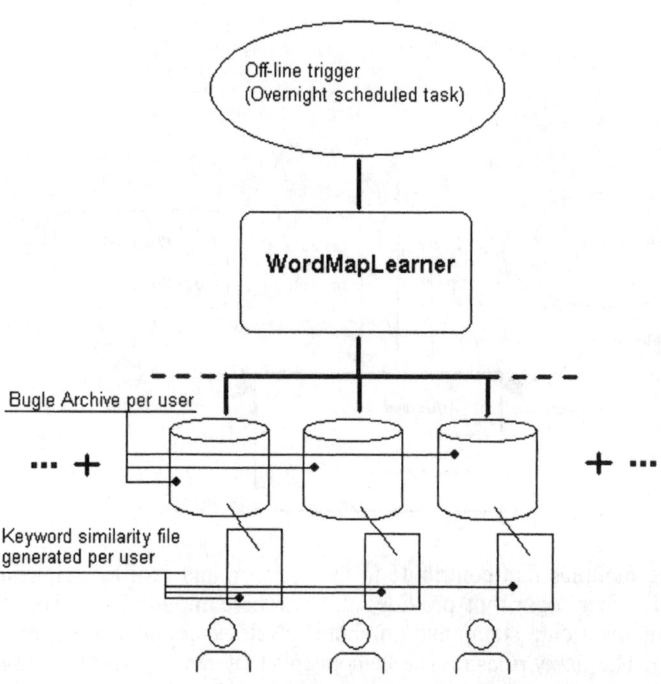

Fig. 6. Individual user profiling based on personalised gathering (archive) of newspapers.

The output of this algorithm is a list of contextually related keywords that are not weighted at this stage. One way of representing this list is by a number of concepts made of keywords: $A1 := \{a11, a12 ...\}, A2 := \{a21, a22,...\},..., An := \{an1, an2,...\}$. Where the upper case letters Ai represent sets of related keywords (referred to as concepts) and lower case aij letters in brackets simply represent keywords.

Since the text repository changes (new documents come in and some documents may be removed), these associations also change. Fig. 1 shows this input into the proposed tool (labelled iArchive in the figure) as input 3. To reflect these changes the system imposes a life span on every concept. The system classifies the concepts into two fuzzy sets: recent and old. The users have control over the size of either set and thus on the life span of concepts; to keep concepts longer the fuzzy set of recent concepts needs to be expanded and the set of old concepts is to be reduced.

Furthermore, two keywords linked to each other (in the context of one concept) have a relevance value that denotes of the strength of their bond. To compute this value the WordMapLearner algorithm runs overnight and the iArchive module compares the new result to the old one (see also Fig. 5 and Fig. 6). If the link between two keywords is confirmed the link is reinforced. For the weighting of concepts the proposed solution consists of normalising the IDF [3] value of every keyword and averaging the value of the concept; this returns a percentage value reflecting the importance of the concept. This value helps us classify the results into three fuzzy sets labelled strong, medium, and weak. Again the users can decide on the size of these data sets, which means that they have control over the concept selection process. Some fuzzy rules help the users manage their profile. For instance if a concept features words with a strong relevance value the system can suggest the users to update their profiles. Or, if the system is about to discard a concept with strong relevance (because its life span has expired) the system can require confirmation from the users. On the whole the system is designed to help the users manage their profile efficiently. Yet, the system can run on its own, without having the users to maintain anything. Users are also allowed to change, add, and remove concepts; they can thoroughly control their profiles. Hence, the system complies with the philosophy of non-obtrusive software application. The fuzzy logic approach gives power to the users if they want to. This system provides fuzzy sets of keywords that contain alternative terms for queries at search time. Other systems either build profiles for pro-active search (e.g. Information Retrieval agents constantly searching for relevant documents) or to expand queries. Here the application gradually builds fuzzy sets of keywords and is able to make helpful suggestions to the users. By giving control to the users with regards to the size of the fuzzy sets they can manage the maintenance of their profiles and they can build more efficient queries.

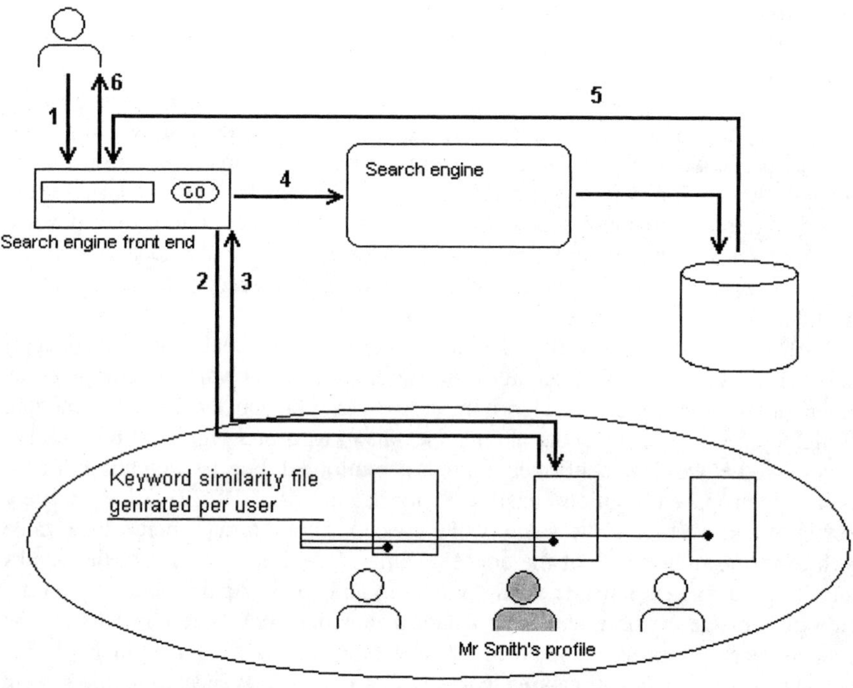

Fig. 7. The on-line searching process, this is divided into three modules: the front-end application (or web based GUI), the user profile and the search engine (the server based application).

Once some concepts have been selected the system can use the vocabulary to expand the users' queries. This can be done either automatically or presented to the users as possible query expansion terms (see also Fig. 7). The latter is the preferred method. The users will be able to distinguish the keywords that are related to their interests and also they can just as easily see the keywords that are not related. Based on this information, the users may decide whether the suggested keywords are to be positive or negative. For instance Fig 2 shows the query: "b2b", Fig 3 shows the system's response. In this profile the keywords associated with "b2b" are "affairs", "relations", and "corporate" it is the users who can decide to add these keywords to the query, or they can use them as negative keywords, or they may decide not to use them at all. In the approach chosen here, the analysis of personal data provides the basis for dynamic Boolean keyword suggestion and fuzzy logic enables the users to build and maintain a profile of keywords that has a temporal nature.

4 Conclusion

The application discussed in this article helps users maintaining a profile of temporary interests. The WordMapLearner algorithm provides the analysis required to extract keywords that are relevant to help the users build an efficient profile. The analysis is based on personal data and therefore the keywords suggested to the users are all adapted to their profiles. The fuzzy rules that help maintaining this profile allow the users to have an informed control over their profile. Fuzzy logic allows the system to identify which are the keywords and concepts that the users need to improve their search. The profile obtained can be used for query expansion. The users can decide if a keyword is of negative or positive relevance to their search. This Information Retrieval solution is implemented in two phases, the first one is the on-line extraction of related keywords with the WordMapLearner algorithm mentioned in the first section. The second phase is the classification of the concepts uncovered into a selection of fuzzy sets. The computation of the related keywords is performed on-line using a set of documents that reflect the users' interests, Fig 6 shows the graphical representation of this process. The search interaction with the users is illustrated in Fig 7. The combination of a search engine (one that can handle Boolean queries) and this technique provides the users with a friendly interface that can help optimise their queries. The classification of concepts into fuzzy sets is shown on Fig 4 and Fig 5.

5 Future Work

This solution allows users to maintain a profile of interests offering some vocabulary to expand their queries. The suggested vocabulary is being extracted from the WordMapLearner algorithm which is able to extract concepts out of the text corpus, the drawback of this type of algorithm is the amount of data it requires before it can be beneficial. Therefore, it might be important to develop algorithms that could provide a closer analysis of semantic keyword contexts. This solution may consist of a hybrid approach that combines two algorithms for small and large text corpora. A small text corpus being the source of recently introduced interests and a large text corpus for sustained interests.

It has been noticed that sometimes the size of concepts is not satisfying, for instance in the example shown, the user had a profile with the keywords b2b, affairs, relations, and corporate. These are keywords that are complementary and that can be used in the same context. Yet, such a profile that it is not quite comprehensive, other suitable keywords are not represented because they have not been found into the text corpus. Therefore, some other techniques to add more vocabulary to the concepts could be introduced. This may include collaborative profiling techniques that are using groups of users with overlapping interests to expand their profiles in a collaborative process. Also some heuristics such as extracting vocabulary out of existing knowledge structures such as semantic trees can be used.

The method discussed in this paper assigns two attributes to a keyword: life span and importance. The importance is worked out from the IDF value of the keywords in the concept. This is likely to disadvantage some concepts that may be relevant but that are ranked down by the IDF approach. Future work in this area will focus on finding a more optimal weighting solution. There could be more categories in which a keyword may belong including its age since it started to appear in a profile (as opposed to the time this keyword has left to belong to a concept) and even the degree of specificity of the keyword. This would allow for more precise profiles and also better cross profile analysis to identify interests that overlap between users.

References

1   MP Thint, B Crabtree, SJ Soltysiak. Adaptive personal agents. *Personal Technologies Journal*, 2(3):141-151, 1998.
2   S Soltysiak, B Crabtree. Identifying and tracking changing interests. *International Journal of Digital Libraries*, 2(1):38 - 53, 1998.
3   K Sparck Jones. Index term weighting. *Information Storage an Retrieval*, (9):313 - 316, 1973.
4   T Kohonen. Self-organized formation of topologically correct feature maps. *Biological Cybernetics*, 43:59-69, 1982.
5   T Kohonen. Self-organising and associative memory. *Springer-Verlag*, 1984.
6   I Koychev. Gradkual Forgetting for Adaptation to Concept Drift. In *ECAI 2000 Workshop Current Issues on Spatio-Temporal Reasoning*, 101 – 106, 2000
7   I Koychev. Tracking changing user interests through prior-leaning of context. In *AH'2002, 2nd International Conference on Adaptive Hypermedia and Adaptive Web Based Systems*, 2002.
8   D Billsus, M Pazzani. Learning and revising user profiles: The identification of interesting web sites. *Machine Learning*, 27:313 - 331, 1997.
9   A Nürnberger. Clustering of document collections using a growing self-organizing map. In *BISC International Workshop on Fuzzy Logic and the Internet (FLINT 2001)*, pages 136-141. ERL, College of Engineering, University of California, Aug 2001.
10 A Nürnberger. Interactive text retrieval supported by self-organising maps. Technical report, BTexact Technologies, IS Lab, 2002.
11 H Ritter, and T Kohonen. Self-organising semantic maps. *Biological Cybernetics*, 61(4):241 - 254, 1989.
12 JJ Rocchio. *Performance Indices for Information Retrieval*. Prentice Hall, 1971. Soft Computing and Information Organisation 11

13   B Crabtree, SJ Soltysiak. Automatic learning of user profiles - towards personalisation of agent services. *BT Technology Journal*, 16(3):110-117, 1998.

14   B Crabtree, SJ Soltysiak. Knowing me, knowing you: Practical issues in the personalisation of agent technology. In *The PAAM'98 Third International Conference on the Practical Application of Intelligent Agents and Multi-Agent Technology. Practical Application Company*, March 23-25 1998.

15   SJ Soltysiak. Intelligent distributed information management systems. Technical report, BTexact Technologies, IS Lab, 1999.

16   D Freitag, J McDermott, D Zabowski, T Mitchel, R Caruana. Experience with a learning personal assistant. *Communications of the ACM*, 7(37):81 - 91, 1994.

Fuzzy Logic in Integrated Management of Networked Systems

Seyed A. Shahrestani

School of Computing and Information Technology
University of Western Sydney
Penrith Campus
Locked Bag 1797
PENRITH SOUTH DC NSW 1797
AUSTRALIA

Email: seyed@ieee.org

Abstract: During the recent past, the networking revolution has been changing the face of computing. Computer networks contain a large number of physical and logical elements that must be managed. However, integrated network management is not concerned with the equipment alone, but a combination of services, applications, and enterprise management concerns drive the solutions. Additionally, as a result of comprehensive monitoring abilities, modern systems result in an overwhelming amount of information. The information may contain incoherent, missing, or unreliable data that need to be filtered and processed. Conventional computer applications provide some degree of automation in processing the data to identify relevant information. But human interplay remains essential, as the data is often incomplete and conflicting. Furthermore, all Internet-based and intranet-based computer systems are vulnerable to security breaches and intrusions by both legitimate users, who abuse their authorities, and unauthorized individuals. With the rapidly increasing dependence of businesses and government agencies on their computer networks, protecting these systems from intrusions is a critical task in integrated network management. These points show the increasing complexities of the networked environments that require radical changes in approaches for their management. In this work, we describe several ways that fuzzy logic can be used to identify or improve the solutions to problems encountered in an integrated network management environment. In particular, we discuss how the utilization of fuzzy logic for representation of imprecise descriptions and uncertainties results in enhanced capabilities in management of networked systems. Some specific application areas that demonstrate the effectiveness of fuzzy logic in improved management of the networks are also discussed. These include advanced help desk, network diagnostic systems, and pro-active management of quality-of-service. Furthermore, the implications and advantages of using fuzzy logic for enhancement of network security, and in particular handling intrusion detection tasks are also reported.

1   Introduction

The ever-increasing complexity of the networks has some profound technical implications for management systems. Modern networked systems result in an overwhelming amount of data and information because of the comprehensive monitoring abilities. Conventional computer applications provide some degree of automation to process and filter the data to identify relevant information, but human interactions remain essential. This is mainly due to the fact that the data is often incomplete and it may reflect on conflicting information. In principle, artificial intelligence (AI) techniques could limit the need for human intervention [12]. In particular, several characteristics of fuzzy logic make it an effective approach for use in an integrated network management environment. For instance, its flexibility in handling uncertainties and its capability to coordinate and manage several models and rules can be mentioned. Such capabilities may result in enhancement of the required solutions in an integrated network management environment.This is particularly the case for some specific application areas that are further discussed in this work.

From a broad point of view, the ability to handle extensive amounts of information is a prerequisite for management of complex systems. For instance, the experience gained on a problem represents the knowledge that can be of value in the future. Obviously, the process of retrieving information relevant to that experience remains a mandatory part. This part will usually pose itself as a difficult problem to solve. Information retrieval (IR) tools are the very bases for any process that deals with large databases. This is the case even if they only support data collection and leaving the actual task of extracting the information to the user or some other software agent. In particular, it can be noted that advanced help desk systems rely heavily on IR. Expressiveness and adaptivity are fundamental features for a data model. The abstraction associated with an object should capture all its peculiarities in an easily manageable representation. In a standard IR context, uncertainty pervades the behavior of both the system and the users. To undertake uncertainty and adaptivity problems simultaneously, fuzzy logic offers excellent solutions.

Among the other functional areas that fuzzy logic can be of great value, fault management can be mentioned. For instance, although the case-based reasoning (CBR) paradigm is reported to give good solutions to alarm correlation problem [9], its high sensitivity to the accuracy of knowledge description should not be ignored. Uncertainty permeates the entire diagnostic process and its management is a fundamental issue in actual diagnostic systems. The information regarding the context of encountered problems and the type of models that can be built to represent them are among crucial aspects of a diagnostic system. While traditionally the main components used in the definition of a context are observations and facts, the data on relevance and confidence may add valuable information. The latter piece of information can be easily amended and handled by fuzzy logic based approaches.

For achieving acceptable quality-of-service (QoS) levels, current network management systems rely heavily on human operators for their correct interpretation of monitored thresholds and triggered alarms. In addition to the problems arising from considering crisp thresholds, such approaches do not scale as the number of thresholds and alarms increase. For a better solution, a network management system should be able to monitor, diagnose and reconfigure application components to ensure that user-level QoS goals are maintained. We discuss how these systems can be significantly enhanced by incorporation of fuzzy logic. Such incorporations will improve the diagnostic rules so that they are more capable of handling ambiguity and incomplete information.

All Internet-based and intranet-based computer systems are vulnerable to intrusions and abuse by both legitimate users (who abuse their authorities) and unauthorized individuals. With the rapidly increasing dependence of businesses and government agencies on their network of computers, protecting these networked systems from intrusions and security breaches is a critical aspect of integrated management. In essence the problem is related to the fact that the personal computer and the Internet have become indispensable parts of everyday lives, while they are exceedingly vulnerable to even simple attacks. The vulnerability of some of these systems stems from the simple fact that they were never intended for a massive interconnection. It is therefore, critical to have the security mechanisms that are able to prevent unauthorized access to system resources and data. However, complete prevention of security breaches does not appear to be practical. Intrusion detection can therefore be regarded as an alternative, or as a compromise to this situation.

Generally speaking, an intrusion detection system (IDS) assumes that an intruder's behavior will be noticeably different from that of a legitimate user. There is also an underlying assumption that many unauthorized actions are detectable one way or another. There are two major approaches for detecting computer security intrusions in real time. These are frequently referred ta as misuse detection and anomaly detection. Misuse detection attempts to detect known (previously identified) attacks against computer systems. Anomaly detection, on the other hand is based on utilization of the knowledge of users' normal behavior to detect attempted attacks. The primary advantage of anomaly detection over misuse detection approaches is the ability to detect novel and unknown intrusion. Successful intrusion detection based on case-based reasoning [6], machine learning [5] and data mining [18] have also been reported. In any of the intrusion detection techniques, the need for exploiting the tolerance for imprecision and uncertainty to achieve robustness and low solution costs is evident.

The main objective of this work is to discuss the ways that fuzzy logic can be used to improve solutions in an integrated network management environment. This is achieved in the remainder of this paper by using the following structure. Section 2 presents the integrated network management environment Section 3 gives an overview of several intrusion detection approaches that is inspired by the principles of natural immune systems. Section 4 focuses on specific applications of fuzzy logic in network management, highlighting how AI and in particular fuzzy logic can be used to improve most other approaches applied to network management problems. The concluding remarks are given in Section 5.

2 Integrated Network Management

The need for distribution of network management functions is already well established. This is evident by the approaches such as definitions of management information base (MIB) for remote monitoring (RMON) or mid-level manager MIB, for example. In general, the integrated network management is concerned with a combination of issues relating to equipment, services, applications, and enterprise management. In this context, various new requirements need to be met by network management solutions. Some of these requirements are mentioned in this section, while some possible enabling approaches for complying with them are discussed in later parts.

Help desk systems play a fundamental role in such activities. These systems are designed to provide customer support through a range of different technology and information retrieval (IR) tools. Efficiency and effectiveness in data retrieval, while crucial for the overall problem solution process, are heavily dependent on the abstraction models. The abstraction associated with an object should capture all its peculiarities in an easily manageable representation [15]. Identification of relevant features of achieving an object abstraction is a complex task and obviously the presence of uncertainties makes this task even harder to be accomplished.

For diagnosis purposes, focusing on case-based reasoning (CBR) paradigm, models that capture the relevance and uncertainty of information in a dynamic manner are essential. This is a requirement for models used in both diagnostic knowledge bases and the underlying processes. Based on such models, a conversational CBR shell implementing nearest-neighbor (NN) retrieval mechanisms for instance, may then be utilized to achieve relatively high precision case-retrieval.

Distributed applications are evolving towards compositions of modular software components with user interfaces based on web browsers and more generally on web technology. Each of these components provides well-defined services that interact with other components via the network. A point that may be easily overlooked though is that the increase in the complexity of distribution, makes it more difficult to manage the end-to-end Quality-of-Service (QoS). The challenge derives in part from the need for interaction of different network and computing domain management scopes. A management system deployed to diagnose QoS degradation should address two major issues. First, to measure the performance of applications, it needs a low-overhead and yet scalable system for quantitative indication of the performance of software components. Second, the performance management system must monitor some of such measurements selectively and diagnose QoS degradation, while being able to adapt to the environment and to integrate with other existing network management systems.

From a management point of view, it is also very critical to have the security mechanisms that are able to prevent unauthorized access to system resources and data. However, complete prevention of security breaches does not appear to be practical. Intrusion detection can be regarded as an alternative, or as a compromise to this situation. In general, an intrusion attempt is defined as the potential possibility of a deliberate unauthorized attempt to access or manipulate information, or render a system unreliable or unusable. An intrusion detection system (IDS) is a tool that attempts to perform intrusion detection. While the complexities of host computers are already making intrusion detection a difficult task, the increasing prevalence of distributed networked-based systems and insecure networks such as the Internet has greatly increased the need for intrusion detection [19]. Given the increasing importance of security management, next section takes a brief look at network intrusion models and detection approaches. The majority of the existing intrusion detection algorithms are mainly dependent on knowledge bases or input/output descriptions of the operation, rather than on deterministic models. Consequently, the utilization of fuzzy logic for representation of imprecise descriptions and uncertainties results in enhanced capabilities of handling intrusion detection through approximate matching. This is further explored in Section4.

3   Intrusion Models and Detection Algorithms

Typically, an IDS would employ statistical anomaly and rule-based misuse models to detect intrusions. The detection in statistical anomaly model is based on the profile of normal user's behavior. It will statistically analyse the parameters of the user's current session and compares them to the user's normal behavior. Any significant deviation between the two is regarded as a suspicious session. As the main aim of this approach is to catch sessions that are not normal, it is also referred to as an 'anomaly' detection model. The second model is dependent on a rule-base of techniques that are known to be used by attackers to penetrate. Comparing the parameters of the user's session with this rule-base carries out the actual act of intrusion detection [3]. This model is sometimes referred to as a misuse detection model, as it essentially looks for patterns of misuse- patterns known to cause security problem.

3.1 Statistical Anomaly Detection

Statistical anomaly detection systems initiate the detection of the security breaches by analysing the audit-log data for identifying the presence of an abnormal usage or system behavior. These systems assume that such abnormal behavior is indicative of an attack being carried out. An anomaly detection system will therefore attempt to recognize the occurrence of 'out of the ordinary' events. For implementation purposes, the first step is concerned with building a statistical base for intrusion detection that contains profiles of normal usage and system behavior. Based on that, these systems can then adaptively expand their statistical base by learning about normal users and system behavior. This model of intrusion detection is essentially based on pattern recognition approaches, i.e. the ability to perceive structure in some data.

3.2 Recognition of Intrusive Patterns

To carry out the pattern recognition act, the raw input data is pre-processed to form a pattern. A pattern is an extract of information regarding various characteristics or features of an object, state of a system, and the like. Patterns either implicitly or explicitly contain names and values of features, and if they exist, relationships among features. The entire act of recognition can be carried out in two steps. In the first step a particular manifestation of an object is described in terms of suitably selected features. The second step, which is much easier than the first one, is to define and implement an unambiguous mapping of these features into class-membership space.

Patterns whose feature values are real numbers (continuous or discrete) can be viewed as vectors in n-dimensional space, where n is the number of features in each pattern. With this representation, each pattern corresponds to a point in the n-dimensional *metric* feature space. In such a space, distance between two points, Euclidean distance being one example, indicates similarities (or differences) of the corresponding two patterns. Generally speaking, the key problem is reduction of the dimensionality of the feature vector (and space). Partitioning the feature space then carries out the actual decision making act (classification) by any of the many available methods; e.g. maximum likelihood, K-nearest neighbors, decision surfaces and discriminate functions. This approach to pattern recognition is generally considered as *statistical* (or decision theoretic).

More specifically for intrusion detection purposes, the statistical analysis attempts to detect any variations in a user's behavior by looking for significant changes in the session in comparison to user's conduct profiles or patterns already saved. The profiles consist of the individual behavior in previous sessions and serve as a means for representing the expected behavior. Obviously, the information content of the patterns that make up the profiles need to be dynamically updated. For intrusion detection purposes, various types of subjects may need to be considered and monitored. These may include users, groups, remote hosts, and overall target systems. Monitoring of groups enables the detection system to single out an individual whose behavior significantly deviates from the overall 'average' group behavior. Detection of system wide deviations in behavior that are not connected to a single user may be achieved by monitoring the target system. For instance, a large deviation in the number of system wide login attempts may be related to an intrusion.

To determine whether the conduct is normal or not, it is characterized in terms of some of its key features. The key features are then applied to individual sessions. While the features employed within different intrusion detection systems may vary substantially, they may be categorised as either a continuous or a discrete feature. A continuous feature is a function of some quantifiable aspect of the behavior such that during the course of the session its value varies continuously. Connection time is an example of this type of feature. This is in contrast to a discrete feature that will necessarily belong to a set of finite values. An example of such a feature is the set of terminal location. For each subject, the maintained profile is a collection of the subject's normal expected behavior during a session described in terms of suitably selected features.

The classification process to determine whether the behavior is anomalous or not is based on statistical evaluations of the patterns stored as profiles specific for each subject. Each session is described by a pattern (usually represented as a vector of real numbers) consisting of the values of the features pre-selected for intrusion detection. The pattern corresponds to the same type of features recorded in the profiles. With the arrival of each audit record, the relevant profiles are solicited and their contents (the patterns they contain) are compared with the pattern (vector) of intrusion detection features. If the point defined by the session vector in the n-dimensional space is far enough from the points corresponding to the vectors stored in the profiles, then the audit record is considered to be anomalous. It can be noted that while the classification is based on the overall pattern of usage (the vector), highly significant deviations of the value of a single feature can also result in the behavior being considered as anomalous.

Important Characteristics

To be useful, the intrusion detection system must maximize the true positive rate and minimize the false positive rate. In most cases (but not all), achieving a very low false positive rate (i.e. a low percentage of normal use classified incorrectly as anomalous) is considered more crucial. This can be achieved by changing the threshold of the distance metric that is used for classifying the session vector. By raising this threshold, the false positive rate will be reduced while this will also lower the true positive rate (i.e. fewer events are considered abnormal).

To increase speed and to reduce misclassification error, particularly when the number of classes (e.g. the number of users) is large or not known, some suggestions have been made for grouping of classes. For example, patterns can be mapped into a generalized indicator vector, on the basis of their similarities. This vector is then used in conjunction with a standard search tree method for identification purposes. Another method first computes a similarity measure- based on distance metric- between each pattern and every other pattern and merges close

samples with each other. Yet another proposed method is to find a pattern prototype (a typical example of certain classes) and use that for establishing the category of a new pattern before comparing it with other exemplars of that category to
recover its specific identity (see [16] for details).

All of the methods described so far start the grouping process by trying to identify similarities between classes and their representing patterns. Obviously, patterns representing the same class of objects should have some features and feature
values in common, while patterns describing members of a different class should
have different values for some or all of these features. In other words, objects are
classified as members of a particular class if they possess some distinctive features, which make them distinguished from other objects present in the universe of
objects. Consequently, one may start the process of grouping of classes on the basis of their evident differences. That is, collect objects (e.g. users), which have
some evident differences from all other objects (classes) into one group. Some efficient algorithms developed for finding the necessary and sufficient conditions
that describe class membership, based on this approach are described in [16].

3.3 Rule-based Misuse Detection

Obviously, attempting to detect intrusions on the basis of deviations from expected behaviors of individual users has some difficulties. For some users, it is
difficult to establish a normal pattern of behavior. Therefor, it will be easy for a
masquerader to go undetected as well. Alternatively, the rule-based detection systems are established on the assumption that most known network attacks can be
characterized by a sequence of events. For implementation purposes, high-level
system state changes or audit-log events during the attacks are used for building
the models that form the rule bases. In a rule-based misuse detection model, the
IDS will monitor system logs for possible matches with known attack profiles
[11]. Rule-based systems generate very few false alarms, as they monitor for
known attack patterns.

There is another situation for which statistical anomaly detection may not be
able to detect intrusions. This is related to the case when legitimate users abuse
their privileges. That is, such abuses are normal behavior for these users and are
consequently undetectable through statistical approaches. For both of these cases,
it may be possible to defend the system by enforcing rules that are meant to describe 'suspicious' patterns of behavior. These types of rules must be independent
of the behavior of an individual user or their deviations from past behavior patterns. These rules are based on the knowledge of past intrusions and known defi-
ciencies of the system security. In some sense, these rules define a minimum
'standard of conduct' for users on the host system. They attempt to define what
can be regarded as the proper behavior whose breaches will be detected.

Most current approaches of detecting intrusions utilize some form of rule-based
analysis. Expert systems are probably the most common form of rule-based intrusion detection approaches; they have been in use for several years [19]. The areas
of KBS, expert systems, and their application to intrusion detection have been and

still are a very active research area. Among the very important aspects of the KBSs, are their knowledge bases and their establishment. This area and related subjects may be considered as a field by itself, referred to as 'knowledge engineering'. Knowledge engineering is the process of converting human knowledge into forms suitable for machines, e.g. rules in expert systems. Some examples of an interdisciplinary approach based for knowledge engineering in computer security systems are described in [17].

For successful intrusion detection, the rule-based sub-system needs to contain knowledge about known system vulnerability, attack scenarios, and other information about suspicious behavior. The rules are independent from the past behavior of the users. With each user gaining access and becoming active, the system generates audit records that in turn are evaluated by the rule-based sub-system. This can result in an anomaly report for users whose activity results in suspicious ratings exceeding a pre-defined threshold value.

Clearly, this type of intrusion detection is limited in the sense that it is not capable of detecting attacks that the system designer does not know about. To benefit from the advantages of both approaches, most intrusion detection systems utilize a hybrid approach, implementing a rule-based component in parallel with statistical anomaly detection. While in general, the inferences made by the two approaches are independent or loosely coupled. The two sub-systems share the same audit records with different internal processing approach [10]. There are arguments and ongoing research in tightening the two together in the hope of achieving a reduced false-positive rate of anomaly detection and eliminating the possibility of multiple alarms.

3.4 Immunology Based Intrusion Detection

This section gives a brief overview of an interesting and somehow different approach to intrusion detection. The design objective for this approach is related to building computer immune systems as inspired by anomaly detection mechanisms in natural immune systems. Such a system would have highly sophisticated notions of identity and protection that provides a general-purpose protection system to complement the traditional systems. The natural immune system tries to distinguish 'self' from the dangerous 'other' or 'nonself' and tries to eliminate the 'other'. This can be viewed as a similar problem in computer security; where 'nonself' might be an unauthorized user, computer viruses or worms, unanticipated code in the form of Trojan horse, or corrupt data.

With fundamental differences between living organisms and computer systems, it is far from obvious how the natural immune systems can be used as models for building competent computer intrusion detection systems. While some of the relevant ideas have been implemented and reported in the relevant literature, many of the appealing parts are still at their theoretical stages. The analogy between computer security problems and biological processes was suggested as early as 1987, when the term 'computer virus' was introduced [19]. But it took some years for the connection between immune systems and computer security to be eventually introduced [8]. This view of computer security can also be of great value for implementing other intrusion detection approaches as well. This type of intrusion detection has been expanded into a distributed, local, and tunable anomaly detection method.

In the immune system, the intrusion detection problem is viewed as a problem of distinguishing self (e.g. legitimate users and authorized actions) from nonself (e.g. intruders). To solve this problem, 'detectors' that match anything not belonging to self are generated. The method relies on a large enough set of random detectors that are eventually capable of detecting all nonself objects. While these systems show several similarities with more traditional techniques in intrusion detection, they are much more autonomous. Such systems present many desirable characteristics [4]. In particular, it can be noted that the detection carried out by the immune system is 'approximate'; the match between antigen (foreign protein) and receptor (surface of the specialized cells in the immune system) need not be exact. This will allow each receptor to bind to a range of similar antigens and vice versa.

One of the main motivations behind these approaches is that the traditional view of computer security is not likely to succeed. Computers are dynamic systems; manufactures, users, and system administrators constantly change the state of the system. Formal verification of such a dynamic system is not practical. Without a formal verification many of the more traditional tools such as encryption, access control, audit trails, and firewalls all become questionable. In turn, this means that perfect implementation of a security policy is impossible, resulting in imperfect system security.

4   Fuzzy Logic and Network Management

Several characteristics of fuzzy logic make it an effective approach for use in an integrated network management environment. In particular, its flexibility in handling uncertainties and its capability to manage several models and rules are of great value. We start this section by giving a broad view of fuzzy logic and AI to establish their particular properties that make them suitable for application in a network management environment. We then proceed to discussing some specific application areas.

4.1 Fuzzy Logic and Artificial Intelligence

The interest in building machines and systems with human-like capabilities has lead to considerable research activity and results. Important features of human capabilities that researchers are interested in implementing in artificial systems include learning, adaptability, self-organization, cognition (and recognition), reasoning, planning, decision-making, action, and the like. All of which are related to intelligence. These research activities form the core of artificial intelligence (AI)

[20]. To achieve higher levels of automation, a number of AI techniques have already been applied to network management problems [12].

Although the focus of this work is on fuzzy logic applications in network management, it can be noted that fuzzy logic can be used to improve most other AI approaches, e.g. knowledge presentation in expert systems. Therefore, while this section gives a more elaborate treatment to fuzzy logic, a very brief overview of some other AI approaches that are found to be of value in network management is also given. This allows for discussing the possible improvements of utilizing fuzzy logic in other AI techniques applied to network management problems in later parts.

Fuzzy Logic

The subject of fuzzy logic is the representation of imprecise descriptions and uncertainties in a logical manner. Many artificial intelligence based systems are mainly dependent on knowledge bases or input/output descriptions of the operation, rather than on deterministic models. Inadequacies in the knowledge base, insufficiency or unreliability of data on the particular object under consideration, or stochastic relations between propositions may lead to uncertainty. In expert systems, lack of consensus among experts can also be considered as uncertainty. In addition, humans (operators, experts...) prefer to think and reason qualitatively, which leads to imprecise descriptions, models, and required actions. Zadeh introduced the calculus of fuzzy logic as a means for representing imprecise propositions (in a natural language) as non-crisp, fuzzy constraints on a variable [21].

Knowledge Based and Expert Systems

Knowledge Based Systems (KBS) are modular structures in which the knowledge is separate from the inference procedure. Knowledge may be utilized in many forms, e.g. collection of facts, heuristics, common sense, etc. When the knowledge is acquired from (and represents) some particular domain expert, the system is considered an expert system. In many cases, production rules or specification of the conditions that must be satisfied for the rule to become applicable represents the knowledge [16]. Also included are the provisions of what should be done in case a rule is activated. Production rules are IF--THEN statements; a `conclusion' is arrived at, upon the establishment of validity of a 'premise' or a number of premises.

Rule-based systems are popular in AI because rules are easy to understand and readily testable. Each rule can be considered independent of the others, allowing for continual updating and incremental construction of the AI programs. Broadly speaking, systems relying on heuristic rules are considered brittle. When a new situation falls outside the rules, they are unable to function and new rules have to be generated. Thus, a very large knowledge base must be created and stored for retrieval purposes. In general, heuristic rules are hard to come up with and are always incomplete. The rules are usually inconsistent; i.e. no two experts come up with the same set.

Artificial Neural Networks

Artificial Neural Networks (ANNs) are dense parallel layers of simple computational nodes. The strengths of the links between the nodes are defined as connection weights. In most cases, one input layer, one output layer, and two internal (hidden) layers will be considered adequate to solve most problems [14]. This is considered as a Multi-Layer Perceptron (MLP) and is widely popular. The connection weights are usually adapted during the training period by back-propagation of errors, which results in a feed-forward network.

Pattern Recognition

Pattern recognition is the ability to perceive structure in some data; it is one of the aspects common to all AI methods. The raw input data is pre-processed to form a pattern. A pattern is an extract of information regarding various characteristics or features of an object, state of a system, etc. Patterns either implicitly or explicitly contain names and values of features, and if they exist, relationships among features. The entire act of recognition can be carried out in two steps. In the first step, a particular manifestation of an object is described in terms of suitably selected features. The second step, which is much easier than the first one, is to define and implement an unambiguous mapping of these features into class-membership space [14]. Patterns whose feature values are real numbers can be viewed as vectors in n-dimensional space, where n is the number of features in each pattern. With this representation, each pattern corresponds to a point in the n-dimensional metric feature space. In such a space, distance between two points indicates similarities (or differences) of the corresponding two patterns. Partitioning the feature space by any of the many available methods, e.g. maximum likelihood, K-nearest neighbors, decision surfaces and discriminate functions then carry out the actual classification.

Case-based Reasoning

Case-based reasoning (CBR) paradigm [1] starts from the assumption that cognitive process is structured as a cycle. The first step is to gather some knowledge, then the knowledge is used to solve a problem and, depending on the result, one may decide to keep track of the new experience. Experience is accumulated either by adding new information or by adapting the existing knowledge. The idea is to solve a problem with the existing skills and, at the same time, improving these skills for future use. From the actual implementation point of view, the focus is on how to aggregate and store the information (cases) and how to retrieve them. The solution of a problem depends on the ability of the system to retrieve similar cases for which a solution is already known. The more common retrieval techniques are inductive retrieval and nearest neighbor.

4.2 Classification of Tasks in Management Layers

Consider the hierarchical model for network management shown in Table 1. As this model in its essential form and the functions of various layers has been discussed in [12], we do not elaborate on them.

Layer	Tasks or Requirements	Information Flow	Control Flow
Business	Decision Support	↑	↓
Service	Information Retrieval (IR)	↑	↓
Network	Resource Management	↑	↓
Element	Fast Control (Connection Admission)	↑	↓

Table 1, Management layers (adapted from [12])

At the highest layer, the problems can be associated with an overwhelming amount of data. The AI techniques should process the data and present only the relevant information by acting as a decision-support tool. At this layer, the response time is important but not critical. This type of task is well suited for techniques that implement search techniques, e.g. genetic algorithm. Also, model-based expert systems can be used to hide the network complexity behind several abstraction levels. In this context, fuzzy logic can be used to handle model/data uncertainties and ambiguities while interpolating between (possibly) several emerging models. The resultant aggregate model will also have some degree of confidence attached to it that will assist the operators in dealing with the presented information.

While in the next section we take a closer look at some of the tasks in the service layer, it can be noted that the above discussion holds for both service and network management layers. For example, AI based network management systems that deal with the problems at network layer, are mostly based upon expert system techniques [13]. At the elements management layer though, the time response becomes the critical factor. It must be noted that fuzzy logic (and ANN) implementations can be hardware-based to achieve fast response (while most other AI approaches are software-based). At this layer, the environment changes rapidly and a slow solution will become irrelevant. The available information is often incomplete and incoherent. The fuzzy logic character in dealing with uncertainties along with its capabilities in handling several sources of information (via interpolations and taking a supervisory role), make it an excellent choice for management support at this layer.

4.3 Advanced Help Desk

In a competitive business environment, customer satisfaction is a vital objective for many companies: high-quality products and high-quality customer service are two strategic aspects. In this context, help desk systems play an important role providing customer support and functions like change, configuration and asset management. The two main components of a help desk system are the front-end and the back-end ones. The former manages the interaction with the customers while the latter deals with information retrieval (IR) issues.

The core functionality is the retrieval of data from a database whose abstraction matches the description of an ideal object, inferred from a query. Implementation issues are critical both for the overall performance of the system and the accuracy of the retrieved information. Customers usually provide data with different degrees of confidence depending on how that information has been collected. Current IR tools do not explicitly model the uncertainty associated with information but they mix the measure of relevance associated to information with the relative measure of confidence. They don't even manage the feedback provided by users about the accuracy and usefulness of the retrieved solutions. An effective use of that information is the key to enable a process of system adaptation. The explicit management of relevance and confidence on information, integrated with an adaptivity process is the key factor for improving the retrieval precision of a help desk system [15]. Fuzzy logic can be used to form an integrated approach to both uncertainty and adaptivity problems. Keywords are still at the base of the abstraction model, but together with relevance information, they will be enriched with information on confidence degree implemented using membership functions.

4.4 Network Diagnostic Systems

The precise identification of the context in which a problem occurs is fundamental in order to diagnose its causes and, eventually to fix it. The more accurate the information on the context, the more precise the diagnosis can be. The goal of a diagnostic system is to maintain and extract from an information base facts, rules and any other type of indications that can help in identifying the problems. The starting point is a set of facts (observations) but the same fact may have different relevance in different contexts. Collecting information on the relevance of facts allows being more precise in the retrieval (matching) process and precision is fundamental when the dimension of the system knowledge base grows.

The problem is that, while observations are hardly disputable, the relevance associated with them may depend on the experience of the observer. Traditionally the confidence and relevance are empirically merged in a single value and this may corrupt the information. A different (or complementary) solution is to explicitly model and manage the uncertainty associated with the observation. The idea is to capture in this way the fact that there is something missing even if we don't know what it is. Certainty may be reinforced or reduced and adaptivity plays a fundamental role in this kind of process. This type of modeling and reinforcement can be best achieved by incorporating fuzzy sets and fuzzy logic.

Furthermore, the association of confidence with the information through fuzzy sets to establish explicit uncertainty models may prove to be beneficial in other respects as well [22]. For example, the fact that confidence values for a symptom are low suggests that a clear understanding of its meaning does not exist and it may need to be investigated more carefully. Looking at the confidence distribution of different symptoms of the same case, we can obtain indications on the reliability of the associated diagnosis proposals: if there is uncertainty on the causes of a problem (case) we may be more careful considering the proposed diagnosis. Qualitative analysis of case descriptors may give indications on the system users, their needs and their problems. This extra layer of information provides a starting point for a more user focused diagnostic system where effectiveness derives not only from technological issues but also from a clearer understanding of the user (human or software agent).

4.5 Quality-of-Service

Distributed applications are increasingly composed of modular off-the-shelf software components and custom code. Current management systems that monitor thresholds and trigger alarms rely on correct interpretation by the operator to determine causal interactions. This approach does not scale as the number of thresholds and alarms increase. For a scalable solution, a management system should be able to monitor, diagnose and apply reconfiguration of application components to ensure that user-level quality-of-service (QoS) goals are maintained. The management system must be pro-active and coordinate with existing network management systems. Emerging problems are corrected before QoS failures occur. The use of knowledge-based systems is ideal for management of these distributed applications. These systems can conceptually be significantly enhanced by incorporation of fuzzy logic. Such incorporation will improve diagnostic rules that are more capable of handling ambiguity and incomplete information.

4.6 Fuzzy Intrusion Detection

Hybrid systems that are claimed to combine the advantages of both statistical and rule-based algorithms, while partially eliminating the shortcomings of each one, are also devised. In general, such systems will use the rule-based approach for detection of previously encountered intrusions and statistical anomaly detection algorithms for checking new types of attacks. An example of this general approach is based on utilization of neural networks that are trained to model the user and system behavior, while the anomaly detection consists of the statistical likelihood analysis of system calls [21]. Another approach is based on state transition analysis [7]. It attempts to model penetrations as a series of state changes that lead from an initial secure state to a target compromised state. A case based reasoning approach to intrusion detection, which alleviates some of the difficulties in acquiring and representing the knowledge is presented in [6]. A data-mining framework for adaptively building intrusion detection models is described in [18]. It utilizes auditing programs to extract an extensive set of features that describe each network connection or host session, and applies data mining approaches to learn rules that accurately capture the behavior of intrusions and normal activities.

For any type of the intrusion detection algorithm, some points need to be further considered. In rule-based (expert) systems administrators or security experts must regularly update the rule base to account for newly discovered attacks. There are some concerns about any system that relies heavily on human operators (or experts) for knowledge elicitation. Some of the crucial ones are:

- Humans, in the course of decision making and reaching a conclusion, might use variables that are not readily measurable or quantifiable.

- Humans might articulate non-significant features. This, among other reasons, can lead to the establishment of inconsistent (from one expert to another) rule bases. Also, the system will be slower than what it should be as some of the rules that make up the knowledge base are of secondary importance.

- Broadly speaking, experts' knowledge is necessarily neither complete nor precise.

For these reasons, it is highly desirable to have systems and algorithms that acquire knowledge from experiential evidence automatically.

The statistical-anomaly detection algorithm will report 'significant' deviations of a behavior from the profile representing the user's normal behavior. While the significant usually refers to a threshold set by the system security officer, in practice it can be difficult to determine the amount that a behavior must deviate from a profile to be considered a possible attack. In the case of distributed anomaly detection based on the mechanisms in natural immune system, it is in fact advantageous to be able to carry out approximate detection.

In any of these algorithms, the need for exploiting the tolerance for imprecision and uncertainty to achieve robustness and low solution costs is evident. This is in fact, the guiding principle of soft computing and more particularly fuzzy logic [23]. The subject of fuzzy logic is the representation of imprecise descriptions and uncertainties in a logical manner. Many intrusion detection systems are mainly dependent on knowledge bases or input/output descriptions of the operation, rather than on deterministic models. Inadequacies in the knowledge base, insufficiency or unreliability of data on the particular object under consideration, or stochastic relations between propositions may lead to uncertainty. Uncertainty refers to any state of affair or process that is not completely determined. In rule-based and expert systems, lack of consensus among experts can also be considered as uncertainty. Also, humans (administrators, security experts...) prefer to think and reason qualitatively, which leads to imprecise descriptions, models, and required actions.

Zadeh introduced the calculus of fuzzy logic as a means for representing imprecise propositions (in a natural language) as non-crisp, fuzzy constraints on a variable [21]. This is 'vagueness': a clear but not precise meaning. That is to say, fuzzy logic started to cover vagueness, but turned out to be useful for dealing with both vagueness and uncertainty. The use of fuzzy reasoning in expert systems is naturally justifiable, as imprecise language is the characteristic of much expert knowledge. In crisp logic, propositions are either true or false, while in fuzzy logic different modes of qualifications are considered.

There seems to be an urgent need for further work on exploring the ways that artificial intelligence techniques can make the intrusion detection systems more efficient. More specifically, intelligent approaches that learn and automatically update user and system profiles need to be investigated. More research to study the implications and advantages of using fuzzy logic for approximate reasoning and handling intrusion detection through approximate matching are definitely required. Additionally, the capabilities of fuzzy logic in using the linguistic variables and fuzzy rules for analysing and summarizing the audit log data need to be investigated.

5 Concluding Remarks

To cope with the increasing complexity of the networks, their management systems have become highly complicated as well. The management system must deal with an overwhelming amount of data that may be incoherent and inconsistent or unreliable. Compared to more conventional techniques, AI approaches are more suitable for this type of tasks. In particular, the capabilities of fuzzy logic in handling vague concepts or systems with uncertainties are of prime significance. We described several ways that fuzzy logic can be used in identifying or improving the solutions to problems encountered in an integrated network management environment. In this work, in addition to a conceptual discussion of this topic, several areas with functional importance are also considered. For instance, it is noted that a key aspect of help desk services is related to information retrieval where uncertainty in data is a major peculiarity.

Another important area is computer security and intrusion detection. Any set of actions that attempt to compromise the integrity, confidentiality, or availability of a resource is defined as an intrusion. Many intrusion detection systems base their operations on analysis of operating system audit trail data. Intrusions can be categorised into two main classes: misuse intrusions and anomaly intrusions. As misuse intrusions follow well-defined patterns they can be detected by performing pattern matching on audit-trail information. Anomalous intrusions are detected by observing significant deviations from normal behavior. Anomaly detection is also performed using other mechanisms, such as neural networks, machine learning classification techniques, and approaches that mimic biological immune systems.

Anomalous intrusions are harder to detect, mainly because are no fixed patterns of intrusion. So, for this type of intrusion detection fuzzy approaches are more suitable. A system that combines human-like capabilities in handling imprecision and adaptive pattern recognition with the alertness of a computer program can be highly advantageous. This is an area that demands further study and research work.

In this work it is argued that a comprehensive solution for uncertainty management can be based on the notion of fuzzy sets in which relevance and confidence is used to enrich the descriptive power of keyword paradigm. In the case of diagnostic systems, fuzzy logic can be used for the explicit modeling of the uncertainty that in turn leads to an actual improvement in terms of case-selection precision.

References

[1] Althoff K, Wess S (1991) Case-based reasoning and expert system development. I in: Contemporary Knowledge Engineering and Cognition, Springer-Verlag, USA.

[2] Benech D (1996) Intelligent agents for system management. In: Proc Distributed Systems: Operation and management.

[3] Denning D (1988) An intrusion-detection model. IEEE Trans Software Engineering 13: 222-228.

[4] D'haeseleer P, Forrest S, Helman P (1997) A distributed approach to anomaly detection1. Technical report, available at http://www.cs.unm.edu/~forrest/papers.html.

[5] Endler D (1998) Intrusion detection. Applying machine learning to Solaris audit data. In: Proc 14th Computer Security Applications Conference, pp 268-279.

[6] Esmaili M, Balachandran B, Safavi-Naini R, Pieprzyk J (1996) Case-based reasoning for intrusion detection. In: Proc 12th Computer Security Applications Conference, pp 214-223.

[7] Ilgun K, R. Kemmerer R, Porras P (1995) State transition analysis: a rule-based intrusion detection approach. IEEE Trans Software Engineering 21: 181–199.

[8] Kephart J (1994) A biologically inspired immune system for computers. In: Brooks R, Maes P (eds) Artificial Life: Proc International Workshop on the Synthesis and Simulation of Living Systems, MA: MIT Press, Cambridge.

[9] Lewis L, Kaikini P (1993) An approach to the alarm correlation problem using inductive modeling technology. Technical note ctron-lml-93-03, Cabletron Systems R&D Center, Merrimack.

[10] Lunt T, Tamaru A, Gilham F, Jagannathan R, Jalali C, Neumann P (1992) A real-time intrusion detection system (IDES). Technical report available at: http://www.sdl.sri.com/nides/reports/9sri.pdf.

[11] Mukherjee B, Heberlein L, Levitt K (1994) Network intrusion detection. IEEE Network 3:26–41.

[12] Muller C, Veitch P, Magill E, Smith D (1995) Emerging AI techniques for network management. In: Proc IEEE GLOBECOM '95, pp 116-120.

[13] Nuansri N, Dillon T, Singh S (1997) An application of neural network and rule-based systems for network management. In: Proc 30th Hawaii International Conference on System Sciences, pp 474-483.

[14] Pao Y (1989) Adaptive Pattern Recognition and Neural Networks. Addison Wesley, USA.

[15] Piccinelli G, Mont M (1998) Fuzzy-set based information retrieval for advanced help desk. Technical report HPL-98-65, HP Laboratories, Bristol. Available at http://www.hpl.hp.com/techreports/98/HPL-98-65.html

[16] Shahrestani S, Yee H, Ypsilantis J (1995) Adaptive recognition by specialized grouping of classes. In: Proc 4th IEEE Conference on Control Applications, Albany, New York, pp 637-642.

[17] Venkatesan R, Bhattacharya S (1997) Threat-adaptive security policy. In: Proc IEEE International Performance, Computing, and Communications Conference, pp 525-531.

[18] Wenke L, Stolfo S, Mok K (1999) A data mining framework for building intrusion detection models. In: Proc. IEEE Symposium on Security and Privacy, pp 120-132.

[19] Wilikens M (1998) RAID'98: Recent advances in intrusion detection- workshop report. Available at http://www.zurich.ibm.com/pub/Other/RAID/

[20] Winston P (1984) Artificial Intelligence. Addison-Wesley, USA.

[21] Zadeh L (1965) Fuzzy sets. In: Information and Control 8: 338-353.

[22] Zadeh L (1983) The role of fuzzy logic in the management of uncertainty in expert systems. In: Fuzzy Sets and Systems 11: 199-228.

[23] Zadeh L (1994) Soft computing and fuzzy logic. In: IEEE Software 11: 48-56.

Smart Homepage-Finder

— A Genetic Fuzzy Neural Agent for Searching Homepages Intelligently

Yuchun Tang and Yanqing Zhang

Department of Computer Science, Georgia State University, Atlanta, GA 30303

Abstract. In this chapter, we propose an intelligent web information search and retrieval model called Web Information Search Task (WIST) based on Computational Web Intelligence (CWI). Homepage-Finder, an intelligent software agent, is designed using Fuzzy Logic (FL), Neural Networks (NN) and Genetic Algorithms (GA) to execute a specific WIST to automatically find all relevant researchers' homepages based on the possibility whether a web page is a personal homepage and the relevance with keywords, given a root URL, some keywords and some structure rules. The simulation results show that Homepage-Finder with non-linear fuzzy reasoning can find more personal homepages (from 797 to 1571) with much higher precision (from 54.4% to 91.0%), higher recall (from 79.0% to 92.5%) than Google and list them in a desired order (average error is 0.082).

1. Introduction

The growth of the WWW is very fast. "The web is growing with a spur of 7 million of pages a day and it has already reached quota of 4 billion of pages." [10]. Finding the desired information is not an easy task because the information available on the WWW is inherently unordered, distributed, and heterogeneous [13]. As a result, "the ability to search and retrieve information from the web efficiently and effectively is a key technology for realizing its full potential." [14]. The task is to retrieve useful information according to some measures on response to user-defined queries which express user's search request.

Expressing a search request is the first thing we need to solve. How to make search interface more expressive? At almost all search scenarios, the desired information is on some web pages (especially HTML/XML formatted). For some similar search requests, the desired web pages share some similar "content characteristics" and/or "structure characteristics".

Traditional search methods let users to submit "keywords" that may be displayed on the desired web pages to express their search requests. So we call them "keyword based search" or "content based search". Obviously, they are not so powerful. There are two problems.

Problem1: Although in some cases a search request can be "crisply" expressed, in many other cases, a search request is inherently fuzzy and thus is hard or even impossible to be expressed "crisply". For example, if 10 pages have the keyword A, 8 pages keyword B, and 5 pages both A and B, so there are 10+8-5=13 pages altogether, suppose the desired pages are the 6 pages including 4 pages with both A and B, 1 page with only A, and 1 page with only B, then how to express the search request by keywords? Whatever the submitted keywords are "A", "B", "A and B", or "A or B", we cannot get the desired pages completely and accurately. Another example is, if a user says "I want to find all personal homepages of the faculties in the department of computer science", how to express the search request? The key is, what is a "personal homepage"? We can not simply submit "homepage" or "home page" as keywords, because some personal homepages don't include these keywords, while some pages which are not personal homepages include these keywords. For example, a class homepage may include a sentence like "welcome to the class homepage". Furthermore, some pages do provide parts of information of the desired faculties, but we can not conclude whether it is a personal homepage or not. "Even though techniques exist for locating exact matches, finding relevant partial match might be a problem". [11]

Problem2: In many, if not all, cases, some keywords provide some structure characteristics while others only provide content characteristics. Traditional search methods do not differentiate the two kinds of keywords. In the aforementioned example "I want to find all personal homepages of the faculties in the department of computer science", if the word "homepage" is displayed on a web page, then maybe the web page is a personal homepage. However, the word "computer science" cannot provide any structure characteristics of the desired web pages. In the following, "keywords" will be referred as those words that only provide content characteristics.

Finding and retrieving information is another thing we need to think about. How to retrieve information more efficiently and more effectively? "Current search engines are known for poor accuracy: they have both low recall (fraction of desired documents that are retrieved) and low precision (fraction of retrieved documents that are desired)." [2]. Furthermore, the most relevant/desired documents are not always displayed at the top of the query result list so that the query result is not listed in a desired order. The focus of current-generation search methods and tools have been on query-processing speed and database size. But recently, some researches have begun to concentrate on providing a short, ranked list of meaningful documents, which requires more effective Information Retrieval (IR); "Ideally, effective IR means high recall and high precision, but in practice it means acceptable compromises." [14] The appeal of FL, NN, and GA, as efficient tools featuring computational intelligence, which is already acknowledged in many areas of Information Technology, plays an important role on addressing this issue [2].

Ivan Ricarte gave a tutorial review of the current state of the art in the area of information retrieval and web search systems and then proposed a reference model to establish the relationships between computational intelligence and information search systems as a means to promote intelligent information search systems development [14].

Because most users of search engines work in a few fairly well-defined knowledge domains, Mori Anvari proposed to design and implement a prototype of an intelligent front-end to enhance the usability and functionality of existing search engines. He also gave an approach to construct ontology (or meta-thesaurus) for a specific knowledge domain and restrict the search to that knowledge domain [1]. Flamenco is another project to develop a general methodology for specifying task-oriented search interfaces across a wide variety of domains and tasks [4].

"A promising way to create useful intelligent agents is to involve both the user's ability to do direct programming along with the agent's ability to accept and automatically create training examples." [3]. In [3], Tina Eliassi-Rad and Jude Shavlik presented and evaluated WAWA's information retrieval system (WAWA-IR) for creating personalized information-finding agents for the web. They also built a "home-page finder" as a case of WAWA-IR agent. Their "home-page finder" is primarily used to find one person's homepages given the person's name, which is also implemented by Ahoy! [5] and HomePageSearch [6], whereas Homepage-Finder here can be used to find a group of people's homepages with some common characteristics such as similar research interests, in the same research group, etc., although our Homepage-Finder can also find a special person's homepages. Another difference is that their "home-page finder" uses back-propagation learning algorithm but our Homepage-Finder uses FNN to model the system and uses GA to optimize the unknown parameters.

Zadeh mentioned: "fuzzy logic may replace classical logic as what may be called the brainware of the Internet" at 2001 BISC INTERNATIONAL WORKSHOP ON FUZZY LOGIC AND THE INTERNET (FLINT2001). The granular fuzzy web search agents are proposed based on granular computing, fuzzy computing and the Internet computing [19]. In general, Computational Web Intelligence (CWI) is a hybrid technology of Computational Intelligence (CI) and Web Technology (WT) dedicating to increasing QoI of e-Business applications on the Internet and wireless networks [18]. Seven major research areas of CWI are (1) Fuzzy WI (FWI), (2) Neural WI (NWI), (3) Evolutionary WI (EWI), (4) Probabilistic WI (PWI), (5) Granular WI (GWI), (6) Rough WI (RWI), and (7) Hybrid WI (HWI). This chapter focuses on the combination of FWI, NWI and EWI.

This chapter is organized as follows. Section 2 proposes an intelligent web information search and retrieval model called Web Information Search Task (WIST). Section 3 presents Homepage-Finder to show how to define and implement a specific WIST to automatically find researchers' homepages. Section

4 gives the simulation results and performance evaluation of Homepage-Finder. Finally, Section 5 concludes this chapter and directs the future works.

2. Web Information Search Task

Many search requests have different content characteristics but share similar structure characteristics. For example, a user wishes to find "all personal homepages of researchers whose research interests are in the field of artificial intelligence and who are in Georgia State University". In this request, the content characteristic is "artificial intelligence", and the structure characteristics can be expressed as "personal homepages". Another request may be "I want to find the information of all faculties of Georgia State University who are members of CWI project", here the content characteristic is "CWI", but the structure characteristics still can be abstracted as "personal homepages".

Based on this observation, we propose an intelligent web information search and retrieval model called Web Information Search Task (WIST). A WIST expresses structure characteristics by simple "structure rules". What is a "structure rule"? Basically, a structure rule is a condition clause defined on web pages' URL, Title, Text, Links, or other related sections. If the condition is satisfied, the target web page may be a desired web page. Otherwise it may be not desired. 2 examples of "structure rules" will be given in Section 3. As a result, all search requests with similar structure characteristics can be "categorized" into a WIST. Then users only need to submit keywords to define content characteristics. Users can also define their own WISTs if necessary.

Design of any new intelligent search engine should be at least based on two main motivations [16, 17]:

- The web environment is, for the most part, unstructured and imprecise. To deal with information in the web environment what is needed is a logic that supports modes of reasoning which are approximate rather than exact. While searches may retrieve thousands of hits, finding decision-relevant and query-relevant information in an imprecise environment is a challenging problem, which has to be addressed.

- Another, and less obvious, is deduction in an unstructured and imprecise environment given the huge stream of complex information.

A WIST is implemented as an intelligent software agent. Essentially, the agent implements a Fuzzy Neural Network (FNN) to infer the possibility whether a web page is desired. So it has the ability of approximate reasoning and deduction to find relevant or partial relevant matches and rank them according to the degree of matching. The agent uses Genetic Algorithms to

- learn to get more accurate parameters of FNN,
- learn to get more suitable structure of FNN, and
- learn to define structure characteristics by adding/modifying structure rules.

Masoud Nikravesh and Tomohiro Takagi proposed an intelligent model called "Fuzzy Conceptual Matching" (FCM) to be used for intelligent information and knowledge retrieval through conceptual matching. [11, 12, 15] The model can be used to calculate conceptually the degree of match to the object or query. In the FCM approach, the "concept" is defined by a series of keywords with different weights depending on the importance of each keyword. "Conceptual Fuzzy Sets" (CFS) are used to describe the "concept". In a CFS, the meaning of a concept is represented by the distribution of the activation values of the other concepts.

Here WIST model may be viewed as an extension of FCM model. Because FCM is an excellent content-based matching method, if we combine the two models together, we will have the advantages of content-based matching and structure-based matching at the same time. For a WIST, the structure characteristics are expressed by structure rules, but the content characteristics can be expressed by "concepts" which are specified by a group of keywords with different weights. Both the structure rules and the concepts can be defined manually according to expert knowledge, or automatically by learning from known web pages. In this way, the aforementioned search request to find "all personal homepages of researchers whose research interests are in the field of artificial intelligence and who are in Georgia State University" can be satisfied better, because the model can also retrieve a homepage, for example, which has not keyword "artificial intelligence" but has keyword "neural networks".

Like many search engines used today, WIST also provides "root URLs" to limit information search only in some special domains. In the above two search requests, the root URL could be defined as "gsu.edu" or "cs.gsu.edu".

3. Homepage-Finder: a case research of WIST

This chapter focuses on solving a specific WIST. It is common for us to look for the information of faculties and/or students with some common characteristics such as similar research interests, in a department, or in some research groups, etc. Browsing their homepages is the way we usually use. Unfortunately, we will quickly find it is time consuming to navigate the university's or department's website link by link because

- some universities or departments' homepage list are not up to date or not complete,
- some faculties have not "normal" homepages but their information is dispersed on many other pages,
- sometimes we want to find all researchers in a research group across departments or even across universities,
- different personal homepages have different content structures (for example, some faculties use <frameset> in their personal index pages).

Searching their homepages by submitting keywords to a current search engine like Google is another possible way. But it is still difficult to get complete and accurate list because current search engines' weak expressiveness and inability to response to a fuzzy search request like "finding homepages".

We implement Homepage-Finder, an intelligent software agent which uses CI technologies including FL, NN, and GA to define and implement the specific WIST to automatically find relevant researchers' homepages based on the possibility whether a web page is a personal homepage and the relevance with keywords, given a root URL, some keywords and some "structure rules". The relevance with keywords is evaluated by Google's PageRank technology [1][7]. Notice here we adopt the exact keyword matching method to show the efficiency of WIST model. If replacing it with FCM method, we will get better result according to the accuracy especially recall.

In the WIST to find homepages, two examples of structure rules are given as follows:

If a web page's URL string includes "/~" and the URL string's last character is "/", then the web page is possible to be a personal homepage

If a web page's Title string includes "homepage" or "home page", then the web page is possible to be a personal homepage

The following will present the methodology and system structure of Homepage-Finder.

3.1 Google Search Engine and Web Services

Google, a popular online web search engine, has recently released the beta version of the Google Web APIs [8]. The Google Web APIs make it possible for software developers to consume their online search engine functions as Web Services.

Homepage-Finder utilizes the Google Search Engine and Web Services to get the web pages' URL and Title matched to the given root URL and keywords, and then uses HTTP to retrieve web pages' Text. As a result, there is no need to develop another agent to navigate in the website referenced by the given root URL so that Homepage-Finder can concentrate on analyzing these retrieved web pages (called "raw data" or "input data") to get ranked list based on the possibility whether they are personal homepages. One more benefit is that the function of Homepage-Finder can be easily integrated into a business successful search engine like Google.

3.2 Fuzzy Inference System (FIS)

After collecting the keyword-matched web pages as input data, Homepage-Finder scores every retrieved web page based on the structure rules defined on the characteristics of its URL, Title, and Text. So a web page gets 3 scores called "URLscore", "Titlescore", and "Textscore", respectively. Homepage-Finder then calculates a "Totalscore" for the web page with a 3-input-1-output TSK fuzzy inference system [9] to infer the possibility whether the web page is a personal homepage. Each of the 4 scores is in the field of [0,1].

Homepage-Finder defines 3 structure rules for URL, 2 structure rules for Title, and 4 structure rules for Text. Each of the 9 structure rules gives a score that is unknown but in the field [0,1]. So there are 9 unknown parameters called "premise parameters" and denoted by url1, url2, url3, title1, title2, text1, text2, text3, text4, respectively. Therefore the two structure rules defined above should be modified as

If a web page's URL string includes "/~" and the URL string's last character is "/", then its URLscore is url1

If a web page's Title string includes "homepage" or "home page", then its Titlescore is title1

Suppose title1=0.7, the second structure rule means "estimated from the title, the possibility the web page is a personal homepage is 70%". It also means "the possibility the web page's Titlescore is HIGH is 70%" and "the possibility the web page's Titlescore is LOW is 30%". The linguistic variables "HIGH" and "LOW" are defined on the 3 input scores:

$$\mu_{HIGH}(URLscore) = URLscore, \quad \mu_{LOW}(URLscore) = 1 - URLscore,$$

$$\mu_{HIGH}(Titlescore) = Titlescore, \quad \mu_{LOW}(Titlescore) = 1 - Titlescore,$$

$$\mu_{HIGH}(Textscore) = Textscore, \quad \mu_{LOW}(Textscore) = 1 - Textscore,$$

"HIGH" means "how much possibility a web page is a personal homepage", "LOW" means "how much possibility a web page is NOT a personal homepage". Essentially, the fuzzification is discrete because Homepage-Finder scores a web page according to the discrete structure rules.

Fuzzy Rules for Homepage-Finder are given below:

FR(1):IF URLscore is LOW and Titlescore is LOW and Textscore is LOW, THEN Totalscore = 0

FR(2):IF URLscore is LOW and Titlescore is LOW and Textscore is HIGH, THEN Totalscore = p_{21} * URLscore + p_{22} * Titlescore + p_{23} * Textscore

FR(3):IF URLscore is LOW and Titlescore is HIGH and Textscore is LOW, THEN Totalscore = p_{31} * URLscore + p_{32} * Titlescore + p_{33} * Textscore

FR(4):IF URLscore is LOW and Titlescore is HIGH and Textscore is HIGH, THEN Totalscore = p_{41} * URLscore + p_{42} * Titlescore + p_{43} * Textscore

FR(5):IF URLscore is HIGH and Titlescore is LOW and Textscore is LOW, THEN Totalscore = p_{51} * URLscore + p_{52} * Titlescore + p_{53} * Textscore

FR(6):IF URLscore is HIGH and Titlescore is LOW and Textscore is HIGH, THEN Totalscore = p_{61} * URLscore + p_{62} * Titlescore + p_{63} * Textscore

FR(7):IF URLscore is HIGH and Titlescore is HIGH and Textscore is LOW, THEN Totalscore = p_{71} * URLscore + p_{72} * Titlescore + p_{73} * Textscore

FR(8):IF URLscore is HIGH and Titlescore is HIGH and Textscore is HIGH, THEN Totalscore = 1

$p_{ij} \in [0,1]$ are called "consequence parameters", $\displaystyle\sum_{j=1}^{3} p_{ij} = 1$, $i \in \{2,3,4,5,6,7\}$.

Homepage-Finder uses product-sum to do fuzzy reasoning:

$$Totalscore = f(\underline{x}) = \sum_{j=1}^{8} (w_j * f_j) / \sum_{j=1}^{8} w_j;$$

$$\underline{x} = (x_1, x_2, x_3) \quad ,$$

x_1, x_2, x_3 $means$ $URLscore, Titlescore$ and $Textscore, respectively;$

$$f_j = FR(j) \quad , \quad j \in \{1,2,3,4,5,6,7,8\};$$

$$w_1 = \mu_{LOW}(x_1) * \mu_{LOW}(x_2) * \mu_{LOW}(x_3);$$
$$w_2 = \mu_{LOW}(x_1) * \mu_{LOW}(x_2) * \mu_{HIGH}(x_3);$$
$$w_3 = \mu_{LOW}(x_1) * \mu_{HIGH}(x_2) * \mu_{LOW}(x_3);$$
$$w_4 = \mu_{LOW}(x_1) * \mu_{HIGH}(x_2) * \mu_{HIGH}(x_3);$$
$$w_5 = \mu_{HIGH}(x_1) * \mu_{LOW}(x_2) * \mu_{LOW}(x_3);$$
$$w_6 = \mu_{HIGH}(x_1) * \mu_{LOW}(x_2) * \mu_{HIGH}(x_3);$$
$$w_7 = \mu_{HIGH}(x_1) * \mu_{HIGH}(x_2) * \mu_{LOW}(x_3);$$
$$w_8 = \mu_{HIGH}(x_1) * \mu_{HIGH}(x_2) * \mu_{HIGH}(x_3);$$

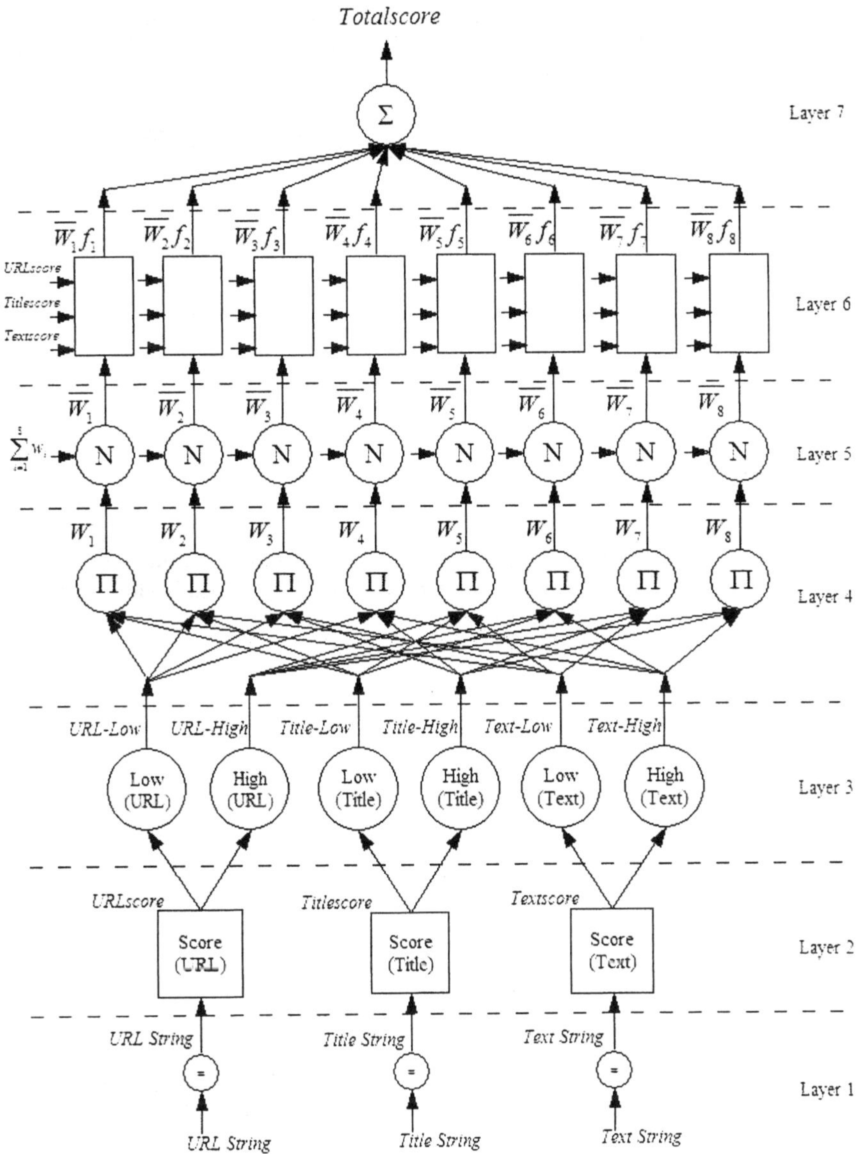

Fig. 1. FNN structure

3.3 Fuzzy Neural Network (FNN)

Hybrid neural networks have useful technical merits because of combination of several relevant techniques [20]. Here, Homepage-Finder uses a fuzzy neural network to do the fuzzy inference. The FNN is functionally equivalent to the 3-input-1-output TSK as stated previously. Fig. 1 shows the architecture of the fuzzy neural network with seven layers.

Layer 1: Input Layer

There are 3 nodes in this layer. Every node is a fixed node whose output is just the same as the input as shown in Fig. 2.

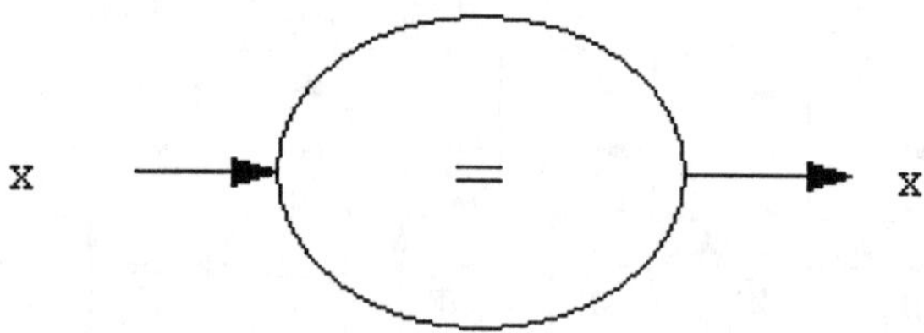

Fig. 2. Input Layer Neuron

Layer 2: Fuzzification Layer

There are 3 nodes in this layer. Every node is an adaptive node with a discrete function as shown in Fig. 3. The input for each of the 3 nodes is URL string, Title string, Text string and the output is URLscore, Titlescore, Textscore, respectively. The discrete functions implement the defined structure rules.

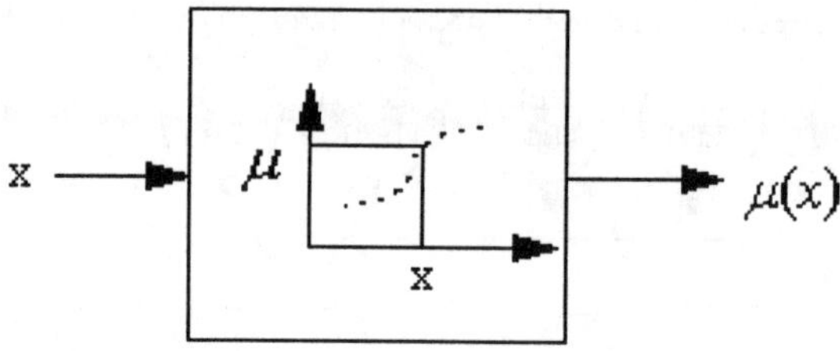

Fig. 3. Fuzzification Layer Neuron

Layer 3: Fuzzification Layer (cont.)

This layer has 6 fixed nodes as shown in Fig. 4 to output URL-High, URL-Low, Title-High, Title-Low, Text-High, and Text-Low, respectively.

$$URL\text{-}High = \mu_{HIGH}(URLscore) = URLscore,$$
$$URL\text{-}Low = \mu_{LOW}(URLscore) = 1 - URLscore,$$
$$Title\text{-}High = \mu_{HIGH}(Titlescore) = Titlescore,$$
$$Title\text{-}Low = \mu_{LOW}(Titlescore) = 1 - Titlescore,$$
$$Text\text{-}High = \mu_{HIGH}(Textscore) = Textscore,$$
$$Text\text{-}Low = \mu_{LOW}(Textscore) = 1 - Textscore,$$

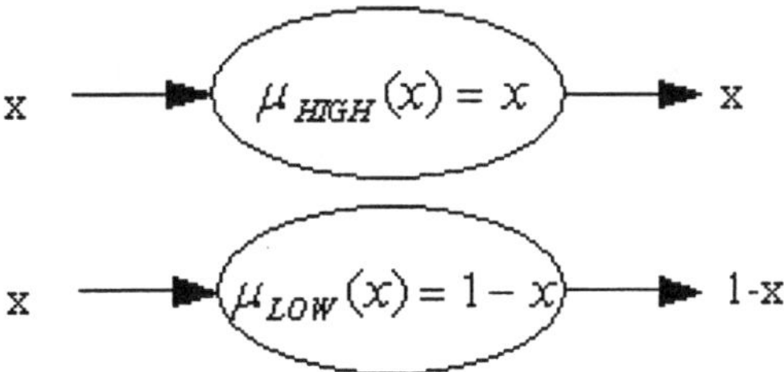

Fig. 4. Fuzzification Layer (cont.) Neuron

Layer 4: Fuzzy-And Layer
There are 8 fixed nodes as shown in Fig. 5 to calculate the firing strengths for fuzzy rules.

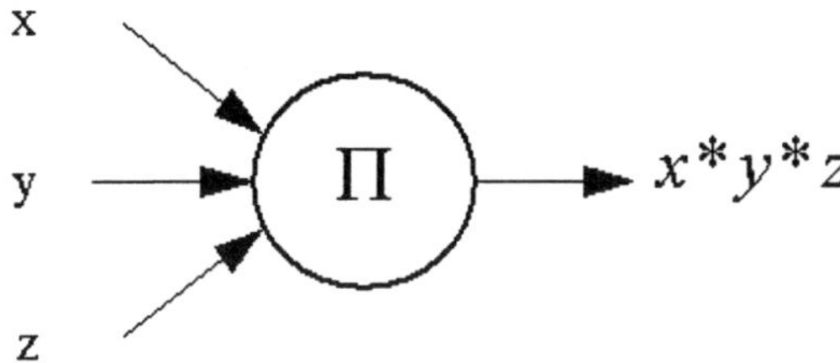

Fig. 5. Fuzzy-And Layer Neuron

Layer 5: Fuzzy-Or Layer
There are 8 fixed nodes as shown in Fig. 6 to normalize the firing strengths.

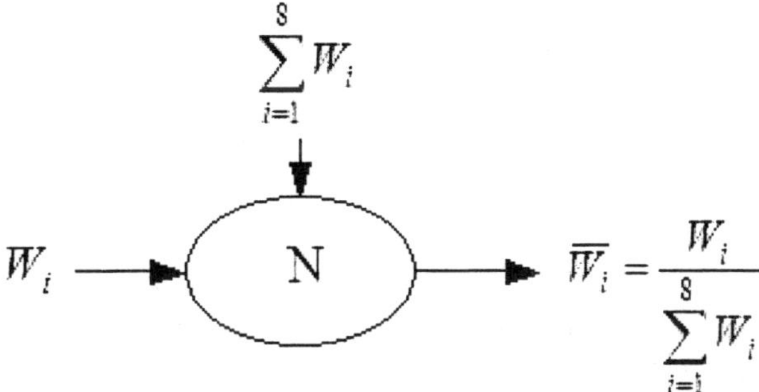

Fig. 6. Fuzzy-Or Layer Neuron

Layer 6: Fuzzy Reasoning Layer
There are 8 adaptive nodes as shown in Fig. 7 to do fuzzy reasoning according to the 8 fuzzy rules, respectively.

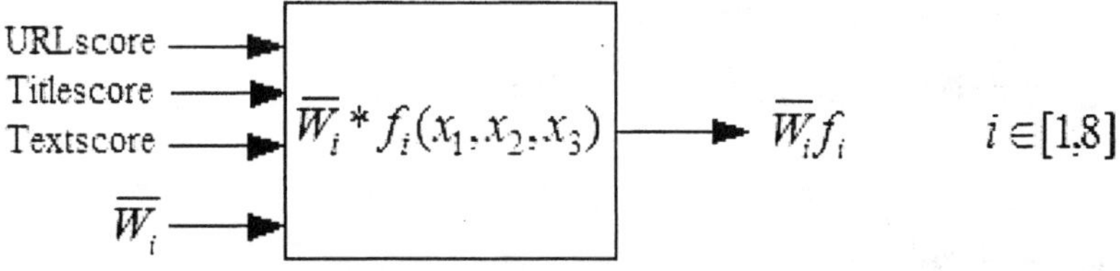

Fig. 7. Fuzzy Reasoning Layer Neuron

Layer 7: Output Layer
There is only one fixed node as shown in Fig. 8 to get the Totalscore.

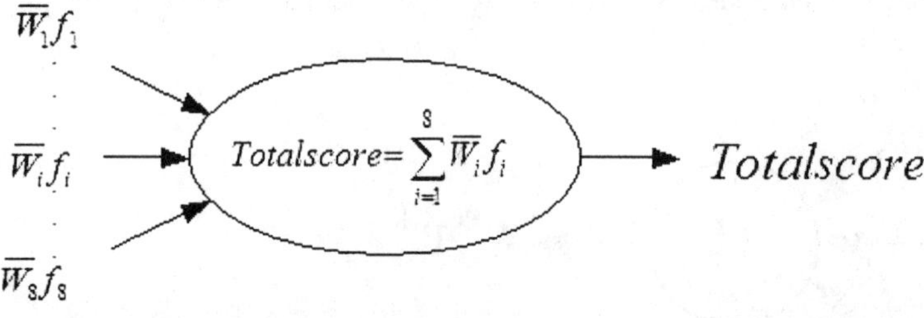

Fig. 8. Output Layer Neuron

3.4 Supervised Learning and Genetic Algorithm

Because Homepage-Finder implements the fuzzy inference with a FNN, so it has the ability to learn to do better job. Homepage-Finder defines the learning objective as minimizing the error function:

$$E = \sum_{i \in training\ data\ set} \frac{1}{2} E_i^2$$

$$E_i = |Totalscore_i - Realscore_i|$$

When building training data set, end users or system administrators need to define the "Realscore" for every web page in the training data set: if a web page is a personal homepage, then its Realscore is 1; otherwise the Realscore is 0.

Traditional back-propagation algorithm is usually adopted, but it is a very time consuming method. The hybrid learning method that combines steepest descent and least squares estimator is a fast learning algorithm [9]. However, the method needs to fix the values of the premise parameters as the prerequisite.

Homepage-Finder uses Genetic Algorithm that is one of the most popular derivative-free optimization methods. GA has many merits [9], but the following two ones are the reasons why Homepage-Finder adopts it:

- GA is parallel. As a result, it can be implemented by using multi-thread technology to massively speed up the learning procedure.
- The performance of GA does not depend on the initial values of the unknown parameters. As a result, the structure rules can be easily defined. For example, end users or system administrators only need to define a structure rule like "If a web page's Title string includes "homepage" or "home page", then its Titlescore is *title*1 " and define *title*1 in the field of [0,1]. What the initial value of *title*1 is does not affect the performance.

Fig. 9 describes the Genetic Algorithm by Java pseudo-code:

```
public boolean beginGATraining() throws Exception{
    int i;
    int generation = 0;
    initialize();
    if ( false == evaluate() ) {
        return false;
    }
    keep_the_best();
    report();
    while ( generation < MAXGENS){
        generation++;
        select();
        crossover();
        mutate();
        if (false == evaluate() ){
            return false;
        }
        elitist();
        report();
    }
    return true;
}
```

Fig. 9. Genetic Algorithms

The function "evaluate( ) " is implemented by multi-thread technology to calculate the error function value, or called the fitness value, of each member in the population. Obviously the minimum value of the error function is 0. But in fact, it's difficult to get so close to 0.

3.5 Multi-threaded Java Application

Homepage-Finder is a multi-thread Java Application so it can efficiently utilize the bandwidth to quickly retrieve web pages. Homepage-Finder also uses multiple

threads to speed up the GA training process. Fig. 10 shows the user interface of
Homepage-Finder.

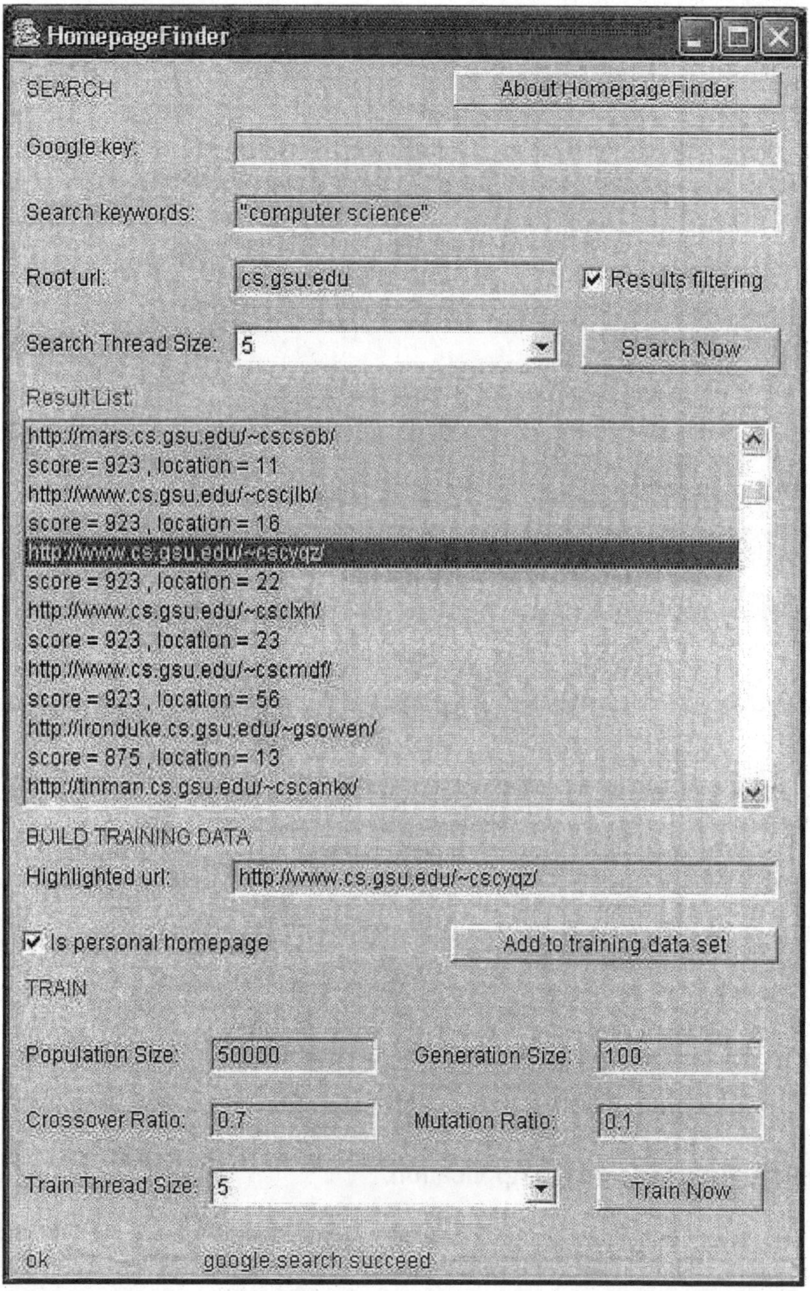

Fig. 10. User Interface of Homepage-Finder

3.6 XML-driven system

Homepage-Finder is an XML-driven system. All of data it uses and produces are XML-formatted, including the GA training data, the NN parameters data, and the query result data, which are validated by corresponding XML Schema. The result XML file is produced per query and is transformed to be HTML-formatted page displayed on XML-Enabled Web Browsers by using XSLT technology. Fig. 11 and Fig. 12 show two sample result lists of Homepage-Finder before XSL transformation and after XSL transformation, respectively.

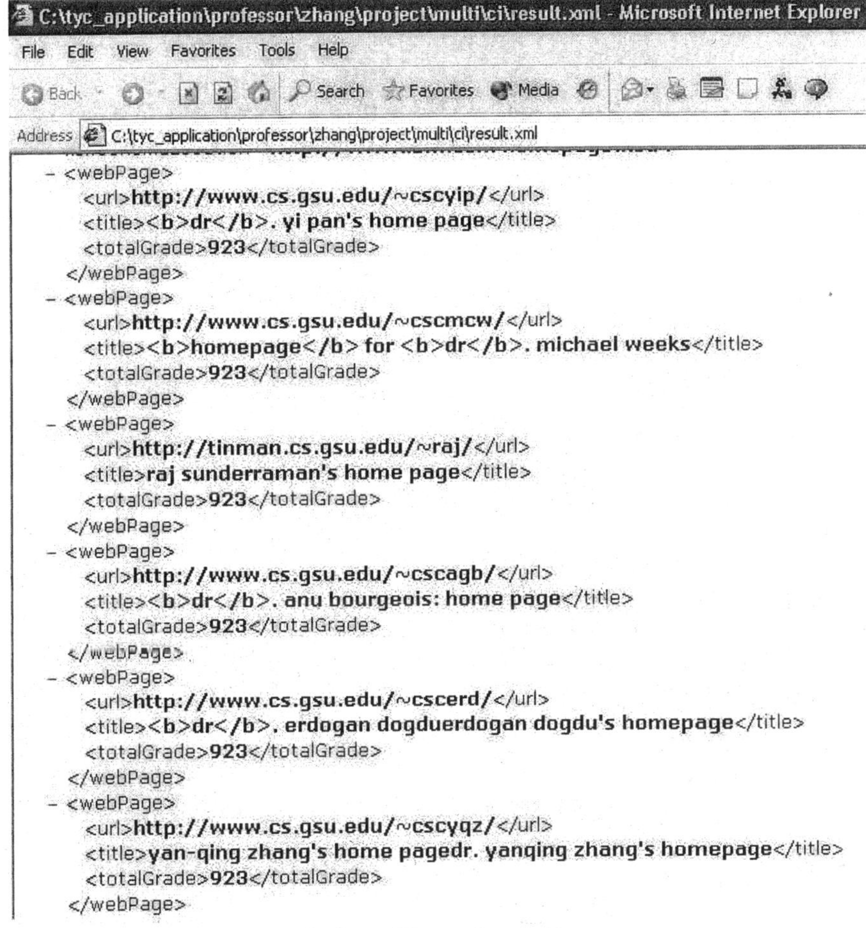

Fig. 11. Result List before Transformation

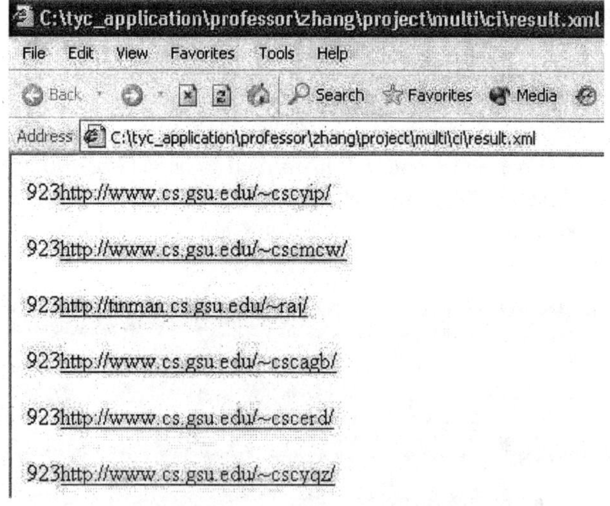

Fig. 12. Result List after Transformation

3.7 Non-linear Fuzzy Inference System

A TSK fuzzy inference system is usually a first-order or linear model. [9] In Homepage-Finder, the FIS is extended to be "nonlinear" by modifying the consequences of fuzzy rules as follows:

FR(1):IF URLscore is LOW and Titlescore is LOW and Textscore is LOW, THEN Totalscore = 0

FR(2):IF URLscore is LOW and Titlescore is LOW and Textscore is HIGH,

THEN Totalscore $= p_{21}*$URLscore$^{q_1} + p_{22}*$Titlescore$^{q_2} + p_{23}*$Textscore^{q_3}

FR(3):IF URLscore is LOW and Titlescore is HIGH and Textscore is LOW,

THEN Totalscore $= p_{31}*$URLscore$^{q_1} + p_{32}*$Titlescore$^{q_2} + p_{33}*$Textscore^{q_3}

FR(4):IF URLscore is LOW and Titlescore is HIGH and Textscore is HIGH,

THEN Totalscore $= p_{41}*$URLscore$^{q_1} + p_{42}*$Titlescore$^{q_2} + p_{43}*$Textscore^{q_3}

FR(5):IF URLscore is HIGH and Titlescore is LOW and Textscore is LOW,

THEN Totalscore $= p_{51}*$URLscore$^{q_1} + p_{52}*$Titlescore$^{q_2} + p_{53}*$Textscore^{q_3}

FR(6):IF URLscore is HIGH and Titlescore is LOW and Textscore is HIGH,

THEN Totalscore $= p_{61}*$URLscore$^{q_1} + p_{62}*$Titlescore$^{q_2} + p_{63}*$Textscore^{q_3}

FR(7):IF URLscore is HIGH and Titlescore is HIGH and Textscore is LOW,

THEN Totalscore $= p_{71}*$URLscore$^{q_1} + p_{72}*$Titlescore$^{q_2} + p_{73}*$Textscore^{q_3}

FR(8):IF URLscore is HIGH and Titlescore is HIGH and Textscore is HIGH,
THEN Totalscore $=1$

$$p_{ij} \in [0,1], \ \sum_{j=1}^{3} p_{ij} = 1, \ q_j \in [0.1, 10], \ i \in \{2,3,4,5,6,7\}, \ j \in \{1,2,3\}.$$

As a result, the orders of URLscore, Titlescore, Textscore are not be fixed to constant 1 but parameters q_1, q_2, q_3 in the field [0.1, 10]. The neurons of layer 6 of FNN are modified to use the new fuzzy rules correspondingly.

4. Performance Evaluations

Four web search systems are listed below for performance comparison:
- Google without Homepage-Finder
- FNN1: linear FNN of Homepage-Finder whose parameters are defined by observation and experience
- FNN2: linear FNN of Homepage-Finder whose parameters are optimized by GA.
- FNN3: non-linear FNN of Homepage-Finder whose parameters are optimized by GA.

FNN2 and FNN3 use the same training data set which consists of 87 web pages selected from the Department of Computer Science of University of Georgia (cs.uga.edu).

4.1 Google Search Strings Definition

The Google search strings are defined by the following example:

If a user wants to find the web pages with the exact phrase "computer science", and if the Root URL is "cs.gsu.edu",
Then the Google search string without Homepage-Finder is
"computer science" professor|dr.|lecturer homepage|"home page" site : cs.gsu.edu
, and the Google search string submitted by Homepage-Finder is
"computer science" professor|dr.|lecturer|homepage|"home page" site : cs.gsu.edu

.

Because Google Web Services limit the maximum number of results received per query to be 1000 [8], Homepage-Finder includes professor|dr.|lecturer|homepage|"home page" in the Google search string to shorten the length of results lists.

4.2 Parameter Identification

There are 9 premise parameters $url1$, $url2$, $url3$, $title1$, $title2$, $text1$, $text2$, $text3$, $text4$ and 21 consequent parameters p_{ij}, q_j, $i \in \{2,3,4,5,6,7\}$, $j \in \{1,2,3\}$. Because of $\sum_{j=1}^{3} p_{ij} = 1$, there are altogether 24 unknown parameters need to be optimized. All parameters and scores except q_j should be in the field [0,1]. For FNN1 and FNN2, $q_j \equiv 1$; for FNN3, $q_j \in [0.1,10]$. But in the actual system, all parameters and scores are multiplied by 1000 and limited to be integers. Realscore is also set to 0 or 1000.

Fig. 13 shows the values of the parameters. The left column is FNN1's parameters, while the middle FNN2, and the right FNN3.

```
- <parameter fitness="6520042">    - <parameter fitness="1771022">    - <parameter fitness="1114444">
    <url1>800</url1>                    <url1>900</url1>                    <url1>900</url1>
    <url2>500</url2>                    <url2>700</url2>                    <url2>700</url2>
    <url3>500</url3>                    <url3>200</url3>                    <url3>100</url3>
    <title1>700</title1>               <title1>900</title1>               <title1>900</title1>
    <title2>500</title2>               <title2>700</title2>               <title2>700</title2>
    <text1>900</text1>                 <text1>900</text1>                 <text1>900</text1>
    <text2>500</text2>                 <text2>700</text2>                 <text2>700</text2>
    <text3>700</text3>                 <text3>800</text3>                 <text3>800</text3>
    <text4>300</text4>                 <text4>600</text4>                 <text4>600</text4>
    <f2p1>200</f2p1>                   <f2p1>200</f2p1>                   <f2p1>300</f2p1>
    <f2p2>200</f2p2>                   <f2p2>0</f2p2>                     <f2p2>100</f2p2>
    <f2p3>600</f2p3>                   <f2p3>800</f2p3>                   <f2p3>600</f2p3>
    <f3p1>200</f3p1>                   <f3p1>500</f3p1>                   <f3p1>500</f3p1>
    <f3p2>400</f3p2>                   <f3p2>100</f3p2>                   <f3p2>0</f3p2>
    <f3p3>400</f3p3>                   <f3p3>400</f3p3>                   <f3p3>500</f3p3>
    <f4p1>300</f4p1>                   <f4p1>0</f4p1>                     <f4p1>0</f4p1>
    <f4p2>200</f4p2>                   <f4p2>100</f4p2>                   <f4p2>100</f4p2>
    <f4p3>500</f4p3>                   <f4p3>900</f4p3>                   <f4p3>900</f4p3>
    <f5p1>300</f5p1>                   <f5p1>700</f5p1>                   <f5p1>500</f5p1>
    <f5p2>100</f5p2>                   <f5p2>0</f5p2>                     <f5p2>200</f5p2>
    <f5p3>600</f5p3>                   <f5p3>300</f5p3>                   <f5p3>300</f5p3>
    <f6p1>400</f6p1>                   <f6p1>800</f6p1>                   <f6p1>700</f6p1>
    <f6p2>100</f6p2>                   <f6p2>0</f6p2>                     <f6p2>0</f6p2>
    <f6p3>500</f6p3>                   <f6p3>200</f6p3>                   <f6p3>300</f6p3>
    <f7p1>300</f7p1>                   <f7p1>900</f7p1>                   <f7p1>900</f7p1>
    <f7p2>300</f7p2>                   <f7p2>0</f7p2>                     <f7p2>0</f7p2>
    <f7p3>400</f7p3>                   <f7p3>100</f7p3>                   <f7p3>100</f7p3>
    <q1>1000</q1>                      <q1>1000</q1>                      <q1>186</q1>
    <q2>1000</q2>                      <q2>1000</q2>                      <q2>4226</q2>
    <q3>1000</q3>                      <q3>1000</q3>                      <q3>173</q3>
    </parameter>                       </parameter>                       </parameter>
```

Fig. 13. Parameter Values

4.3 Results

The performance is evaluated in 8 Computer Science Departments. Homepage-Finder only retrieves the web pages with $Totalscore \geq 600$. The desired web pages in precision are decided manually, that is, a person decides every retrieved web page whether it is a personal homepage or whether it provides some information of a faculty or a group of faculties. The desired web pages in recall are defined as all personal homepages whose URLs are in the sites referred by Root URLs, includes the search keywords, and are displayed on the "list pages" as shown in the following 8 tables. The titles of the 8 tables are formatted as "search KEYWORDS in ROOT URLs". The first columns of the tables list the numbers of the retrieved web pages; the second columns give precision, that is, how many retrieved web pages are desired; the third columns give recall, that is, how many desired web pages are retrieved; the fourth columns give the average error of the desired web pages in recall which are retrieved. The higher a web page's Totalscore is, the lower the web page's error, and thus the higher the web page is ordered in the result list. For example, in the TABLE 1, FNN1 retrieves 25 web pages, in which 23 are desired; in the 17 desired web pages defined in recall, 14 are retrieved by FNN1; and the average error of the 14 web pages is 0.150, which means that Totalscore is averagely 850 for each of the 14 web pages.

	length	precision	recall	avg(Ei)
Google	32	22/32=68.8%	11/17=64.7%	N/A
FNN1	25	23/25=92%	14/17=82.4%	0.150
FNN2	40	37/40=92.5%	15/17=88.2%	0.067
FNN3	54	48/54=88.9%	15/17=88.2%	0.035

Table 1. Search "computer science" in cs.gsu.edu

list page: http://www.cs.gsu.edu/people/faculty.html

	length	precision	recall	avg(Ei)
Google	25	17/25=68%	0/16=0%	N/A
FNN1	46	45/46=97.8%	16/16=100%	0.365
FNN2	59	58/59=98.3%	16/16=100%	0.164
FNN3	65	63/65=96.9%	16/16=100%	0.090

Table 1. Search "computer science" in cs.caltech.edu

list page: http://www.cs.caltech.edu/people.html

	length	precision	recall	avg(Ei)
Google	156	112/156=71.8%	28/29=96.6%	N/A
FNN1	188	178/188=94.7%	29/29=100%	0.235
FNN2	241	220/241=91.3%	29/29=100%	0.114
FNN3	278	252/278=90.7%	29/29=100%	0.053

Table 3. Search "computer science" in cs.virginia.edu

list page: http://www.cs.virginia.edu/people/faculty.html

	length	precision	recall	avg(Ei)
Google	101	45/101=44.6%	1/14=7.1%	N/A
FNN1	36	35/36=97.2%	0/14=0%	N/A
FNN2	86	83/86=96.5%	14/14=100%	0.308
FNN3	114	103/114=90.4%	14/14=100%	0.161

Table 4. Search "computer" in csee.usf.edu

list page: http://www.csee.usf.edu/faculty_pages/faculty_list.html

	length	precision	recall	avg(Ei)
Google	125	77/125=61.6%	33/45=73.3%	N/A
FNN1	109	95/109=87.2%	28/45=62.2%	0.306
FNN2	282	229/282=81.2%	40/45=88.9%	0.208
FNN3	365	302/365=82.7%	41/45=91.1%	0.109

Table 5. Search "computer science" in cs.washington.edu

list page: http://www.cs.washington.edu/people/faculty/

	length	precision	recall	avg(Ei)

	length	precision	recall	avg(Ei)
Google	408	242/408=59.3%	52/52=100%	N/A
FNN1	134	133/134=99.3%	42/52=80.8%	0.385
FNN2	192	171/192=89.1%	43/52=82.7%	0.250
FNN3	216	195/216=90.3%	43/52=82.7%	0.116

Table 6. Search "computer science" in cs.uiuc.edu

list page: http://www.cs.uiuc.edu/people/faculty/index.html

	length	precision	recall	avg(Ei)
Google	247	161/247=65.2%	42/45=93.3%	N/A
FNN1	175	173/175=98.9%	33/45=73.3%	0.215
FNN2	307	301/307=98.0%	41/45=91.1%	0.085
FNN3	352	340/352=96.6%	45/45=100%	0.059

Table 7. Search "computer science" in cs.berkeley.edu

list page: http://www.eecs.berkeley.edu/Faculty/Lists/CS/

	length	precision	recall	avg(Ei)
Google	371	121/371=32.6%	32/34=94.1%	N/A
FNN1	169	167/169=98.8%	27/34=79.4%	0.141
FNN2	219	207/219=94.5%	30/34=88.2%	0.078
FNN3	283	268/283=94.7%	30/34=88.2%	0.044

Table 8. Search "computer sciences" in cs.wisc.edu

list page: http://www.cs.wisc.edu/faculty_staff.html

The results show that:
- FNN1, FNN2 and FNN3 have much higher precision than Google.
- FNN2 and FNN3 have higher recall than Google and FNN1.
- FNN2 can find more personal homepages than Google with much higher precision, higher recall and list them in a desired order.
- FNN3 can find more personal homepages and list them in a more desired order than FNN2, while keep precision and recall at the same level.

	length	precision	recall	avg(Ei)
Google	1465	797/1465=54.4%	199/252=79.0%	N/A
FNN1	882	849/882=96.3%	189/252=75%	0.267
FNN2	1426	1306/1426=91.6%	228/252=90.5%	0.159
FNN3	1727	1571/1727=91.0%	233/252=92.5%	0.082

Table 9. Simulation results including the above 8 departments

For any search request, FNN3 should have at least equal recall to Google according to the search string definitions. But in fact, because of Google Web Service's limitation, FNN3 can only access at most 1000 results [8]. As a result, the recalls of FNNs are lowered. That is why the recalls of FNN3 are lower than Google in Tables 6 and Table 8.

5. Conclusions

Finding desired information from the web efficiently and effectively is a key technology to realize its full potential and is not an easy task. [13, 14] In this chapter, we propose an intelligent web information search and retrieval model called Web Information Search Task (WIST) to address this issue, especially when a search request cannot or is difficult to be expressed in a crisp way. Firstly, similar search requests are abstracted as a WIST according to their structure characteristics defined by simple "structure rules". The WIST then retrieves the desired web pages according to some keywords and/or some root URLs submitted by a user. After gradually learning, the WIST can express the search requests much better, retrieve the desired web pages with high precision and high recall, and list the query result in a desired order.

This chapter also presents Homepage-Finder, an intelligent software agent which uses CI technologies including FL, NN, and GA to define and implement a specific WIST to automatically find all relevant researchers' homepages based on the possibility whether a web page is a personal homepage and the relevance with keywords, given a root URL, some keywords and some structure rules. The results show that Homepage-Finder with non-linear fuzzy reasoning can execute the WIST with high precision and high recall, and list the query result from higher score to lower score.

Homepage-Finder adopts the exact keyword matching method. In the future, we want to replace it with fuzzy concept matching method to make Homepage-Finder and other WIST-based "finders" more powerful and more intelligent.

6. References

[1] Mori Anvari, "Search Engines: Key to Knowledge Acquisition," In Proceedings of the 2001 BISC International Workshop on Fuzzy Logic and the Internet, pp. 25-29, August, 2001.

[2] Soumen Chakrabarti, "Data Mining for Hypertext: A Tutorial survey," SIGKDD: SIGKDD Explorations: Newsletter of the Special Interest Group (SIG) on Knowledge Discovery & Data Mining, ACM 1(2): 1-11, 2000

[3] Tina Eliassi-Rad, Jude Shavlik, "A System for Building Intelligent Agents that Learn to Retrieve and Extract Information," International Journal on User Modeling and User-Adapted Interaction, Special Issue on User Modeling and Intelligent Agents, 2001.

[4] Marti Hearst, "Using Dynamic Metadata to Improve Search User Interfaces," In Proceedings of the 2001 BISC International Workshop on Fuzzy Logic and the Internet, pp. 2, August, 2001.

[5] http://www.cs.washington.edu/research/projects/WebWare1/www/ahoy/
Jonathan Shakes, Marc Langheinrich & Oren Etzioni, "Dynamic Reference Sifting: A Case Study in the Homepage Domain," In Proceedings of the Sixth International World Wide Web Conference, pp. 189-200, 1997

[6] http://hpsearch.uni-trier.de/

[7] http://www.google.com.

[8] http://www.google.com/apis/api_faq.html#tech8.

[9] J.-S. R. Jang, C.-T. Sun, E. Mizutani, "Neuro-Fuzzy and Soft Computing, A Computational Approach to Learning and Machine Intelligence," Prentice Hall, Upper Saddle River, NJ, 1st edition, pp.81-84, 1996.

[10] Vincenzo Loia, Masoud Nikravesh, and Lotfi A. Zadeh, "Foreword: Fuzzy Logic and the Internet," Soft Computing, Vol. 6, pp. 285-286, Springer-Verlag, Aug. 2002.

[11] M. Nikravesh, T. Takagi, et al., "Web Intelligence: Conceptual-based Model," UC Berkeley Electronics Research Laboratory, Memorandum No. UCB/ERL M03/19, June 2003.

[12] M. Nikravesh, et al., "Perception-Based Decision processing and Analysis," UC Berkeley Electronics Research Laboratory, Memorandum No. UCB/ERL M03/21, June 2003.

[13] Sankar K. Pal, "Web Mining in Soft Computing Framework: Relevance, State of the Art and Future Directions," IEEE Transactions on Neural Networks, vol 13, no. 5, pp. 1163-1177, 2002.

[14] Ivan Ricarte, "A Reference Model for Intelligent Information Search," In Proceedings of the 2001 BISC International Workshop on Fuzzy Logic and the Internet, pp. 80-85, August, 2001.

[15] T. Takagi and M.Tajima, "Proposal of a Search Engine based on Conceptual Matching of Text Notes," IEEE International Conference on Fuzzy Systems FUZZ-IEEE'2001, 2001.

[16] L. A. Zadeh, "The Problem of Deduction in an Environment of Imprecision, Uncertainty, and Partial Truth," in M. Nikravesh and B. Azvine, FLINT 2001, New Directions in Enhancing the Power of Internet, UC Berkeley Electronics Research Laboratory, Memorandum No. UCB/ERL M01/28, August 2001.

[17] L. A. Zadeh, "A Prototype-Centered Approach to Adding Deduction Capability to Search Engines – The Concept of Protoform," BISC Seminar, Feb 7, 2002, UC Berkeley, 2002.

[18] Y.-Q. Zhang, T. Y. Lin, "Computational Web Intelligence (CWI): Synergy of Computational Intelligence and Web Technology," Proc. of FUZZ-IEEE2002 of World Congress on Computational Intelligence 2002: Special Session on Computational Web Intelligence, pp. 1104-1107, May 2002.

[19] Y.-Q. Zhang, S. Hang, T.Y. Lin, and Y.Y. Yao, "Granular Fuzzy Web Search Agents," Proc. of FLINT2001, pp. 95-100, Aug. 14-18, 2001.

[20] Y.-Q. Zhang and A. Kandel, "Compensatory Genetic Fuzzy Neural Networks and Their Applications," Series in Machine Perception Artificial Intelligence, Vol. 30, World Scientific, 1998.

Druck: Strauss Offsetdruck, Mörlenbach
Verarbeitung: Schäffer, Grünstadt